RECENT ADVANCES IN BIOPOLYMERS

Edited by **Farzana Khan Perveen**

Recent Advances in Biopolymers
http://dx.doi.org/10.5772/60630
Edited by Farzana Khan Perveen

Contributors

Valentina Dinca, Livia Elena Sima, Laurentiu Rusen, Thomas Lippert, Maria Dinescu, Maria Farsari, Anca Florina Bonciu, Pratima Parashar, Lourdes Ballinas, Samuel Oluwatobi Oluwafemi, Divya Bhatnagar, Miriam Rafailovich, Marcia Simon, Ion N. Mihailescu, Emanuel Axente, Felix Sima, Carmen Ristoscu, Natalia Serban, Olayemi Fakayode, Adewale Odunayo Oladipo, Sandile Songca, Agostinho Antunes, Masanao Imai, Keita Kashima, Peter Papoh Ndibewu, Taki Netshioswi, Tina Lefakane, Prince Ngobeni

Published by InTech

Janeza Trdine 9, 51000 Rijeka, Croatia

Publishing Process Manager Iva Simcic

Technical Editor InTech DTP team

Cover InTech Design team

First published March, 2016

Printed in Croatia

Additional hard copies can be obtained from orders@intechopen.com

Recent Advances in Biopolymers, Edited by Farzana Khan Perveen

p. cm.

ISBN 978-953-51-2255-5

For EU product safety concerns:
IN TECH d.o.o., Prolaz Marije Krucifikse Kozulić 3, 51000 Rijeka, Croatia,
info@intechopen.com or visit our website at intechopen.com.

Meet the editor

Dr Farzana Khan Perveen (Gold Medallist) is born in Karachi, obtained her BSc(Hons) and MSc (Zoology:Entomology) from the University of Karachi, MAS (Monbush Scolar; Agriculture:Agronomy) from the Nagoya University, Japan, and PhD (Research and Course works from the Nagoya University; Toxicology) degree from the University of Karachi. She is Founder/Chairperson of the Department of Zoology (DOZ) and Ex-Controller of Examinations at Shaheed Benazir Bhutto University and Ex-Founder/Ex-Chairperson of DOZ, Hazara University and Kohat University of Science & Technology. She is author of 125 high-impact research papers; 122 abstracts; 40 books; and 4 chapters and editor of 3 books and supervises BS, MSc, MPhil, MS, and PhD students. She has organized and participated in numerous international and national conferences and received multiple-awards and fellowships. She is a member of research societies, editorial boards of Journals, and World Commission on Protected Areas, International Union for Conservation of Nature. Her fields of interest are Zoology, Biotechnology, Biochemistry, and Bioengineering.

Contents

Preface

Polymer is an organic or inorganic, natural or synthetic material composed of macromolecules of high molecular weight whose structure consists of repeating units, called monomers, joined in sequence linked by covalent bonds. They are widely present in the environment. They are essential to life and also produced by various species of biological systems; therefore, they are called *biopolymers*.

The residues coming from the woodlands and agricultural exploitation constitute the most abundant biomass available in earth. Its importance as source of renewable energy has grown in addition to their environmental impact. Biomass waste is a lignocellulosic feedstock, which contains three main biopolymers: cellulose, hemicellulose, and lignin. It is used for the production of numerous value-added products because of their chemical composition, but it is necessary to recover efficiently from the valuable biopolymer as intact as possible by different processing alternatives. Synthetic polymers are used as supports for the migration of protein and cell growth factors. In addition to serving as support for tissue must be biocompatible, it should not cause immunogenic reactions and mechanical properties of supportable tensions. Polymer-coated noble metal nanoparticles are currently of particular interest to investigators in the fields of nanobiomedicine and fundamental biomaterials. These materials not only are exhibiting imaging properties in response to stimuli but can also efficiently deliver various drugs and therapeutic genes.

To reduce the use of non-renewable resources and to minimize the environmental pollution caused by synthetic materials, the quest for utilizing biomaterials for a wide range of biomedical and industrial applications is increased. Biopolymers are polymeric materials produced in nature by many biochemical processes induced by plants or microorganisms. The recent developments and trends of biopolymers have grown especially in the field of nanotechnology. An overview of recent advances in the field of laser-based synthesis of biopolymer thin films is presented here for biomedical applications. The biopolymer thin films have been used in tissue engineering, cell instructive environments, and drug delivery systems. Their applications have been used in fabrication of organic and hybrid organic-inorganic coatings. Matrix-assisted pulsed laser evaporation (MAPLE) and combinatorial-MAPLE have been introduced and compared with other conventional methods for thin films assembling on solid substrates. Regenerative medicine has been prepared with materials that are biodegradable, biocompatible, and structurally and chemically stable and that can mimic the properties of the native extracellular matrix (ECM). Hydrogels are hydrophilic three-dimensional networks that have long received attention in the field of regenerative medicine due to their unique properties. Silver nanoparticles (AgNps) have attracted much interest in biomedical engineering, since they have excellent antimicrobial properties. Silver nano-polymer composites have application in biochemical sensors, antimicrobial activity, and drug delivery system. Silver nanoparticles are more effective than ionic homologues (Ag^+) for their antimicrobial activity. Antimicrobial properties of silver nanoparticle are used by their incorporation into medical devices, tissues, and other health-related products for skin pathologies to reduce the risk of contamination and to promote higher preventive infection control.

Recent advances in biopolymers, including functional biomaterials for the manufacturing of personal protective garments (PPGs) or equipment (PPE), have dramatically improved their efficiency and performance. Good and acceptable permeation characteristics, mechanical strength, and durability are common attributes of these materials. Membranes prepared from oceanic biopolymers have a high potential in membrane separation processes and water purification. They more bio-compatible and lower-cost materials compared with artificial polymers. They have excellent performance in the separation engineering and regulation factors controlling membrane properties. The controlled interfacial properties of materials and modulated behaviours of cells and biomolecules on their surface are a requirement in the development of new generation of high-performance biomaterials for regenerative medicine applications. Roughness, chemistry, and mechanics of biomaterials are all sensed by cells. Organization of the environment at nano- and micro-scale as well as chemical signals trigger specific responses with further impact on cell fate. Particularly, human mesenchymal stem cells (hMSCs) are of great promise both in basic developmental biology studies and in regenerative medicine, as progenitors of bone cells. A new numerical-cum-graphical method called TI2BioP (Topological Indices to BioPolymers) has been developed to estimate topological indices (TIs) from two-dimensional (2D) graphical approaches for the natural biopolymers DNA, RNA, and proteins, in which the long biopolymeric sequences turns into 2D artificial graphs such as Cartesian and Four-color maps; however, it also reads other 2D graphs from the thermodynamic folding of DNA/RNA strings inferred from other programs. Its version 2.0 is freely available at http://ti2biop.sourceforge.net/.

Dr Farzana Khan Perveen
Founder Chairperson and Associate Professor
Department of Zoology
Ex-Controller of Examinations
Shaheed Benazir Bhutto University (SBBU), Khyber Pakhtunkhwa
Pakistan

Section 1

Synthesis of Biopolymers

Chapter 1

Biopolymers from Waste Biomass — Extraction, Modification and Ulterior Uses

Lourdes Ballinas-Casarrubias, Alejandro Camacho-Davila, Nestor Gutierrez-Méndez, Víctor Hugo Ramos-Sánchez, David Chávez-Flores, Laura Manjarrez-Nevárez, Gerardo Zaragoza-Galán and Guillermo González-Sanchez

Additional information is available at the end of the chapter

http://dx.doi.org/10.5772/61855

Abstract

The residues coming from woodlands and agricultural exploitation constitute the most abundant biomass available on earth. Its importance as a source of renewable energy has grown in addition to the environmental impact. Biomass waste is a lignocellulosic feedstock which contains three main biopolymers: cellulose, hemicellulose and lignin. It could be utilized for the production of a number of value-added products due to their chemical composition, but it is necessary to efficiently recover the valuable biopolymer as intact as possible by different processing techniques.For different applications, the principal objective of pre-treatment is to keep the cellulose intact, meanwhile hemicellulose and lignin are removed. The yields of the fractions depend on the pre-treatment method, which is the most expensive step in biomass conversion. Traditionally, cellulose is obtained by kraft, sulphite and soda treatments. These methods are non-environmentally friendly and generate huge quantities of toxic wastes. Recently developed models considering the environmental laws encourage the sustainable processing of biomass into value-added products. The use of ionic liquids as new solvents for biomass waste and organosolv processes is reviewed, which are used to obtain cellulose. One of the possible applications of cellulose is membrane synthesis, which has been reported for other biomass materials, such as sugarcane bagasse, mango seed and newspaper. In this chapter, some green pre-treatment methods, different sustainable routes for cellulose modification and some of the results obtained on membrane development based on waste biomass are discussed.

Keywords: Biomass, Waste, Pre-treatment, Membranes

1. Introduction

An alarming environmental and economic situation is created by agro-industrial and forestry wastes, which are produced in large amounts annually. For example, in Mexico in 2011, sawmills produced 3.9 million m^3 of wood which generated 100 tons of waste. Chihuahua has the second place in timber production with an 18.3%, in 2011 at the national level. At the state level, *Pinus* sp. (955, 654 m^3 roll) and oak (51, 170 m^3 roll) are widely used, of which 71.6% are used by sawmills. In 2008, there were 19, 246 m^3 of waste in the genus *Quercus* (oak). Although fluctuations are reported each year and production of oak has reduced since 2007, 20–25% of the generated waste at sawmills corresponds to this genus [1]. Generally, these residues are exposed to the environment causing problems due to the suspension of particles of sawdust in the air. These residues are used for the production of pellets; however, the amount of bioaccumulation is very large and hence this technique cannot be applied. Generally, sub-products are used directly as a fuel or to manufacture agglomerated materials, but this usage is not enough to overcome the problems of its disposition.

Another interesting subproduct is the agave bagasse. The raw material to produce tequila with 'Denominación de Origen' (denomination of origin) is agave tequilana. Bagasse is the final waste produced after the head of the plant is cooked and pressed to extract the sugar liquor which will produce tequila by fermentation of *Saccharomyces cerevisiae*. According to the National Consortium of Tequila, 350, 000 tons of bagasse waste on a dry basis is produced annually compared with 14.1 million litres of tequila produced in 2012. The use of agave bagasse as a resource helps to reduce environmental degradation that is caused by the accumulation of the tequila waste. These compounds are accumulated and disposed in large volumes and are resistant to degradation, and they are stored for 6–8 years [2]. For instance, using the bagasse from agave tequilana is very helpful, from an economic and environmental point of view.

Biomass waste, as a lignocellulosic feedstock, contains three main biopolymers: cellulose, hemicellulose and lignin. Due to their chemical composition based on carbohydrates, they are used for the production of a number of value-added compounds. Many processing alternatives are being studied to efficiently recover biomass polymers. The main purpose of pre-treatment is to remove as much hemicellulose and lignin as possible, while simultaneously keeping enough cellulose undamaged [3]. Fractions with a high yield of cellulose depend on the pre-treatment method, which still remains the most expensive and time-consuming step in biomass conversion. Furthermore, it is very difficult to generalize the process conditions even for a similar type of biomass due to the broad variability of the raw material composition. Cellulose is currently separated from lignocellulosic materials using non-environmentally friendly processes, such as kraft, sulphite and soda treatments, with the consequent generation of toxic wastes. In the last few years, new rules have been framed due to the growing restrictions in the environment, based on the sustainable processing of biomass into value-added products [4]. In this chapter, traditional and environmentally friendly methods for biomass pre-treatment are reviewed, with special emphasis on the use of green technologies [5, 6]. A brief description about nanostructured composites used as membrane materials is provided. Some results obtained using agave bagasse and sawdust are also highlighted.

http://dx.doi.org/10.5772/61855

2. Materials of cellulosic origin: pre-treatment and new treatments of biomass for obtaining cellulose

Residues from biomass consist of three macromolecular components: cellulose, lignin and polyoses as well as small quantities of extractives and mineral salts, having a lamellar structure in which the components are distributed.

The cellulose and hemicellulose are predominant in the region of the cell wall, and lignin is distributed throughout the structure, presenting a maximum concentration at the middle lamella. The transformation processes of lignocellulosic fractions are limited by the structure of the biomass.

Generally, the breakdown of the complex cellulose–hemicellulose–lignin or its fractionation by pre-treatment techniques and delignification are required [7]. Cellulose (Figure 1) is a polymer formed by long linear chains of glucose residues linked by beta-D-glucose bonds, where the glucose molecules are linked through carbons 1 and 4 to form celobiose; 1–4 hydroxide link occurs at the β-position (equatorial), resulting in the formation of a linear polymer. Hemicellulose is approximately 25% of the biomass, and it is a linear polysaccharide polymer similar to cellulose, except that it is primarily composed of xylans, glucomannans and arabinose. Another difference is that the resistance is lower due to its degree of polymerization (70–200 units). Like cellulose, it is a polar molecule and therefore contributes to the polarity of the wood. Hemicelluloses contain various residues of sugars, generally from two to four or up to six.

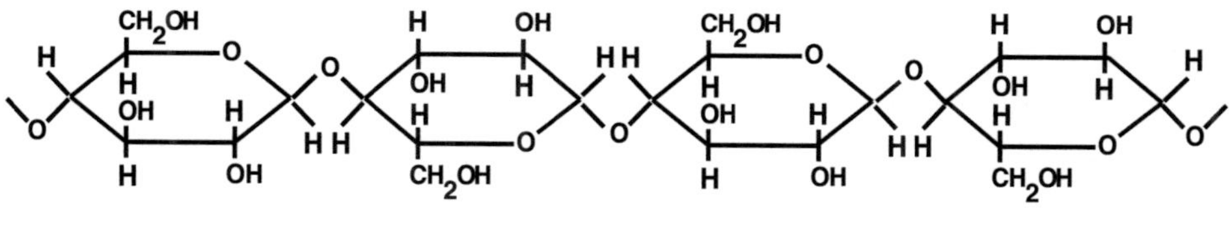

Figure 1. Structure of cellulose; central part of the molecular chain [8].

Lignin is a highly stable macromolecule, being the second most abundant biopolymer, after cellulose. Lignin is primarily composed of ϱ-phenyl propane units forming an amorphous tridimensional macromolecule, representing 20–30% of the total mass of the lignocellulosic material. The structure of lignin is attributed to its biosynthetic mechanism, which is processed via radicals from precursors such as sinapylic, coniferylic and p-coumarylic alcohols. The following are the most abundant types of bonds formed between the phenylpropane units: β–O-4 e α – O-4 (50–65%), β-5 (6–15 %), β-1 (9–15%), 5 (5–9 %) y β–β (2–5%).

Cellulose is one of the most common biopolymers representing about 1.5×10^{12} tons of total annual production of biomass. It is regarded as an inexhaustible raw material for the production of eco-friendly and biocompatible products. The wood pulp is the most important for processing cellulose, which is aimed at the production of paper and paperboard source.

Approximately 2% was used for the production of regenerated cellulose fibres, films and for synthesizing many cellulose esters and ethers [9]. These derivatives, produced on an industrial scale, are used in optical films, portion means, additives, pharmaceuticals, cosmetics and coating, among others [7]. In its polymeric form, it can be used for papermaking or converted into derivatives such as acetyl cellulose, nitrocellulose, ethanol, artificial fibres, colloids with medicinal uses, celluloid, various plastic materials and also in obtaining esters and cellulose ethers.

Several studies have shown that it is possible to obtain cellulose from agro-industrial waste, such as waste from fibre, bagasse and wood [10, 11]. However, the cellulose obtained from this resource has viscosity values unsuitable for papermaking. One option for using it is to chemically derive it for compounds that can be commercially exploited. To obtain cellulose from agro-industrial waste, the separation of macromolecular components should be carried out by chemical, physical or enzymatic processes.

The modification of cellulose has been widely studied due to the possibility of obtaining compounds with a wide industrial application. These modifications include the production of cellulose esters, with applications ranging from auxiliaries in large-scale polymerization, the viscosity thickeners in cosmetics, foods, drugs and membranes. Also, the study of the use of lignin has both economic and environmental importance due to the wide availability of the same from agro-industrial wastes.

There are several processes of delignification in the treatment of biomass. Most of them are based on the previously optimized methods for the delignification of wood. The main industrial processes of chemical delignification of lignocellulosic materials are the kraft, soda and sulphite processes, which release large amounts of toxic chemicals to the environment [12]. The complex cellulose–hemicellulose–lignin must be fractionated by pre-treatment to remove maximum lignin, by not weakening too much the cellulose fibres.

The kraft method stands out among all other chemical pulping processes due to its high efficiency in lignin removal, without damaging much of the cellulose fibres. Of all the methods, this method is used in 60% worldwide [8]. The kraft process uses sodium hydroxide and sodium sulphate in large digesters to which splintered wood is introduced ; this process removes maximum lignin leaving a brown paste; however, it produces toxic emissions such as sulphur dioxide (primary contributor to acid rain). A black liquor is obtained that burns modern plants, generating vapours that can be used for generating electric energy. Waste water is also treated, purified and recycled to regenerate sodium sulphide and caustic soda.

Another method used is the 'sulphite', where a clearer, weak and soft paste is produced. Similar to the kraft method, the chemicals used can be recycled, but the recovery efficiency of sulphuric acid is lower, and also a higher amount of sulphur dioxide is released per ton of the pulp produced.

Unlike traditional industrial pulping methods, processes using techniques of steam explosion and organic/water solvent mixtures (organosolv) or their combination allow the use of not only the cellulose but also the lignin and polyoses for the production of various chemical inputs, also being environmentally friendly routes [7].

http://dx.doi.org/10.5772/61855

Organosolv process has been investigated to extract lignin from pulp for paper production. In this process, an organic or aqueous organic solvent is mixed with a catalyst (usually an inorganic acid such as HCl or H_2SO_4) to break the bonds of the lignin and hemicellulose [13]. Organic acids as catalysts can also be used, such as oxalic, salicylic and acetylsalicylic acids. Table 1 shows some of the pre-treatment processes, their advantages and limitations.

Pretreatment	Advantage	Limitations
Steam explosion	Causes degradation of hemicellulose and lignin transformation; effective cost.	Incomplete disruption of lignin-carbohydrate matrix.
Fibre explosion with aqueous ammonia	Increase the accessibility area, removes a large amount of lignin and hemicellulose, does not produce inhibitors for subsequent processes.	It is not efficient with biomass with a high content of lignin.
Explosion with CO_2	Increases the accessibility area, does not produce inhibitors for subsequent processes.	Unchanged lignin or hemicellulose
Ozonólisis	Reduces the lignin content; don´t produce toxic waste.	Large amounts of ozone are required, it is an expensive process
Biological	Degrades lignin and hemicellulose; low energy requirement	The hydrolysis speed is very low
Organosolv	Hydrolyzed lignin and hemicelluloses.	Solvents need to be drained from the reactor, evaporated, condensed, and recycled: high cost.

Table 1. Pre-treatment processes. Advantages and limitations.

The objective of pre-treatment is to remove the maximum amount of lignin and hemicellulose as possible, while maintaining the cellulose intact. Typically used solvents are methanol, ethanol, acetone, ethylene glycol, triethylene glycol and phenol. The name ethanosolv refers to an organosolv process with ethanol. The ethanosolv method is more convenient than other pre-treatments because it is safer, since ethanol is less toxic than other solvents [13].

In addition to the advantages presented in Table 1, organosolv pre-treatment has the following benefits: 1) organic solvents are easily recovered by distillation and can be recycled back to the pre-treatment process; 2) after pre-treatment, the recovered lignin can be used for various co-products (vanillic acid, vanillin and p-coumaryl); and 3) pre-treatment contemplates the use of all components of the biomass, making it viable for biorefinery of lignocellulosic biomass. The ethanosolv process is currently one of the best pre-treatments of biomass residues from an environmental perspective [14–21].

At the top of the development of innovative methods for pulping, it was found that the ionic liquids (ILs) were the most studied solvent in the last decade [7, 12, 22–26].

The first report of ionic liquids as non-derivatizing solvents of cellulose was published in 1934 by Charles Graenacher. He applied the N-alkylpyridinium salts for dissolving cellulose and a homogeneous reaction medium [24]. However, it was in 2002 when Swatloski reported the use of comparable salts with boiling points below 100 ° C, currently known as ionic liquids, provoking a new start in the research of cellulose. It has been observed that the most promising ILs for the modification of cellulose are salts of 1-alkyl-3-methylimidazolium [27–29].

To date, much is known about ILs: organic salts with low melting point, of which those containing ammonium, pyridinium and imidazolium are the most suitable for dissolving cellulose, in other words, those organic salts with asymmetric cations to interact appropriately with the skeleton of cellulose.

A generally accepted theory to explain the interaction between the ionic liquid and polymers has not been proposed. Such solvents are considered non-derivatizing solvents, i.e. with non-covalent bonds, as shown in studies of ^{13}C nuclear magnetic resonance (NMR). These studies indicate that the chloride interaction is higher for the breaking the hydrogen bond, than in the cation, as in the DMAc/LiCl systems.

The most widely used ionic liquids, with melting temperatures below 55°C, and extremely low viscosities, include the AmimCl (1-allyl-3-methylimidazolium chloride) and EMIMAc (ethyl 1-ethyl-3-methylimidazolium acetate). These are considered as non-toxic, non-corrosive and biodegradable. Therefore, the use of ILs for extracting cellulose from waste is of particular interest for the development of green methods for obtaining this vegetal biopolymer.

Another method for biomass waste treatment processes is solid-state fermentation, in which lignolytic fungi and/or the enzymes that they produce are used [14, 30–33]. These are better known as techniques of biopulping or biobleaching, if xylanases or laccases are used. A fermentation process in solid phase is defined as the one occurring in the absence or almost nule presence of water, using natural and inert substrates as solid supports. The substrate supplies nutrients to microbial growth, and it serves as a means of attachment of cells. Among the most important factors for the growth of microorganisms are the particle size, the level of water and humidity. White rot fungi, such as *Phanerochaete chrysosporium* and *Ceriporiopsis subvermispora,* are the most important [14, 30–33]. However, the main limitations are the long incubation times thereof, two to four weeks, and the yield loss, because the fungus attacks the polysaccharides simultaneously.

Therefore, there is a current area of research on the biotechnological processes using enzymes instead of fungus [34]. The enzymes added directly, such as the lignin peroxidase complex, into the biomass showed better pulping yields.

3. Chemical modification of cellulose

By the chemical treatment of cellulose, different compounds can be obtained with new and particular features, including hydrophobic compounds, water-soluble products, thickeners, ion-exchange materials resistant to the action of microorganisms, etc. In these chemical modifications, cellulose ethers, such as methylcellulose (MC), hydroxyl propyl methyl cellulose (HPMC) and cellulose acetate, are mainly included. The esterification of the cellulose in acetic acid is not possible and therefore cannot be used as an esterifying agent. Typically, acetic anhydride is used as an acetylating agent. This acetylation reaction of cellulose with acetic anhydride can be represented by the following equation:

$$C_6H_7O_2(OH)_3 + 3\,(CH_3CO)_2\,O \rightarrow C_6H_70(OCOCH_3)_3 + 3\,CH_3COOH$$

If the reaction is only performed with acetic anhydride, esterification is very slow; therefore a catalyst is needed. The most commonly used are sulphuric or perchloric acid. Acid-catalyzed treatments have many limitations. First, the acid itself is not environmentally friendly reactive, and its discharge can cause environmental pollution. Second, its corrosivity requires expensive construction materials, which increase the cost of the process [35].

The esterification of cellulose using ionic liquids is an area of current development [10, 36, 37].

The main advantage of the process is that a regioselective esterification can be obtained. The direct esterification with highly sterically demanding acylating reactives has been studied with substances such as pivaloyl chloride and adamantyl and 2,4,6-trimethylbenzoyl of cellulose in solvents such as DMAc/LiCl, DMSO/TBAF and AmimCl [27].

In 2011, Wang reported the direct extraction of cellulose with AmimCl, with an 85% yield of cellulose initially contained in the fibre, using the DMSO/water system as the precipitant [36]. Esterification in this medium can be performed in one step from the cellulose fibre, obtaining cellulose acetates with 2.16–2.63 degrees of substitution (DS) of. The major advantage of the process is that it involves a one-step preparation. In a homogeneous medium, the DS can be easily controlled by adjusting the time of acetylation, as acetylated products may be obtained according to the needs of synthesis.

4. Preparation and characterization of membranes

The membranes have overcome many shortcomings of conventional separation methods because of features such as ease of operation, energy advantages, selectivity and low operating costs. Through selective complexation reactions (reactive membranes) and processes of ultra-, micro- and nanofiltration (inert membranes), we have been able to make highly specific separations [38]. Currently, they are the basic materials that stimulate scientific research and technological development.

The membrane is clearly the most important part of the separation system. A membrane is a permeable or semi-permeable phase, which divides two fluid phases: feeding (F) and recovery (S). It is essentially a barrier that allows the selective passage of the chemical species, controlling the relative speeds of transport, which ideally produces a phase (F) without the transported components and a phase (S) enriched with them. The operation of a membrane, therefore, will depend on its structure, largely determining the separation mechanism and its application. For this reason, the science of membranes has clearly focused on the development of materials with different properties such as synthetic polymers, modified natural products, inorganic membranes, ceramic, metal and liquids [39–41].

The flow across the membranes is possible when a directive force is applied, as concentration gradients, temperature, and pressure or electrical potential.

This parameter is used as a criterion for the classification of membrane processes that exist; for example, ultra-, micro- and nanofiltrations are carried out thanks to a hydrostatic pressure gradient, distillation by membrane is due to a temperature gradient, and due to concentration gradients processes of pervaporation, dialysis and reactive liquid membranes transportation are achieved.

In general, the membranes are classified into three: inert, reactive and composite. An inert membrane is one in which any chemical transformation is not seen involved in the membrane during separation, whereas in a reactive membrane, the chemical interaction between the chemical of interest and the reactive component of the membrane determines the process transport. In a composite membrane at least two types of materials are identified; one of them can be reactive and the other inert, that is the reason why a 'hybrid' transport occurs in the system.

In the processes of ultra-, micro- and nanofiltration, inert membranes of certain physicochemical properties are generally used. Depending on the application, it is necessary to control the distribution and pore size, because the transport process is related to a molecular sieving.

The filtration membranes can be easily obtained by phase inversion method. This is a process by which a polymer is transformed in a controlled way (at fixed conditions of temperature, humidity and pressure) from a liquid to a solid state. The transition to solid can be adjusted so that the morphology of the membrane can be determined. A widely used method for the synthesis of membranes is precipitation by controlled solvent evaporation [42, 43].

Generally, the cost of operating a membrane process depends on the material that the membrane contains. For this reason, it is extremely important to find new synthesis routes that are economically feasible and environmentally friendly, to obtain membranes. This study is intended to produce composite membranes from various lignocellulosic materials, in a novel way, particularly from agro-industrial wastes.

Acetylated cellulose can be regenerated by ILs solutions by precipitation in a wide variety of non-solvents, including water, alcohols and acetone. If the membrane casting is made simultaneously, an adjustment of the topographical and morphological characteristics can be attained. A large amount of work has been reported using ILs as an alternative to the traditional methods of obtaining membranes [10, 11, 37, 44–48].

The formation of cellulose acetate membranes via phase inversion using ionic liquids such as (thiocyanate 1-butyl-3-methylimidazolium) and solvents such as [BMIM] SCN, 1-n-butyl-3-methylimidazolium thiocyanate [49] has been studied. The process is compared to phase inversion in water using N-methyl-2-pyrrolidone as a solvent and acetone. It is found that the

http://dx.doi.org/10.5772/61855

ionic liquid has distinct effects on the process of phase inversion and in the morphology of the membranes obtained by comparing the process with NMP,N-Methyl-2-pyrrolidone and acetone due to its viscous nature.

5. Recent advances in membrane synthesis using biomass waste

As mentioned above, agave bagasse and sawdust are biomass wastes which can be revalorized by obtaining new membrane materials. Results on the organosolv treatment of *Pinus* sp. sawdust have been recently reported [42]. The following graph relates the rheological behaviour of the membranes obtained using the acetylated cellulose by the organosolv pre-treatment of sawdust.

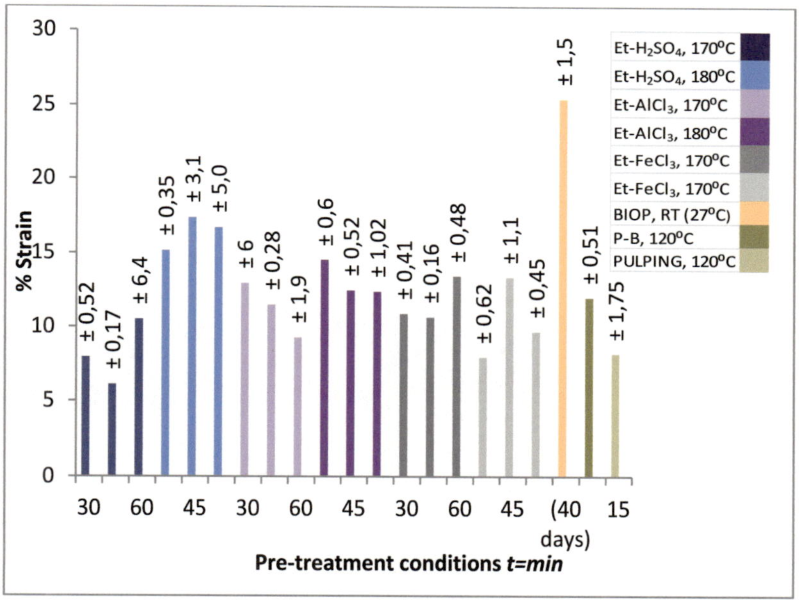

Figure 2. Percentage strain of membranes of acetylated cellulose. Cellulose was recovered by organosolv pre-treatments stated in the figure, such as ethanosolv–H_2SO_4, i.e. using sulphuric acid as catalyst and so on; BIOP refers to biopulping using *Phanerochaete chrysosporium*, P–B refers to pulping and bleaching and pulping to the traditional method using 10% sulphuric acid at 120°C.

Strain was recorded at an angular velocity of 6, 286 rad/s. Strain was modified by the cellulose recovery profile. Biopulping, the longest pre-treatment performed, gave the best cellulose for membrane resistance. Materials obtained were mainly dense (Figure 3). These can resist up to 9.1 bar, addressing a micro- and ultra-filtration with the rejection of calcium in the range of 9.51–89.75%.

Similar pre-treatments were applied for agave bagasse. Membranes were obtained as shown in Figure 4. Cellulose was obtained by a traditional method using 10% sulphuric acid at 120°C as mentioned elsewhere [48]. Acetylation was also performed using the Fischer method, and acetylated cellulignin was recovered with dichloromethane. This solution was the medium for casting at different evaporation temperatures and humidity. These materials attain nanofiltration pressure and flow conditions.

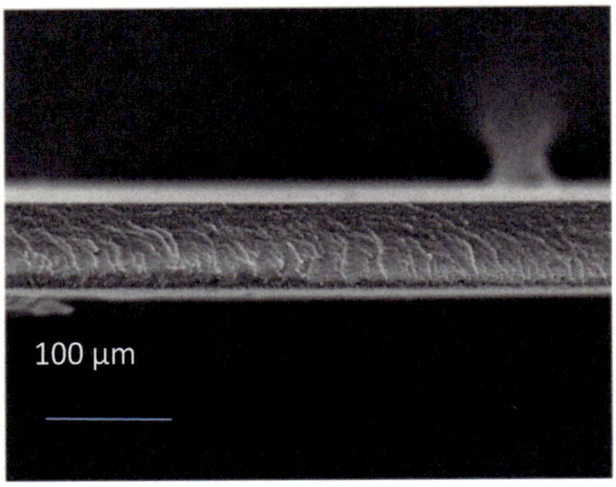

Figure 3. Scanning electron micrograph of membranes manufactured by acetylated cellulose of *Pinus* spp. saw dust.

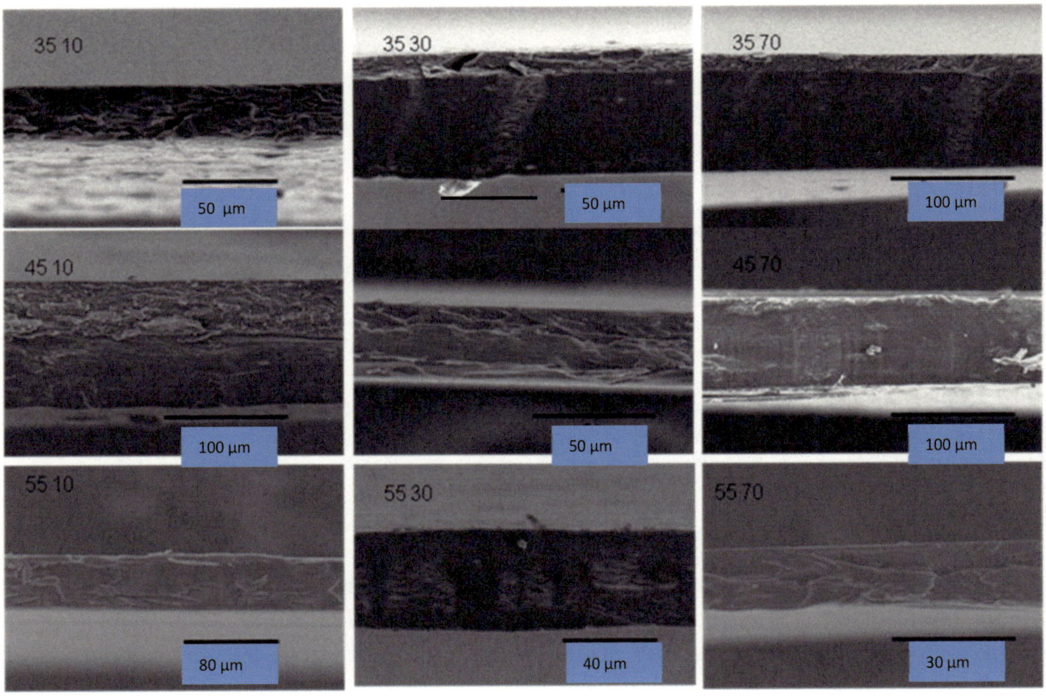

Figure 4. Acetylated cellulignin membranes from agave bagasse. Numbers on the left corners represent the following: The first two correspond to temperature (°C), i.e. 34, 45, 55; the second two correspond to humidity (%), i.e. 10, 30, 70.

http://dx.doi.org/10.5772/61855

6. Conclusion

Biomass waste provides an extraordinary source of biopolymers susceptible to transformation and revalorization, such as cellulose acetates, which can be further used as raw materials for membrane production. With the application of new environmental regulation for green technologies, these can even be obtained by the use of ionic liquids.

Biomass pre-treatment is a fundamental step for biomass utilization. Organosolv processes release good-quality cellulose, which can be further functionalized for other applications such as the generation of acetylated cellulignin. It could also be a good alternative to implement ILs or biological routes for cellulose recovery.

Author details

Lourdes Ballinas-Casarrubias[1*], Alejandro Camacho-Davila[1], Nestor Gutierrez-Méndez[1], Víctor Hugo Ramos-Sánchez[1], David Chávez-Flores[1], Laura Manjarrez-Nevárez[1], Gerardo Zaragoza-Galán[1] and Guillermo González-Sanchez[2]

*Address all correspondence to: lourdes.ballinas@gmail.com

1 Facultad de Ciencias Químicas, Universidad Autónoma de Chihuahua, Circuito Universitario, Chihuahua, Chih., México

2 Centro de Investigación en Materiales Avanzados, S.C., Chihuahua, Chih., México

References

[1] Ministry of Environment and Natural Resources. Statistical Yearbook of Forestry Production. 1st edition, Mexico: Mexico DF, 2011.

[2] Iñiguez-Covarrubias G, Lange S E, Rowell R M. Utilization of byproducts from the tequila industry: Part 1: Agave bagasse as a raw material for animal feeding and fiberboard production. Bioresource Technology, 2001; 77: 25–32.

[3] Poletto M, Júnior H, Zattera A. Native cellulose: Structure, characterization and thermal properties. Materials (Basel) 2014; 7: 6105–6019.

[4] Farzana P. Biogas for both: Generating energy and waste management. In: Advances in Energy Research and Development. Editors: Chern, M J., Lanin, V., Sarker, M., Vaxiri, N., ORIC Publications, Arkansas, USA; 2014. Online: URL: www.oricpub.com Manuscript ID 1110011.

[5] Gity Mir M S and Sayaf M. Compostable polymers and nanocomposites—A big chance for planet earth. In: Recycling Materials Based on Environmentally Friendly Techniques. Editor: Achilias D.S., Intech; 2015. DOI: 10.5772/58503.

[6] Novak A C, Sydney E B, Soccol C R. Biotransformation of Waste Biomass into High Value Biochemicals. 1st edition, NY, USA: Springer, 2014; 389–411. DOI: 10.1007/978-1-4614-8005-1.

[7] Lemm D, Heublein B, Fing H.P, Bohn A. Cellulose: fascinating biopolymer and sustainable raw material. Angewandte Chemie, International Edition. 2005; 44, 3358–3393.

[8] Fengel D, Wegener G. Wood Chemistry, Ultrastructure, Reactions. 1984; Berlin: Walter de Gruyter, 613 p.

[9] Tuck C O, Perez E, Horvath I T, Sheldon R A, Poliakoff M. Valorization of Biomass: Deriving more value from waste. Science. 2012; 337, 695–699.

[10] Ass B, Ciacco G, Frollini E. Cellulose acetates from linters and sisal: Correlation between synthesis conditions in DMAc/LiCl and product properties. Bioresource Technology. 2006; 97, 1696–1702.

[11] Liu C, Sin R C, Zhang A P, Ren J L, Wang A, Qin M H, Chao Z N, Luo W. Homogeneous modification of sugarcane bagasse cellulose with succinic anhydride using a ionic liquid as reaction medium. Carbohydrate Research, 2007, 342, 919–926.

[12] Pinkerte A, Marsh K, Pang S, Staige M. Ionic liquids and their interaction with cellulose. Chemical Review. 2008, 109, 6712–6728.

[13] Kim Y Y, Yu A, Han M, Choi G, Chung B. Ethanosolv pretreatment of barley straw with iron(III) chloride for enzymatic saccharification. Journal Chemical Technology and Biotechnology. 2010; 85, 1494–1498.

[14] Pandey A, Soccol C, Mitchell D. New developments in solid state fermentation: Bioprocesses and products. Process Biochemistry. 2000; 35, 1153–1169.

[15] Poletto M, Pistor V, Zeni M, Zattera A. Crystalline properties and decomposition kinetics of cellulose fibers in wood pulp obtained by two pulping processes. Polymer Degradation and Stability. 2011; 96, 679–685.

[16] López F, Pérez A, García J C, Feria M J, García M M, Fernández M. Cellulosic pulp from Leucaena diversifolia by soda-ethanol pulping process. Chemical Engineering Journal. 2011; 166, 22–29.

[17] Díaz M J, Alfaro A, García M M, Eugenio M E, Ariza J, López F. Ethanol pulping from tagasaste. A new promising source for cellulose pulp. Industrial & Engineering Chemistry Research. 2004; 43, 1875–1881.

[18] Oliet M, García J, Gilarranz M A. Solvent effects in autocatalyzed alcohol-water pulping. Comparative study between ethanol and methanol as delignifying agents. Chemical Engineering Journal. 2002; 87, 157–162.

[19] Jiménez L, Pérez I, García J C, Rodríguez A. Influence of process variables in the ethanol pulping of olive tree trimmings. Bioresource Technology. 2001; 78, 63–69.

[20] Baptista C, Robert D, Duarte A P. Effect of pulping conditions on lignin structure from maritime pine kraft pulps. Chemical Engineering Journal. 2005; 121, 153–158.

[21] Xu Y, Li K, Zhang M. Lignin precipitation on the pulp fibers in the ethanol-based organosolv pulping. Colloids and Surfaces A: Physicochemistry Engineering Aspects. 2005; 301, 255–263.

[22] Lovell C S, Walker A, Robin A D, Radhi A, Tanner S F, Budtova T, Ries M E. Influence of cellulose on ion diffusivity in 1-ethyl-3-methyl-imidazolium acetate cellulose solutions. Biomacromolecules. 2010; 11, 2927–2935.

[23] Liebert T, Heinze T. Tailored cellulose esters synthesis and structure determination. Biomacromolecules. 2005; 6, 333–340.

[24] Liebert T. Cellulose solvents – Remarkable history, bright future. In: Cellulose solvents: For analysis, shaping and chemical modification, ACS Symposium Series American Chemical Society, Washington DC USA, 2010. Volume 1033, pp. 3–64.

[25] Sashina E S and Kashirskii D A. Pyridinium-based ionic liquids — Application for cellulose processing. In: Ionic Liquids-Current State of the Art. Intech; 2015. DOI: 10.5772/59286

[26] Lan W, Liu C F, Yue F S, Sun R C. Rapid dissolution of cellulose in ionic liquid with different methods. In: Cellulose - Fundamental Aspects. Intech; 2013. DOI: 10.5772/52517

[27] Fox S C, Li B, Xu D, Edgar K J. Regioselective esterification and etherification of cellulose: A review. Biomacromolecules. 2011; 12, 1956–1972.

[28] Kocherbitov V, Ulvenlund S, Kober M, Jarring K, Arnebrant T. Hydration of microcrystalline cellulose and milled cellulose studied by sorption calorimetry. Journal of Physical Chemistry. B 2008; 112, 3728–3734.

[29] Huang K, Wang B, Cao Y, Li H, Wang J, Lin W, Mu C, Lia D. Homogeneous preparation of cellulose acetate propionate (CAP) and cellulose acetate butyrate (CAB) from sugarcane bagasse cellulose in ionic liquid. Journal of Agricultural and food Chemistry. 2010; 3, 45–47.

[30] Ferraz A, Guerra A, Mendoca R, Masarin F, Vicentim M, Aguilar A, Pavan P. Technological advances and mechanistic basis for fungal biopulping. Enzyme and Microbial Technology. 2008; 43, 178–185.

[31] Tanaka H, Koike K, Itakura S, Enoki, A. Degradation of wood and enzyme production by *Ceriporiopsis subvermispora*. Enzyme and Microbial Technology. 2009; 45, 384–390.

[32] Giles R, Galloway E, Elliot G, Parrow M. Two stage fungal biopulping for improved enzymatic hydrolysis of wood. Bioresource Technology. 2011; 102, 8011–8016.

[33] Zeng J, Singh D, Chen S. Biological pretreatment of wheat straw by *Phanerochaete chrysosporium* supplemented with inorganic salts. Bioresource Technology. 2011; 102, 3206–3214.

[34] Zhao J, Li X, Qu Y. Application of enzymes in producing bleached pulp from wheat straw. Bioresource Technology. 2006; 97, 470–1476.

[35] Sikorski P, Wada M, Heux L, Shintani H, Stokke B. Crystal structure of cellulose triacetate. Macromolecules. 2004; 37, 4547–4553.

[36] Wang, X, Li H, Tang Q. Cellulose extraction from wood chip in an ionic liquid 1-allyl-3-methilimidazolium chloride (AmimCl). Bioresource Technology. 2011; 102, 7959–7965.

[37] Cao Y, Zhang J, He J, Li H, Zhang Y. Homogeneous acetylation of cellulose at relatively high concentrations in an ionic liquid. Product Engineering and Chemical Technology. 2010; 18(3), 515–522.

[38] Feeman B, Pinnau I. Gas and liquids separations using membranes: An overview. In Advanced Materials for Membrane Separations, 2004, ACS Symposium Series, Washington D. C., USA.

[39] Rajesh S, Shobana K, Anitharaj S, Mmohan D. Preparation morphology, performance, and hydrophilicity studies of polyamide-imide incorporated cellulose acetate ultrafiltration membranes. Industrial Engineering and Chemistry Research. 2011; 20, 5550–5564.

[40] Xing D, Peng N, Chung T. Formation of cellulose acetate membranes via phase inversion using ionic liquid [BMIM]SCN as the solvent. Industrial Engineering and Chemical Research. 2010; 49, 8761–8769.

[41] Meireles C, Rodrigues G, Fernandes M, Cerqueira D, Nascimiento R, Mundim E, Poletto, P, Zeni M. Characterization of asymmetric membranes of cellulose acetate from biomass: Newspaper and mango seed. Carbohydrate polymers. 2010; 80, 954–961.

[42] Ballinas Casarrubias L, Saucedo Acosta T, Mac Donald Pizaña K, Ruiz Cuilty K, Nevárez-Moorillón G V, Gutierréz-Méndez N, Torras Font C, Chávez Flores D, González-Sánchez G. Organosolv pretreatment for cellulose recovery from sawdust for its ulterior use in membrane synthesis and operation. Desalination and Water treatment, 2015.

[43] Ma H, Hsiao B, Chu B. Thin film nanofibrous composite membranes containing cellulose of chitin barrier layers fabricated by ionic liquids. Polymer. 2011; 52, 2594–2599.

[44] Stefanescu C, Daly W, Negulescu I. Biocomposite films prepared from ionic liquid solutions of chitosan and cellulose. Carbohydrate Polymers. 2011; 87(1), 435–443.

[45] Xing D Y, Peng N, Chung T S. Formation of cellulose acetate membranes via phase inversion using ionic liquid, [BMIM]SCN, as the solvent. Industrial Engineering and Chemical Research. 2010; 49, 8761–8769.

[46] Ballinas L, Manjarrez L, Hermosillo Valdez R, Gonçalves A, Moraes-Rocha G. González-Sánchez, G. Study of microscopic structure of cellulose and lignin based membranes. Microscopy and Microanalysis 14 (Suppl 2). 2008; 1206–1207 DOI:10.1017/s1431927608085863.

[47] Lia D. Homogeneous preparation of cellulose acetate propionate (CAP) and cellulose acetate butyrate (CAB) from sugarcane bagasse cellulose in ionic liquid. Journal of Agricultural and food Chemistry. 2010; 3, 45–47.

[48] Cao Y, Wu J, Meng T, Zhang J, He J, Li H, Zhang Y. Acetone soluble cellulose acetates prepared by one step homogeneous acetylation of cornhusk cellulose in an ionic liquid 1-allyl-3-methilimidazolium chloride (AmimCl). Carbohydrate Polymers. 2007; 69, 665–672.

Chapter 2

Biopolymer-mediated Green Synthesis of Noble Metal Nanostructures

Olayemi J. Fakayode, Adewale O. Oladipo, Oluwatobi S. Oluwafemi and Sandile P. Songca

Additional information is available at the end of the chapter

http://dx.doi.org/10.5772/62127

Abstract

Polymer-coated noble metal nanoparticles are currently of particular interest to investigators in the fields of nanobiomedicine and fundamental biomaterials. These materials not only exhibit imaging properties in response to stimuli but also efficiently deliver various drugs and therapeutic genes. Even though a large number of polymer-coated noble metal nanoparticles have been fabricated over the past decade, most of these materials still present some challenges emanating from their synthesis. The metal nanoparticles when encapsulated in a polymer and taken up by human cells might show a lower degree of toxicity; however, the degree of toxicity for some of the starting materials and precursors has raised serious concerns. Hence, there is a need to implement the principle of green chemistry in the synthesis of nanomaterials. The use of environmentally benign materials for the synthesis of metal nanoparticles provides numerous benefits ranging from biocompatibility, availability, cost-effectiveness, amenable scale-up to eco-friendliness. The biopolymer-based nanovehicles have been found to be more suitable in the field of nanotechnology owing to their high reproducibility, ease of manufacture, functional modification and safety (they are not carcinogenic). Unlike synthetic polymers where the raw material can be derived from petrochemicals or chemical industrial processes, biopolymers are produced from renewable resources such as plant and/or living organism. They are degradable by natural processes down to elemental entities that can be resorbed in the environment. Furthermore, they can also be modified to serve a particular purpose which explains the myriad of their potential applications. The macromolecular chain of these biopolymers possesses a large number of hydroxyl groups which can easily complex with metal ions. Additionally, these biopolymers also contain supramolecular structures that can lead to new functionalities of their composites with metal and semiconductor nanoparticles. In this chapter, a comprehensive discussion on different biopolymers, green synthesis of noble metal nanostructures, mechanisms, characterization and application in various fields is presented.

Keywords: Biopolymers, Nanoparticles, Green synthesis, Bionanocomposites, Noble metals

1. Introduction

In our modern day, nanotechnology has continued to play a vital role in a plethora of biomedical and biotechnological applications especially in sensing, imaging and treatment of various diseases. A variety of noble metal nanostructures encapsulated or coated with biopolymers have been studied by researchers in the field of nanobiomedicine. This is due to the interesting properties and wide-spectrum application of biopolymer-based nanomaterials. These materials have been shown to combine both the intrinsic features of noble metal nanostructures and the biological features presented by renewable source polymers [1–3]. They have been found not only to exhibit excellent imaging properties but also to efficiently deliver various drugs and therapeutic genes [4–7]. Noble metal nanostructures used in therapy and diagnosis must be non-toxic, stable in biological media and should be specific for the target [8,9]. However, the requirement of these three factors has hindered the use of many noble metal nanostructures in a variety of biomedical applications. Hence, there is a need for the conjugation of these materials with functional biological molecules to necessitate and improve their efficacy for biomedical functionalities. Bionanocomposites exhibit better biomedical values than their naked nanoparticle counterparts. They demonstrate colloidal stability, maintain plasmonic properties and show little or no effect on cell viability in the biological cell system [10,11].

Generally, the synthesis of most noble metal nanostructures requires a high concentration of surfactant, which directs the asymmetric geometry. However, studies have shown that surfactants tend to degrade biological membranes, thus raising significant concern about the cytotoxicity of these materials [12,13]. The cytotoxicity of these surfactant-bound nanostructures can be reduced by minimizing the surfactant concentration below the critical micellar concentration. This reduction is effected at the expense of the stability of the nanomaterials solution, and as a result, their unique optical properties in biological environments are compromised.

There have been calls for the development of nanomaterials based on the principles of green chemistry and, consequently, a myriad of studies have emerged. Some of the proposed solutions are based on the substitution of toxic reagents with more eco-friendly materials. As a result, a lot of research has been done on developing more environmentally benign synthetic methods for noble metal nanomaterials owing to their wide area of applications. In addition, there have been calls for the development of sustainable process and practices in order to define products as green based.

Raveendran and co-workers [14] report three main steps to be considered in the green synthesis of metallic nanoparticles. These are (i) the choice of solvent medium used for the synthesis, (ii) use of environmentally benign reducing agent and (iii) the choice of non-toxic material as stabilizing agent. The use of environmentally benign materials for the synthesis of metal nanoparticles provides numerous benefits ranging from biocompatibility, renewable feedstocks, cost-effectiveness, waste prevention, amenable scale-up, synthetic steps reduction, use of safer solvents and increase of energy efficiency to eco-friendliness [15–17].

1.1. Significance and limitations of nanomaterials

Nanomaterials possess phenomenal ability to create better materials as they are currently being used in numerous products and industrial applications (Figure 1). Most of these materials present evolutionary development of existing technologies. The use of biomaterials such as ascorbic acid, glutathione, sugars, glycerol, orange peel, plant extracts and yeast, as alternative raw materials for synthesizing nanomaterials presents an even more fascinating approach, thus impacting their significance [18,19]. In the area of nanomaterial research, noble metal nanostructures have received tremendous attention due to their wide range of applications in electronics, sensing, catalysis, photonics, environmental clean-up, imaging, water purification, cancer therapy, labelling and drug delivery [20–23]. Most notable among the breakthrough applications of biopolymer-based noble metal nanostructures include: the use of biopolymer-based biodegradable food-packaging materials in the place of non-biodegradable plastics [24], nanocrystal metal–chitosan granular composite material for water purification [25], selective tumour ablation using polymer-coated gold nanorods and metal nanostructure for sensing applications [26–28].

Nevertheless, with noble metal nanostructures having such a phenomenal advantage, there also exist limitations to their application. There have been increasing concerns relating to their toxicity arising from their bioaccumulation in human body. It has been reported that they preferentially accumulate in the liver and spleen [29,30]. As a result, they cause many dysfunctions to these organs causing disruption to their body activities. The nanoparticles size, shape and surface functional groups play a vital role in their toxicity effect. There exist a challenge of conjugating nanoparticles to the surface of metal nanoparticles with negligible or no toxicity effect. As the nanoparticles become conjugated, the surface chemistry, size and shape tend to change, thus influencing their cell and protein interface interactions. This inevitably leads to unspecific and non-selective activities in the body [31].

1.2. Noble metal nanoparticles

Nanoparticles of gold, silver and platinum are classified as noble metal nanostructures. They are nanosized metals with at least one dimension within the nanometre size range of 1–100 nm. Unlike bulk metals, which are typically ductile and possess high thermal and electrical conductivity, metal nanoparticles completely differ in such physical properties. The properties exhibited by metal nanoparticles are completely different from those exhibited by their corresponding bulk metals due to the absence of electron delocalization [32]. Noble metal nanoparticles have very large surface-area-to-volume ratio when compared to their bulk counterparts, thus making them attractive for a wide array of many applications. Their enhanced properties can be tailored by controlling their shape, size and composition [33,34]. A new generation of hybrid nanostructured materials presents an emerging field in nanotechnology. Nanobiocomposite over the past few years has become a term used to designate composites, which contain naturally occurring polymers (biopolymers) and an inorganic structure. The development of nanocomposites comprising a biopolymer matrix with noble metal nanostructures has been extensively studied and considerable efforts are now being directed at biopolymer-based nanocomposites with improved properties [25,26].

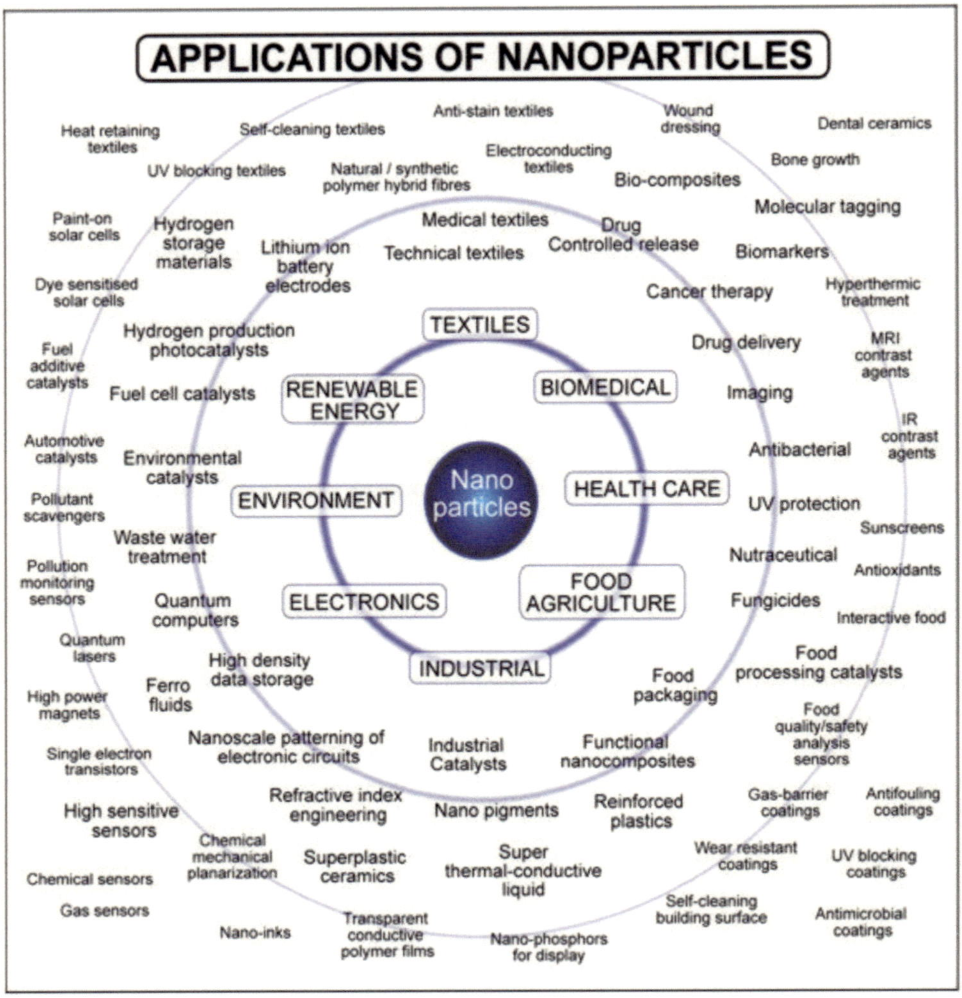

Figure 1. Applications of nanoparticles [134].

2. Synthesis and characterization of noble metal nanoparticles

Nanoparticle synthesis over the past decades has received enormous attention due to their extensive applications. Today, researchers have been able to controllably synthesize noble metal nanoparticles of various shapes and dimensions. These include zero-dimensional, one-dimensional, two-dimensional and three-dimensional nanostructures in a remarkable fashion [35–40]. Noble metal nanoparticles have been synthesized using various methods. Some of these methods include the electrochemical method [41], photochemical method [42], chemical reduction method [33], sonochemical method [43] and the biosynthesis method [44]. A universal reflection on the environmental risks or hazards and sustainability arising from most of these methods other than biosynthetic routes prompted the need for greener approaches in the manufacture of these materials. As a result, considerable attention has been focused on the biosynthesis as the method of choice for the synthesis of these nanoparticles. Metal nanoparticles have been embedded in a host biopolymer matrix in a variety of ways, which include *in situ* chemical synthesis, plasma polymerization combined with metal evaporation, ion implantation and vacuum deposition on viscous flow polymer.

2.1. Synthesis using non-polysaccharide reductants

Interesting biopolymers have been used to synthesize gold, silver and platinum nanoparticles via direct *in situ* reduction [45]. This method enables rapid nucleation and growth of noble metal nanoparticles embedded in a biopolymer matrix using non-polysaccharide reductants. The result is either nanoparticle composites or more complex nanostructures, depending on the reaction conditions used. The non-polysaccharide reductant can be either strong like sodium borohydride or mild like sodium citrate and tetrakis(hydroxymethyl)phosphonium chloride (THPC).

2.1.1. Sodium citrate

The use of sodium citrate in the synthesis of gold and silver nanoparticles is perhaps the first method to be developed [46] and it is still relevant today. It is a simple process for reducing metal salts to nanoparticles with modest monodispersed spherical shape. The particle diameters are usually within the range of 10–20 nm. Colloidal gold and silver nanoparticles are produced by this technique because the citrate ions act as both capping and reducing agents. Following this trend, spherical silver nanoparticles with monodispersed and controlled sizes have also been effectively synthesized. In addition, monodispersed gold nanoparticles with controlled size have also been reported. This is achieved by altering the concentration of gold chloride ions with respect to the citrate [47]. Furthermore, Bastus and co-workers [48] report the enhanced citrate-reducing ability in combination with tannic acid. The tannic acid is instrumental to the formation of the seeds and subsequent high yield of the final spherical particles.

2.1.2. Sodium Borohydride ($NaBH_4$)

The most common chemical reduction method for preparing noble metal nanoparticles involves the reduction of their metal salts or precursor (from M^{+n} to M^0) using sodium borohydride as the reducing agent. Sodium borohydride is a strong reducing agent in organic and inorganic chemistry, and it is used to reduce most transition metal ions to zero-oxidation metal nanoparticle state M(0) in the presence of a colloidal stabilizer [49]. The stabilizing agent prevents the aggregation of the nanoparticles after formation. $NaBH_4$ reduces metal precursors by hydride transfer. However, single-electron transfer is also a possibility, given the electron-rich nature of borohydride anion. Sodium borohydride can also act as bi- or tridentate ligands [50]. It has also been demonstrated that $NaBH_4$ has the ability to control the size and morphology of nanoparticles to a large extent using appropriate synthetic techniques. As a result, nanoparticles with very large surface area relative to volume have become feasible. This is particularly important because the surface plasmon resonance (SPR) band of nanoparticles is a function of morphology [51,52]. The concentration of $NaBH_4$ has been found to affect the size, shape, dispersity and even hydrodynamic diameter of the metal nanoparticles. As the concentration of borohydride increases, the aspect ratio (length divided by width) of the nanorods formed decreases. Moreover, at a certain higher concentration, shape deformation occurs. Also, ageing the sodium borohydride solution prior to use has been found to influence the shape formation of the nanorods [53]. Monodispersed Ag-NPs have also been synthesized using borohydride reduction of silver salt. The borohydride anions are adsorbed onto silver nanoparticles and aggregation is prevented by polyvinyl pyrrolidone (PVP) addition. A large excess of the reductant is also needed to stabilize the nanoparticle product [49] since they are adsorbed onto silver nanoparticles after they are formed.

2.1.3. Tetrakis (Hydroxymethyl) Phosphonium Chloride (THPC)

THPC is an exceptionally strong reducing agent capable of synthesizing smaller gold or silver nanoparticles usually used as seeds for the formation of nanoshells [54] and gold nanoclusters employed for fluorescence imaging [55,56]. Typically, it serves as a good reducing agent in alkaline solution. However, to achieve good nanoparticle stability, a stabilizer is often added alongside with THPC during synthesis. Nonetheless, Hueso and co-workers have reported the use of THPC as both reducing agent and stabilizing ligand in the synthesis of ultra-small noble metal nanoparticles. This is achieved in a single-step room-temperature method with a wide variety of monometallic and bi/tri nano alloys as products [57]

2.2. Synthesis using polysaccharide reductants

Polysaccharides are long polymers of monosaccharide sugars and their derivatives. Unlike proteins or nucleic acids, these polymers can either be straight chained or branched. They can be of one type of monosaccharide (homopolysaccharides), or more than one (heteropolysaccharides) in repeating units. Polysaccharides can also be divided into groups according to their two major functions: contribution to structural components of cells and energy storage. Cellulose is a structural polysaccharide while starch is mainly an energy storage polysaccharide. Starch is the main energy storage in plants while glycogen is the main energy storage in animals. Both starch and cellulose are polymers of glucose but they differ based on the configuration of C–O bonding [58]. They are generally insoluble in cold water, although at around 100 °C, starch is water-soluble while cellulose needs a temperature higher than 300 °C to be water-soluble [59]. Upon hydrolysis with acids or enzymes, they eventually yield their constituent monosaccharide sugars. Polysaccharides have the ability to coordinate metal ions and thus act as stabilizing agents. Furthermore, they have been known to exhibit reducing properties in the synthesis of metal nanostructures. This dual role has enabled the production of metal nanoparticles with improved properties and functionalities such as different shapes and sizes, hydrophilicity, biocompatibility, specificity and non-toxicity. Due to their non-toxic nature, polysaccharides are recognized as green reducing agents for nanostructure synthesis. Their structural hydroxyl groups provide them with strong reducing ability and solubility in water. Other natural polysaccharides such as chitosan have also gained attention in the preparation of nanoparticles [60]. Under this section, we will review the use of starch, cellulose and dextran for the synthesis of noble metal nanoparticles.

2.2.1. *Starch*

Chemically, starch consists of long chains of D-glucose (a monosaccharide) units joined by 1,4-glycosidic bonds. They are formed in plants during photosynthesis. They are present in many plant-based food sources such as root vegetables, for example, potatoes and cereals. The eco-friendly biosynthesis of metal nanoparticles using starch has been well reported. Most of them have been performed through reduction processes. It is a simple method which yields high amount of nanoparticles and to some extent allows for control of particle size. El-Rafie and co-workers [61] report the use of alkali-treated maize starch as both reducing agent and stabilizer for the production of silver nanoparticles. In their report, a redox reaction occurs between $AgNO_3$ and alkali-treated starch to obtain the nanoparticles. The nanosilver obtained is spherical in shape with particle size ranging between 4 and 6 nm. Valencia and co-workers [62] also report the use of starch as reducing agent at 90 °C in the presence of NaOH. Spherical-shaped Ag-NPs with particle diameter between 10 and 30 nm were produced. Furthermore, the role of pH in the green-mediated synthesis of starch-capped Ag-NPs has been investigated. Ag-NPs are synthesized using two green materials: glucose as the reducing agent and starch as the stabilizing agent. The presence of an accelerator (NaOH) is found to affect the size, size distribution and the pH of the solution of the as-synthesized Ag-NPs. At mildly acidic conditions, occurrence of starch hydrolysis affects the particle size. The insufficient protection of the particle surface led to the formation of larger particles but with defined sizes. However, less starch hydrolysis results in nanoparticle aggregation leading to a blue shift and splitting of the spectra [63]. Spherical nanoparticles and nanowires of gold and silver with various sizes and shapes have also been prepared using a monosaccharide (D-glucose) and a polysaccharide (soluble starch) [64].

Gold nanoparticles coated with different sugars are successfully synthesized using a non-toxic, water-soluble phosphine amino acid (THPAL) as a reducing agent. The sugars used are glucose (monosaccharide); sucrose, maltose or lactose (disaccharides); raffinose (trisaccharide) and starch (polysaccharide) [65]. The capping ligand plays a vital role in transforming the shapes of the nanoparticles formed. Platinum nanoparticles (Pt-NPs) with uniform particle size (2–4 nm) have also been synthesized using soluble starch. The soluble starch performs a dual role as both reducing and stabilizing agents. Under alkaline treatment, the degraded intermediate species with reducing potentials (i.e., small molecules of aldehydes and hydroxyl ketones) generated *in situ*, completely reduces the platinum ions. The hydroxyl group in the starch structure complexes with the platinum ions and prevents aggregation or precipitation, thus sufficiently stabilizing the platinum nanoparticles formed [66]. Furthermore, the use of glucose as a reducing agent and starch as stabilizing agent to protect Pt-NPs core in buffer solution have also been reported [67].

2.2.2. *Cellulose*

Cellulose chemically consists of long chains of six-membered ring glucose (a monosaccharide) molecules. The repeating units comprise two anhydroglucose rings $(C_6H_{10}O_5)$ *n*, where *n* = 10,000–15,000. They comprise the fibre and are indigestible by humans but digestible by some other animals [68–70]. Celluloses are ideal materials upon which new biopolymer composites have thrived. For instance, carboxymethyl cellulose (CMC) obtained from cotton have been used in the synthesis of Ag-NPs at different conditions. The degree of polymerization of CMC is shown to have significant impact on the reduction capacity, particle size and stability of the silver nanoparticles formed [71]. Also, synthesis of Ag-NPs has been carried out with cellulose playing a dual role of both reducing and capping agent in aqueous medium. The cellulose is extracted from an environmentally problematic aquatic weed, water hyacinth. Particle shape and size evolutions of the silver nanoparticles are achieved by varying the pH of the solution and reaction time [72]. Metal nanoparticles of gold, silver and platinum have also been prepared using cellulose gel. Filter paper pulp is used to generate the cellulose solution which was allowed to coagulate to form a gel. The nanoparticles of gold, silver and platinum formed are well dispersed and stabilized by the cellulose network. The nanoporous structure of the cellulose hydrogels also prevents aggregation of the nanoparticles formed [73]. Some bacteria are also capable of producing cellulose via digestion. Cellulose of this nature is different from cellulose obtained from trees and cotton because it is free from lignin and hemicellulose. Colloidal silver nanoparticles could be prepared using cellulose of bacterial origin. Bacterial cellulose (BC) is immersed in the silver nitrate solution and reduced by different reducing agents (ascorbic acid, hydrazine and hydroxylamine) under the influence of gelatine and polyvinyl pyrrolidone (PVP), as colloidal stabilizers. The particle size and size distribution of the BC–Ag^0 composite depend on the reductant type and stabilizer used. Best results are obtained for ascorbic acid-reduced/gelatine-protected Ag composite [74].

2.2.3. Dextran

Dextran are water-soluble branched polysaccharides of glucose (dextrose) produced by lactic acid bacterial action on sucrose. The linear chains are linked by α-1,6 glycosidic bond between glucose molecules while the branches begin from α-1,3 linkages. In this method, metal nanostructures are prepared using an environmentally benign solvent and dextran as capping agent, or in some cases dextran serves as both reducing and capping agents. Wang and coworkers [75] report the synthesis of well-dispersed, uniform and biocompatible gold nanoparticles at room temperature using dextran as both reducing and stabilizing agent. The as-synthesized dextran-coated Au-NPs are stable in high ionic strength medium, thereby making them important materials for fabricating novel nanostructures.

Environmentally friendly copolymers of dextran have also been used to synthesize gold-copolymer and silver-copolymer nanoparticles. The simple green synthetic method makes use of graft copolymers, dextran-graft-poly(ε-caprolactone) and dextran-graft-poly(δ-valerolactone) as reductants and stabilizers in the synthesis of the noble metal nanoparticles. The amphiphilic nature of the copolymer is crucial during the process, as this allows interaction of the copolymer with the metal surface and the aqueous medium. The OH end groups of the grafted blocks of copolymers are responsible for the reduction of gold and silver ions [76,77].

2.3. Characterization of polysaccharide-stabilized gold and silver nanoparticles

2.3.1. *UV–Visible spectrophotometry*

UV–Visible spectrophotometry is an essential powerful tool for characterizing the optical properties of various materials at molecular and atomic levels. Typically, gold and silver nanoparticles exhibit a UV–Visible characteristic absorption band peak around 520 and 400 nm, respectively, due to their SPR properties [78–81]. The SPR is due to the collective oscillation of light-excited free-moving conduction band electrons present at the surface of the noble metals. Stabilizing these metals with polysaccharides usually causes a change in the electron density of the metals and thus a shift in the positioning of their SPR peaks. However, the direction and the actual positioning of SPR peak depend on the type of nanoparticle [79,80], particle size and structure of polysaccharide involved [82–84].

2.3.2. *Fourier Transform Infrared Spectroscopy (FTIR)*

FTIR is used to identify the functional groups present in the polysaccharide used to stabilize metal nanoparticles [80]. In conjunction with UV–Visible, dynamic light scattering and Zeta potential data, FTIR can be used to evaluate the extent of polysaccharide stabilization of metal nanoparticles (NPs). Stereotypically, polysaccharide-stabilized gold or silver NPs usually show absorption wavenumber bands for O–H (3500–3200 cm^{-1}: alcohols; 3300–2500 cm^{-1}: acids), C–H (3000–2850 cm^{-1}: saturated alkanes) and C–O (1320–1000 cm^{-1}: alcohols and ethers) groups associated with general carbohydrate structure [79,80,83,85]. To ascertain the effectiveness of stabilization of nanoparticles by the polysaccharide, shifts in the affected groups are usually observed. The hydroxyl, hemiacetal and aldehyde groups play significant roles in the synthesis and stabilization of polysaccharide-capped silver [85,86] and gold nanoparticles [79,83].

2.3.3. *Zeta potential*

The zeta potential measures the overall charge on polymer-stabilized noble metal nanoparticle system [87] as either negative or positive in millivolts. As mentioned earlier, it may be used in conjunction with FTIR to assess the extent of nanoparticle stabilization by polysaccharides. When stabilized with carboxylic acid-terminated polysaccharides, gold and silver NPs are usually surrounded by negative charges. The negatively charged polysaccharide molecules interact with one another via electrostatic repulsion and thus help to prevent agglomeration. Similarly, positively charged polysaccharide-stabilized gold and silver nanoparticles are stabilized via electrostatic repulsion of the polysaccharide's molecules. The higher the magnitude of the zeta potentials, the greater is the extent of charge-to-charge electrostatic repulsion and thus the greater is the stability of nanoparticles [82] and vice versa. The negatively or positively charged polysaccharide-stabilized silver nanoparticles are important for efficient antimicrobial efficacy of silver nanoparticles against gram-positive and gram-negative microbes, respectively [82].

2.3.4. Dynamic Light Scattering (DLS)

The dynamic light scattering, also known as the photon correlation spectroscopy, uses light scattered by moving particles in a dispersible solution system to determine the hydrodynamic size distribution of various particles present in the solution. DLS particle-size estimation is based on measurement of the intensities of scattered lights by various particles present in the solution, provided the particles' movement is Brownian in nature. Therefore, it is used to evaluate the homogeneity of a solution and thus the aggregation of nanoparticle in the solution. For a reliable DLS measurement, a polydispersity index of between 0.05 and 0.7 may be preferable. Also, the refractive indices (n) and absorption values (k) of both the material and solvent must be accurately known for a reliable conversion of intensity data to volume or number size distribution. By default, particle sizes are obtained from the extrapolation from the intensity's data. However, data can be converted to obtain particle sizes relative to volume (size based on concentration of particles) or number of primary particles (size based on percentage number of primary particles) using Mie formula. The most important use of DLS's characterization of polysaccharide-stabilized gold and silver nanoparticles is to evaluate the extent of size characteristics after stabilization. For example, it has been reported that particle size decreases with increasing concentration of polysaccharide [80,83] up to a certain extent and then increases at higher concentration due to aggregation [80]. Thus, DLS can be used to obtain the optimal concentration of the stabilizing agent that favours non-aggregation of particles in a given solution [83]. DLS can also be used to evaluate the effect of the swelling of polymers on the size of nanoparticles when the temperature of the medium is increased.

2.3.5. Transmission Electron Microscopy (TEM)

TEM is a two-dimensional technique used to probe the shape and morphology of a nanoparticle system [83,88]. The shapes of nanoparticles affect the optical, chemical and biological properties of metal nanoparticles. For example, silver nanoparticles have been shown to evolve different shapes as temperature increases when synthesized and stabilized by methyl cellulose [89]. This change in shape morphology of stabilized silver nanoparticles was observed using the TEM technique (Figure 2). Furthermore, TEM can be used to estimate the particle-size distribution of nanoparticle systems [86]. The particle-size data obtained from TEM are usually compared to the information obtained from other particle-size measurements such as DLS and X-ray diffractometry for analytical reliability and accuracy. Furthermore, TEM can be used in conjunction with FTIR, UV–Vis and other probing techniques to elucidate the mechanism of evolution of nanoparticles during the course of the synthesis (Figure 3).

2.3.6. Scanning Electron Microscopy (SEM), Inductively coupled Plasma Spectroscopy (ICP-MS) and powder X-Ray Diffractometry (XRD)

SEM is similar to TEM but differs in its magnifications and image outputs. SEM is a three-dimensional imaging probe, which can be used to visualize the three-dimensional images of stabilized nanoparticles [80,90,91] (Figure 2). ICP-MS is mainly used to determine the concentration of metals in a material. Thus, it can be used to estimate the concentration of gold and silver nanoparticles [92] within the surrounding structure of polysaccharide stabilization. The

http://dx.doi.org/10.5772/62127

amount of nanoparticles present within a stabilizing matrix is essential for such sensitive applications such as environmental sensing, industrial catalysis and biomedical diagnostics and therapies. The powder XRD technique is used to probe the extent of the crystallinity of both unstabilized and stabilized nanoparticles. Particles size can also be estimated using XRD technique. Usually, a face-centred cubic structure is observed for polysaccharide-stabilized gold [79,83,93] and silver nanoparticles [82,90].

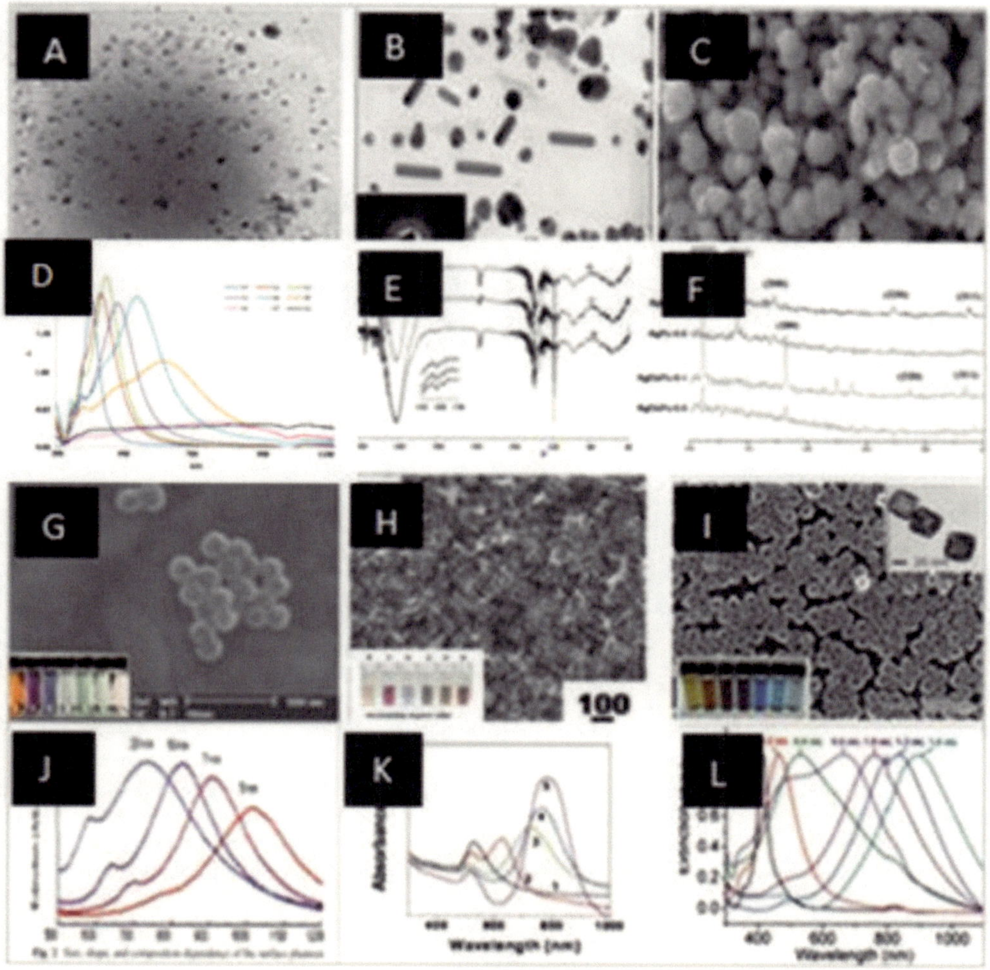

Figure 2. A: TEM image of methyl cellulose-capped silver nanoparticles at 60 °C [89]; B: TEM image of methyl cellulose-capped silver nanoparticles at 90 °C [89]; C: SEM image of starch-stabilized silver nanoparticles [90]; D: UV–Vis for starch-capped silver nanoparticles [90]; E: FTIR of starch-capped silver nanoparticles [90]; F: XRD for pullulan/oxidized pullulan-stabilized silver nanoparticles [82]; Some gold nanostructures (Top): nanoshells (G) [115], nanorods (H) [114] nanocages (I) [91] and their corresponding SPR characteristics (J–L) (Below).

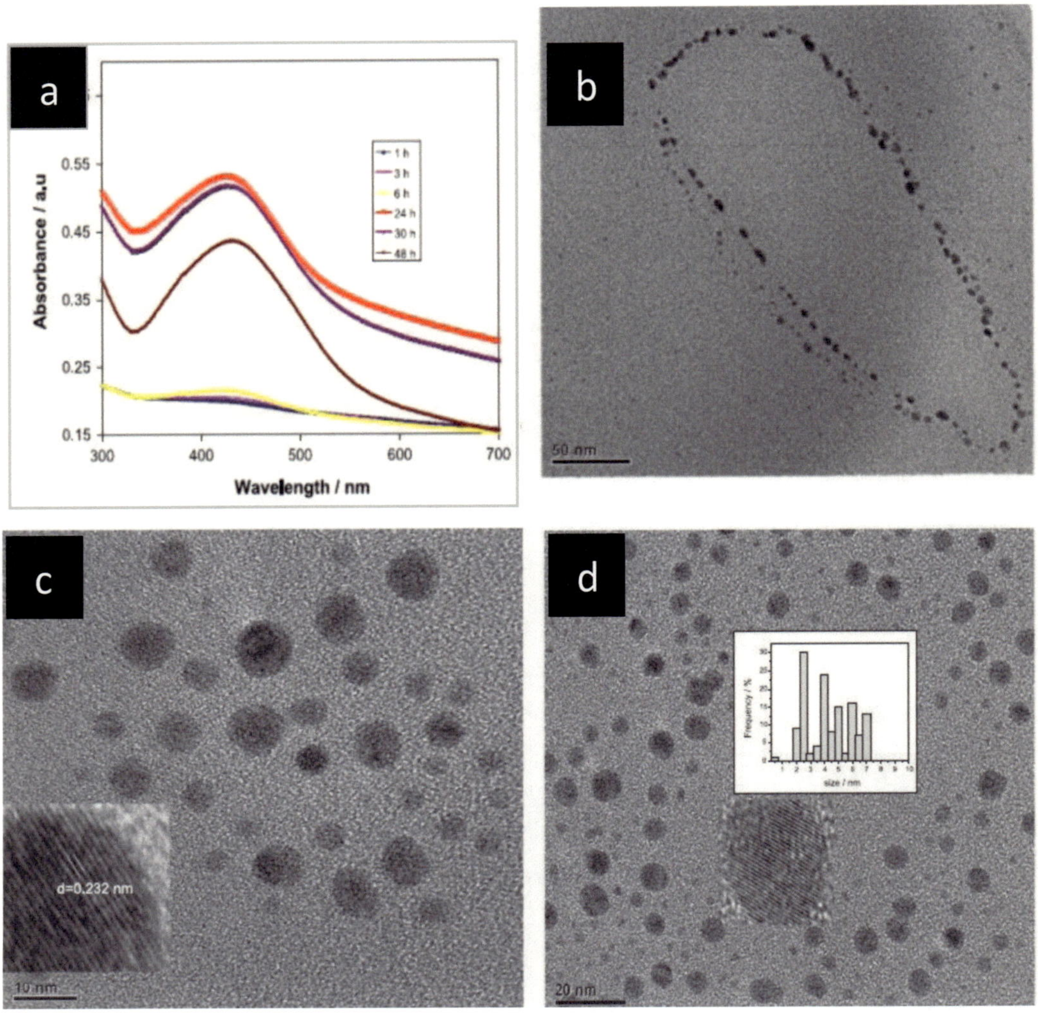

Figure 3. Elucidation of a reaction mechanism using UV–Vis and TEM. a: UV–Vis absorption spectra of Ag-NPs at different reaction times during the synthesis of Ag-NPs. b: TEM image of Ag-NPs (at 1 h reaction time) showing necklace arrangement. c: TEM image at 24 h reaction time, scale bar, 10 nm (inset shows high-resolution image). d: TEM image at 30 h reaction time, scale bar, 20 nm (inset shows high-resolution image and particle-size distribution) [81].

3. Biomedical applications of polysaccharide-stabilized gold and silver nanoparticles

3.1. Gold nanoparticles

Gold nanoparticles are essential nanomaterials for many biomedical applications such as targeted biological sensing [94], site-specific drug delivery, diagnostic imaging [84,95,96] and photothermal therapy [95,97]. They are essentially utilized for these applications due to their unique SPR property, inertness [84,96], non-toxicity, enhanced permeability retention effect, conjugatable surfaces for the attachment of targeting [98] and therapeutic agents [99], near-infrared light absorption [97], high light absorption [95] and scattering properties and generation of therapeutic heat after light irradiation [95]. Recent advances in the use of gold nanoparticles have focused on the use of polysaccharide-stabilized gold nanoparticles. One of the reasons for this is the better efficiency of polysaccharides in synthesizing and stabilizing gold nanoparticles compared to other stabilizing agents such as oligosaccharides or mono-saccharides [100]. Other reasons are their low toxicity, easily conjugatable functional groups [101] and good biocompatibility [102,103]. Some of the recent biomedical applications of gold nanoparticles are provided in the next section.

3.1.1. *Targeted plasmonic biosensing detection*

Various structures of gold nanoparticles such as the nanospheres, nanorods and nanocages have been shown to exhibit different SPR characteristics [78,103], which can be used to detect biomarkers [104] and pathogens in body fluid samples. In targeted plasmonic-assisted biosensing, stabilized gold nanoparticles are conjugated to some specific biomolecule targeting agents such as carbohydrates [102,105] peptides [106], antibodies and oligonucleotides [102] which react specifically with some disease biomarkers or microbes. This interaction often leads to the aggregation of gold nanoparticles [104] which causes shifting of the surface plasmon resonance peak. This shift effect can be detected by visual colour change, spectrophotometric measurement or dynamic light scattering techniques. Targeted plasmonic biosensing could be successfully employed for the rapid improvement of various diagnostic techniques such as homophase immunoassay, DNA assay and enzymatic assay [104].

3.1.2. *Targeted bioimaging enhancement*

Gold nanoparticles are plasmonic entities which absorb and scatter light from the visible to near-infrared optical region where tissues are relatively transparent to light [106,107]. These properties enable them to be used as contrasting agents in many biomedical imaging applications such as magnetic resonance imaging and computed tomography [108]. When excited with near-infrared light, gold nanoparticles absorb and scatter light more than the surrounding tissues leading to a better detection of their sites of localization. In targeted magnetic resonance imaging (MRI) technique, stabilized gold NPs are conjugated to an active targeting material such as a specific site-targeting carbohydrate [105], peptides [106,109,110], aptamers [108] or external magnetic field-driven iron oxide nanoparticles (IONPs) [111] in order to achieve active site-targeting, detection and better imaging. Superparamagnetic iron oxide nanoparticles (SPIONs) are excellent MRI contrast agents [112,113]. Thus, utilizing stabilized gold-SPION hybrids, such as SPION-gold core–shell, promotes greater improvement of the MRI technique. Furthermore, gold NPs may be coupled with conventional imaging dyes or contrasting agents to improve their imaging efficacies.

3.1.3. Targeted fluorescence detection

In targeted fluorescence detection, stabilized gold nanoparticles are either bonded directly [98] or conjugated to a fluorescence molecule [95] bonded to an active site-targeting biomolecule [110]. The technique makes use of (i) the synergetic properties of the selective localization of the gold nanoparticles in disease tissues, (ii) interaction of the active targeting biomolecules with their corresponding disease tissue receptors and (iii) emitted fluorescence light from the gold nanoparticles [98] or conjugated fluorophore to improve detection of disease tissues [110]. For diagnosis purposes, the active site targeting gold nanoparticle solution may be injected via intravenous means and the body exposed to the fluorescence spectroscopy imaging technique. The diseased tissues containing the gold nanoparticle fluorophore conjugates will emit light of specific wavelength while the healthy tissue remains non-fluorescent [95,110].

3.1.4. Targeted Photothermal Therapy (PTT)

Targeted PTT involves the use of active site targeting stabilized gold nanoparticles (Au-NPs) for specific tumour cell destructions. The Au-NPs generate therapeutic heat after irradiation with light of appropriate wavelength and dose. The site targeting agent interacts with the tumour cell surfaces in order to pave way for the nanoparticles to gain entrance into the cell where the therapeutic function will take place. Especially suitable for this purpose are the gold nanorods [78,114] and nanoshells [78,115] whose aspect ratio and the ratio of the shell to core diameter can be tuned to absorb light in the near-infrared regions where biological tissues are transparent to light. Gold nanorods have been extensively applied for the eradication of cancer cells via PTT [102]. Other gold nanoparticle structures that can be used for PTT are gold nanocages [78,91]. However, all gold nanostructures exhibit different PTT efficacies due to their different structures and surface plasmon resonance properties.

3.2. Silver nanoparticles

3.2.1. Diagnostic enhancement

Stabilized silver nanoparticles (Ag-NPs), being a plasmonic entity also exhibit biological imaging characteristic [96,116] and plasmonic and fluorescence detection of biomolecules [116,117,118] and toxic heavy metals [119,120]. Like gold nanoparticles, it could be used to improve image contrast of disease tissues when functionalized or employed as nanocarriers of specific site-targeting agents such as peptides, antibodies and carbohydrates which show some specific affinity towards certain molecules present in disease tissues. As nanocarriers of these biomolecules, silver nanoparticles may cause enhanced biological image contrast via their light absorption and scattering properties. However, the choice of silver nanoparticles for fluorescence imaging function may be more for silver nanoclusters (Ag-NCs) compared to other silver morphologies due to their small size-dependent fluorescence effect. It has been reported that as the size of silver nanoparticles approaches the dimension of the Fermi wavelength of an electron, they emit fluorescence light which can be tuned from the visible to near-infrared region after suitable light irradiations [116]. Moreover, Ag-NCs have also been found to be useful as an excellent nanobiosensor agent for the detection of biotoxins [118] and toxic inorganic heavy metals [120]. Nonetheless, like gold nanoparticles, the interaction of functionalized, stabilized silver nanoparticles with disease biomarkers such as proteins may cause aggregation and thus a shift in the position of their SPR band which may be used to detect these biomarkers.

3.2.2. *Targeted antimicrobial activity*

Silver nanoparticles are especially known for their antibacterial activities [90,121–125]. However, in recent years, reports about the antiviral effects of silver nanoparticles are beginning to emerge [126–128]. In their study, Gusseme and co-workers [128] employed biogenic silver nanoparticles to remove murine norovirus 1 (MNV-1) from drinking water. The biogenic silver was mixed with a water filter to effect the viral removal. Also, the use of silver nanoparticles in preventing the binding of HIV-1 to human host cells via binding of nanoparticles to the virus' glycoproteins *in vitro* has been reported [129]. The sulphur group of the virus' gp120 glycoprotein has been suggested to be the binding target site [129,130]. Furthermore, the applications of silver nanoparticles such as antifungal [131] and antiplasmodium agents [132] have also been reported.

3.2.3. *Anti-angiogenesis activity*

Silver nanoparticles could also function as anti-angiogenesis agents by preventing the formation of new blood vessels responsible for the development of many human diseases such as cancer, macular degeneration and inflammatory diseases [133]. In their study, Gurunathan and co-workers [133] suggested that silver nanoparticles readily interfere with the angiogenesis pathway by inhibiting the activation of p13k/akt.

4. Conclusion

Biopolymers have continued to revolutionize the advancements of green nanotechnology. Nanoscale derivatives of polymers like starch, cellulose and dextran can be synthesized on a large scale and can be used for the manufacture of bionanocomposites. They present promising alternatives to the environmental challenging non-biodegradable plastics via industrially viable process. Most of these biopolymers also offer a route for completely green synthesis of several noble metal nanoparticles performing the dual role of stabilizing and reducing agents in aqueous medium. Biopolymer-mediated synthesis offers numerous advantages which include efficient solubility in water, non-cytotoxicity and biocompatibility in biomedical applications. Polysaccharide-capped silver and gold nanoparticles are emerging as excellent nanomaterials in many fields, including biomedical and environmental. This is due to the result of the excellent functionalities inferred on them by the polysaccharide molecules. These properties include long-term stability and high solubility in aqueous solution, specific biomolecule targeting which induces specific therapeutic functions and toxicity reduction.

Author details

Olayemi J. Fakayode[1,2], Adewale O. Oladipo[1,2], Oluwatobi S. Oluwafemi[1,2*] and Sandile P. Songca[3]

*Address all correspondence to: oluwafemi.oluwatobi@gmail.com

1 Department of Applied Chemistry, University of Johannesburg, Doornfontein, Johannesburg, South Africa

2 Centre for Nanomaterials Science Research, University of Johannesburg, Johannesburg, South Africa

3 Department of Chemistry, Walter Sisulu University, Private bag X1, Mthatha, South Africa

References

[1] Djoković V, Božanic DK, Vodnik VV., Krsmanović RM, Trandafilovic LV., Dimitrijević-Branković S. Structure and optical properties of noble-metal and oxide nanoparticles dispersed in various polysaccharide biopolymers. Phys Chem Interfaces Nanomater. 2011;18(1):8098–98.

[2] Dos Santos DS, Goulet PJG, Pieczonka NPW, Oliveira ON, Aroca RF. Gold nanoparticle embedded, self-sustained chitosan films as substrates for surface-enhanced Raman scattering. Langmuir. 2004;20(23):10273–7.

[3] Porel S, Venkatram N, Rao DN, Radhakrishnan TP. In situ synthesis of metal nanoparticles in polymer matrix and their optical limiting applications. J Nanosci Nanotechnol. 2007;7(6):1887–92.

[4] B. Goudoulas T. Polymers and biopolymers as drug delivery systems in nanomedicine. Recent Pat Nanomed. 2012;2(1):52–61.

[5] Nitta SK, Numata K. Biopolymer-based nanoparticles for drug/gene delivery and tissue engineering. Int J Mol Sci. 2013;14(1):1629–54.

[6] Liu Z, Jiao Y, Wang Y, Zhou C, Zhang Z. Polysaccharides-based nanoparticles as drug delivery systems. Adv Drug Deliv Rev. Elsevier B.V.2008;60(15):1650–62.

[7] Zhang L, Chan JM, Gu FX, Rhee JW, Wang AZ, Radovic-Moreno AF, et al. Self-assembled lipid-polymer hybrid nanoparticles. A robust drug delivery platform. ACS Nano. 2008;2(8):1696–702.

[8] Johnston H, Brown D, Kermanizadeh A, Gubbins E, Stone V. Investigating the relationship between nanomaterial hazard and physicochemical properties: Informing

the exploitation of nanomaterials within therapeutic and diagnostic applications. J Control Release. Elsevier B.V. 2012;164(3):307–13.

[9] Farokhzad OC, Langer R. Impact of nanotechnology on drug delivery. ACS Nano. 2009;3(1):16–20.

[10] Tebbe M, Kuttner C, Männel M, Fery A, Chanana M. Colloidally stable and surfactant-free protein-coated gold nanorods in biological media. Appl Mater Interfaces. 2015;7:5984–91.

[11] Li Z, Huang P, Zhang X, Lin J, Yang S, Liu B, et al. RGD-conjugated dendrimer-modified gold nanorods for in vivo tumor targeting and photothermal therapy. Mol Pharm. 2010;7(1):94–104.

[12] Alkilany AM, Nagaria PK, Hexel CR, Shaw TJ, Murphy CJ, Wyatt MD. Cellular uptake and cytotoxicity of gold nanorods: Molecular origin of cytotoxicity and surface effects. Small. 2009;5:701–8.

[13] Lewinski N, Colvin V, Drezek R. Cytotoxicity of nanoparticles. Small. 2008;4(1):26–49.

[14] Raveendran P, Fu J, Wallen SL. Completely "green" synthesis and stabilization of metal nanoparticles. J Am Chem Soc. 2003;125(46):13940–1.

[15] Wurm FR, Weiss CK. Nanoparticles from renewable polymers. Front Chem. 2014;2(July):49.

[16] Rajan M, Raj V. Potential drug delivery applications of chitosan based nanomaterials. Int Rev Chem Eng. 2013;5(2):145–55.

[17] Wu G, Li P, Feng H, Zhang X, Chu PK. Engineering and functionalization of biomaterials via surface modification. J Mater Chem B. 2015;3(10):2024–42.

[18] Cinelli M, Coles SR, Nadagouda MN, Błaszczyński J, Słowiński R, Varma RS, et al. A green chemistry-based classification model for the synthesis of silver nanoparticles. Green Chem. 2015;17(5):2825–39.

[19] Cabrera FC, Mohan H, dos Santos RJ, Agostini DLS, Aroca RF, Rodríguez-Pérez MA, et al. Green synthesis of gold nanoparticles with self-sustained natural rubber membranes. J Nanomater. 2013; 1–10.

[20] Huang X, Neretina S, El-Sayed MA. Gold nanorods: From synthesis and properties to biological and biomedical applications. Adv Mater. 2009; 21(48):4880–910.

[21] Nath D, Banerjee P. Green nanotechnology - a new hope for medical biology. Environ Toxicol Pharmacol. Elsevier B.V. 2013;36(3):997–1014.

[22] Xie J, Zhang Q, Lee JY, Wang DIC. The synthesis of SERS-active gold nanoflower tags for in vivo applications. ACS Nano. 2008;2:2473–80.

[23] Sahadev A Shankarappa, Manzoor Koyakutty, Shantikumar V Nair. Efficacy versus toxicity - The ying and yang in translating nanomedicines. Nanomater Nanotechnol. 2014;4(23): 1–9.

[24] Tang XZ, Kumar P, Alavi S, Sandeep KP. Recent advances in biopolymers and biopolymer-based nanocomposites for food packaging materials. Crit Rev Food Sci Nutr. 2012;52(5):426–42.

[25] Sankar MU, Aigal S, Maliyekkal SM, Chaudhary A, Anshup, Kumar AA, et al. Biopolymer-reinforced synthetic granular nanocomposites for affordable point-of-use water purification. Proc Natl Acad Sci USA. 2013;110 (21):8459–64.

[26] Csáki A, Thiele M, Jatschka J, Dathe A, Zopf D, Stranik O, et al. Plasmonic nanoparticle synthesis and bioconjugation for bioanalytical sensing. Eng Life Sci. 2015; 15(3): 266-275.

[27] Joy NA., Janiszewski BK, Novak S, Johnson TW, Oh SH, Raghunathan A, et al. Thermal stability of gold nanorods for high-temperature plasmonic sensing. J Phys Chem C. 2013;117(22):11718–24.

[28] Sung HK, Oh SY, Park C, Kim Y. Colorimetric detection of Co^{2+} ion using silver nanoparticles with spherical, plate, and rod shapes. Langmuir. 2013;29(28):8978–82.

[29] Siemianowicz K, Likus W, Markowski J. Medical aspects of nanomaterial toxicity. In: Soloneski S, Larramendy ML, editors. Nanomaterials - Toxicity and Risk Assessment. InTech; 2015. p.161–75

[30] González-muñoz M, Díez P, González-gonzález M, Dégano RM, Ibarrola N, Orfao A, et al. Evaluation strategies of nanomaterials toxicity. In: Soloneski S, Larramendy ML, editors. Nanomaterials - Toxicity and Risk Assessment. InTech; 2015. p. 23–37.

[31] Clichici S, Filip A. In vivo assessment of nanomaterials toxicity. In: Soloneski S, Larramendy ML, editors. Nanomaterials - Toxicity and Risk Assessment. InTech; 2015. p. 93–121

[32] Campelo JM, Luna D, Luque R, Marinas JM, Romero AA. Sustainable preparation of supported metal nanoparticles and their applications in catalysis. Chem Sus Chem. 2009;2(1):18–45.

[33] Niu W, Zhang L, Xu G. Seed-mediated growth of noble metal nanocrystals: crystal growth and shape control. Nanoscale. 2013;5(8):3172–81.

[34] Burda C, Chen X, Narayanan R, El-Sayed M A. Chemistry and properties of nanocrystals of different shapes. Chem Rev. 2005;1025–102.

[35] Babu VJ, Vempati S, Uyar T, Ramakrishna S. Review of one-dimensional and two-dimensional nanostructured materials for hydrogen generation. Phys Chem Chem Phy. 2015;17(5):2960–86.

[36] Tiwari JN, Tiwari RN, Kim KS. Zero-dimensional, one-dimensional, two-dimensional and three-dimensional nanostructured materials for advanced electrochemical energy devices. Prog Mater Sci. 2012;57(4):724–803.

[37] Murphy CJ, Gole AM, Hunyadi SE, Orendorff CJ. One-dimensional colloidal gold and silver nanostructures. Inorg Chem. 2006;45(19):7544–54.

[38] Song Y, Yang Y, Medforth CJ, Pereira E, Singh AK, Xu H, et al. Controlled synthesis of 2-D and 3-D dendritic platinum nanostructures. J Am Chem Soc. 2004;126(2):635–45.

[39] Zhang Y, Grady NK, Ayala-Orozco C, Halas NJ. Three-dimensional nanostructures as highly efficient generators of second harmonic light. Nano Lett. 2011;11(12):5519–23.

[40] Kuchibhatla SVNT, Karakoti AS, Bera D, Seal S. One dimensional nanostructured materials. Prog Mater Sci. 2007;52(5):699–913.

[41] Mohl M, Kumar A, Mohana Reddy AL, Kukovecz A, Konya Z, Kiricsi I, et al. Synthesis of catalytic porous metallic nanorods by galvanic exchange reaction. J Phys Chem C. 2010;114:389–93.

[42] Abdelrasoul GN, Cingolani R, Diaspro A, Athanassiou A, Pignatelli F. Photochemical synthesis: Effect of UV irradiation on gold nanorods morphology. J Photochem Photobiol A Chem. 2014;275:7–11.

[43] Zhu Y, Wang X, Guo W, Wang J, Wang C. Sonochemical synthesis of silver nanorods by reduction of silver nitrate in aqueous solution. Ultrason Sonochem. 2010;17(4):675–9.

[44] Sathishkumar M, Pavagadhi S, Mahadevan A, Balasubramanian R. Biosynthesis of gold nanoparticles and related cytotoxicity evaluation using A549 cells. Ecotoxicol Environ Saf. 2014;114:232–40.

[45] Ayyad O, Muñoz-Rojas D, Oró-Solé J, Gómez-Romero P. From silver nanoparticles to nanostructures through matrix chemistry. J Nanopart Res. 2010;12(1):337–45.

[46] Turkevich J, Stevenson PC, Hillier J. A study of the nucleation and growth processes in the synthesis of colloidal gold. Discuss Faraday Soc. 1951;11(c):55–75.

[47] Zhao L, Jiang D, Cai Y, Ji X, Xie R, Yang W. Tuning the size of gold nanoparticles in the citrate reduction by chloride ions. Nanoscale. 2012;4(16):5071.

[48] Bastús NG, Merkoçi F, Piella J, Puntes V. Synthesis of highly monodisperse citrate-stabilized silver nanoparticles of up to 200 nm: Kinetic control and catalytic properties. Chem Mater. 2014;26(9):2836–46.

[49] Mavani K, Shah M. Synthesis of silver nanoparticles by using sodium borohydride as a reducing agent. Int J Eng Res Technol. 2013;2(3):1–5.

[50] Deraedt C, Salmon L, Gatard S, Ciganda R, Hernandez R, Ruiz J, et al. Sodium borohydride stabilizes very active gold nanoparticle catalysts. Chem Commun. Royal Society of Chemistry. 2014;50(91):14194–6.

[51] El-Sayed MA. Small Is different: shape-, composition-dependent properties of some colloidal semiconductor nanocrystals. Acc. Chem. Res., 2004; 37(5):326–333.

[52] Lee KS, El-Sayed MA. Dependence of the enhanced optical scattering efficiency relative to that of absorption for gold metal nanorods on aspect ratio, size, end-cap shape, and medium refractive index. J Phys Chem B. 2005;109(43):20331–8.

[53] Samal AK, Sreeprasad TS, Pradeep T. Investigation of the role of $NaBH_4$ in the chemical synthesis of gold nanorods. J Nanopart Res. 2009;12(5):1777–86.

[54] Yong KT, Sahoo Y, Swihart MT, Prasad PN. Synthesis and plasmonic properties of silver and gold nanoshells on polystyrene cores of different size and of gold-silver core-shell nanostructures. Colloids Surf A Physicochem Eng Asp. 2006;290:89–105.

[55] Zhang Z, Xu L, Li H, Kong J. Wavelength-tunable luminescent gold nanoparticles generated by cooperation ligand exchange and their potential application in cellular imaging. RSC Adv. 2013;3(1):59–63.

[56] Kumar R, Korideck H, Ngwa W, Berbeco RI, Makrigiorgos GM, Sridhar S. Third generation gold nanoplatform optimized for radiation therapy. Transl Cancer Res. 2013;2(4):228–39.

[57] Hueso JL, Sebastián V, Mayoral Á, Usón L, Arruebo M, Santamaría J. Beyond gold: rediscovering tetrakis-(hydroxymethyl)-phosphonium chloride (THPC) as an effective agent for the synthesis of ultra-small noble metal nanoparticles and Pt-containing nanoalloys. RSC Adv. 2013;3(26):10427.

[58] Atsushi F, Dhepe PL. Sustainable green catalysis by supported metal nanoparticles. Chem Rec. 2009;9(4):224–35.

[59] Sasaki C, Sumimoto K, Asada C, Nakamura Y. Direct hydrolysis of cellulose to glucose using ultra-high temperature and pressure steam explosion. Carbohydr Polym. 2012;89(1):298–301.

[60] Venkatesham M, Ayodhya D, Madhusudhan A, Veera Babu N, Veerabhadram G. A novel green one-step synthesis of silver nanoparticles using chitosan: catalytic activity and antimicrobial studies. Appl Nanosci. 2012;4(1):113–9.

[61] Ahmed HB, Zahran MK. Facile precursor for synthesis of silver nanoparticles using alkali treated maize starch. Int Sch Res Not. 2014:1–12.

[62] Ayala Valencia G, Cristina de Oliveira Vercik L, Ferrari R, Vercik A. Synthesis and characterization of silver nanoparticles using water-soluble starch and its antibacterial activity on Staphylococcus aureus. Starch - Stärke. 2013;65(11–12):931–7.

[63] Singh M, Sinha I, Mandal RK. Role of pH in the green synthesis of silver nanoparticles. Mater Lett. 2009;63(3–4):425–7.

[64] Shervani Z, Yamamoto Y. Carbohydrate-directed synthesis of silver and gold nanoparticles: Effect of the structure of carbohydrates and reducing agents on the size and morphology of the composites. Carbohydr Res. 2011;346(5):651–8.

[65] Katti KK, Kattumuri V, Bhaskaran S, Katti KV, Kannan R. Facile and general method for synthesis of sugar coated gold nanoparticles. Int J Green Nanotechnol Biomed. 2012;29(6):997–1003.

[66] Tongsakul D, Wongravee K, Thammacharoen C, Ekgasit S. Enhancement of the reduction efficiency of soluble starch for platinum nanoparticles synthesis. Carbohydr Res. 2012;357:90–7.

[67] Engelbrekt C, SØrensen KH, Lübcke T, Zhang J, Li Q, Pan C, et al. 1.7 nm platinum nanoparticles: Synthesis with glucose starch, characterization and catalysis. Chem Phys Chem. 2010;11(13):2844–53.

[68] Habibi Y, Lucia LA, Rojas OJ. Cellulose nanocrystals: chemistry, self-assembly, and applications. Chem Rev. 2010;110:3479–500.

[69] Moon RJ, Martini A, Nairn J, Simonsen J, Youngblood J. Cellulose nanomaterials review: structure, properties and nanocomposites. Chem Soc Rev. 2011;40(7):3941–94.

[70] Siqueira G, Bras J, Dufresne A. Cellulosic bionanocomposites: A review of preparation, properties and applications. Polym (Basel). 2010;2(4):728–65.

[71] Hebeish AA., El-Rafie MH, Abdel-Mohdy FA, Abdel-Halim ES, Emam HE. Carboxymethyl cellulose for green synthesis and stabilization of silver nanoparticles. Carbohydr Polym. 2010;82(3):933–41.

[72] Mochochoko T, Oluwafemi OS, Jumbam DN, Songca SP. Green synthesis of silver nanoparticles using cellulose extracted from an aquatic weed; water hyacinth. Carbohydr Polym. 2013;98(1):290–4.

[73] Cai J, Kimura S, Wada M, Kuga S. Nanoporous cellulose as metal nanoparticles support. Biomacromolecules. 2009;10(1):87–94.

[74] De Santa Maria LC, Santos ALC, Oliveira PC, Barud HS, Messaddeq Y, Ribeiro SJL. Synthesis and characterization of silver nanoparticles impregnated into bacterial cellulose. Mater Lett. 2009;63(9–10):797–9.

[75] Wang Y, Zhan L, Huang CZ. One-pot preparation of dextran-capped gold nanoparticles at room temperature and colorimetric detection of dihydralazine sulfate in uric samples. Anal Methods. 2010;2(12):1982.

[76] Saldias C, Leiva A, Valderas J, Radić D. Environmentally friendly synthesis of noble metal nanoparticles assisted by biodegradable dextran-graft-lactone copolymers. Polym Adv Technol. 2014;25(4):372–9.

[77] Bezugly M, Kutsevol N, Rawiso M, Upr C, Cedex S, Bezugla T. Water-soluble branched copolymers dextran-polyacrylamide and their anionic derivates as matrices for metal nanoparticles in situ synthesis. Chemik. 2012;66(8):862–7.

[78] Huang X, El-Sayed, MA. Plasmonic photo-thermal therapy (PPTT). Alexandria J Med. 2011;47(1):1–9.

[79] Hussain ST, Iqbal M, Mazhar M. Size control synthesis of starch capped-gold nanoparticles. 2009;11:1383–91.

[80] Mandal A, Sekar S, Chandrasekara, N. Vibrational spectroscopic investigation on interaction of sago starch capped silver nanoparticles with collagen: a comparative physicochemical study using FT-IR and FT-Raman. RSC Adv. 2015;5:15763–71.

[81] Oluwafemi SO, Vuyelwa N, Scriba M, Songca, SP. Green controlled synthesis of monodispersed, stable and smaller sized starch-capped silver nanoparticles. Mater Lett. 2013;106:332–6.

[82] Coseri S, Spatareanu A, Sacarescu L, Rimbu C, Suteu D, Spirk S, Harabagiu V. Green synthesis of the silver nanoparticles mediated by pullulan. Carbohydr Polym. 2015;116:9–17.

[83] Tagad CK, Rajdeo KS, Kulkarni A, More P, Aiyer RC, Sabharwal, S. Green synthesis of polysaccharide stabilized gold nanoparticles: chemo catalytic and room temperature operable vapor sensing application. RSC Adv., 2014; 4, 24014–24019.

[84] Vijayakumar S, Ganesan, S. In vitro cytotoxicity assay on gold nanoparticles with different stabilizing agents. J Nanomater. 2012:1–9.

[85] Mochochoko T, Oluwafemi OS, Jumbam DN, Songca SP. Green synthesis of silver nanoparticles using cellulose extracted from an aquatic weed; water hyacinth. Carbohydr Polym. 2013;98(1):290–4.

[86] El-Sheikh MA, El-Rafie SM, Abdel-Halim ES, El-Rafie MH. Green synthesis of hydroxyethyl cellulose-stabilized silver nanoparticles. J Polym. 2013; 1–11.

[87] Gupta NR, Prasad BLV, Gopinath CS, Badiger, MV. A nanocomposite of silver and thermo-associating polymer by a green route: a potential soft–hard material for controlled drug release. RSC Adv. 2014;4:10261–8.

[88] Aramwit P, Bang N, Ratanavaraporn J, Ekgasit, S. Green synthesis of silk sericin-capped silver nanoparticles and their potent anti-bacterial activity. Nanoscale Res Lett. 2014;9(1):1–7.

[89] Bhui DK, Pyne S, Sarkar P, Bar H, Sahoo GP, Misra, A. Temperature controlled synthesis of silver nanostructures of variable morphologies in aqueous methyl cellulose matrix. J Mol Liq. 2011;158(3):170–4.

[90] Cheng F, Betts JW, Kelly SM, Hector, AL. Green synthesis of highly concentrated aqueous colloidal solutions of large starch-stabilised silver nanoplatelets. Mat Sci Eng C. 2015;46:530–7.

[91] Hu M, Chen J, Li ZY, Au L, Hartland GV, Li X, Marquez M, Xia Y. Gold nanostructures: engineering their plasmonic properties for biomedical applications. Chem Soc Rev. 2006;35(11):1084–94.

[92] Fabricius AL, Duester L, Meermann B, Ternes TA. ICP-MS-based characterization of inorganic nanoparticles—sample preparation and off-line fractionation strategies. Anal Bioanal Chem. 2014; 406(2):467–79.

[93] Hien NQ, Phu DV, Duy NN, Quoc LA. Radiation synthesis and characterization of hyaluronan capped gold nanoparticles. Carbohydr Polym. 2012;89(2):537–41.

[94] Khlebtsov N, Bogatyrev V, Dykman L, Khlebtsov B, Staroverov S. Analytical and theranostic applications of gold nanoparticles and multifunctional nanocomposites. Theranostics. 2013;3(3):167–80.

[95] Jang B, Park J, Tung C, Kim I, Choi Y. Gold nanorod-photosensitizer complex for near-infrared fluorescence imaging and photodynamic/photothermal therapy in vivo. ACS nano. 2011; 5(2):1086–94.

[96] Wolfbeis OS. An overview of nanoparticles commonly used in fluorescent bioimaging. Chem. Soc. Rev., 2015: 44, 4743-4768.

[97] Huang X, Jain PK, El-Sayed IH, El-Sayed MA. Plasmonic photothermal therapy (PPTT) using gold nanoparticles. Lasers Med Sci. 2008;23(3):217–28.

[98] He H, Xie C, Ren J. Nonbleaching fluorescence of gold nanoparticles and its applications in cancer cell imaging. Anal Chem. 2008; 80(15):5951–7.

[99] Lee HJ, Liu Y, Zhao J, Zhou M, Bouchard RR, Mitcham T, Wallace M, Stafford RJ, Li C, Gupta S, and Melancon MP. In vitro and in vivo mapping of drug release after laser ablation thermal therapy with doxorubicin-loaded hollow gold nanoshells using fluorescence and photoacoustic imaging. J Control Release. 2013;172(1):1–17.

[100] Kavita KK, Vijaya K, Sharanya B, Kattesh VK, Kannan, R. Facile and general method for synthesis of sugar coated gold nanoparticles. Int J Green Nanotechnol and Biomed. 2012;29(6):997–1003.

[101] Sathiyanarayanan G, Vignesh V, Saibaba G, Vinothkanna A, Dineshkumar K, Viswanathana MB, Selvin J. Synthesis of carbohydrate polymer encrusted gold nanoparticles using bacterial exopolysaccharide: a novel and greener approach. RSC Adv. 2014;4:22817–27.

[102] Locatelli E, Monaco I, Franchini MC. Surface modifications of gold nanorods for applications in nanomedicine. RSC Adv. 2015;5:21681–99.

[103] Biju V. Chemical modifications and bioconjugate reactions of nanomaterials for sensing, imaging, drug delivery and therapy. Chem Soc Rev. 2014;43(3):737–962.

[104] Valentini P, Pompa PP. Gold nanoparticles for naked-eye DNA detection: smart. RSC Adv. 2013; 3:19181–90.

[105] Adak AK, Lin H, Lin C, Lin H. Biomolecular chemistry studying carbohydrate – protein interactions. Org Biomol Chem. 2014;12:5563–73.

[106] Oyelere AK, Chen PC, Huang X, El-Sayed IH, El-Sayed MA. Peptide-conjugated gold nanorods for nuclear targeting. Bioconjug Chem. 2007;18(5):1490–7.

[107] Zhou F, Xing D, Ou Z, Wu B., Resasco DE, Chen WR. Cancer photothermal therapy in the near-infrared region by using single-walled carbon nanotubes. J Biomed Opt. 2009;14(2):021009.

[108] Kim D, Jeong YY, Jon S. A drug-loaded aptamer gold nanoparticle bioconjugate for combined CT imaging and therapy of prostate cancer. ACS Nano. 2010;4(7):3689–96.

[109] Chen WH, Chen JX, Cheng H, Chen CS, Yang J, Xu XD, Wang Y, Zhuo RX, Zhang XZ. A new anti-cancer strategy of damaging mitochondria by pro-apoptotic peptide functionalized gold nanoparticles. Chem Commun (Cambridge, England). 2013;49(57):6403–5.

[110] Meyers JD, Cheng Y, Broome, AM, Agnes RS, Schluchter MD, Margevicius S, Wang X, Kenney ME, Burda C, Basilion JP. Peptide-targeted gold nanoparticles for photodynamic therapy of brain cancer. Particle & Particle Systems Characterization. 2015;32:448–57.

[111] Wagstaff AJ, Brown SD, Holden MR, Craig GE, Plumb JA, Brown RE, Schreiter N, Chrzanowski W, Wheate NJ. Cisplatin drug delivery using gold-coated iron oxide nanoparticles for enhanced tumour targeting with external magnetic fields. Inorganica Chim Acta. 2012;393:328–33.

[112] Thomas R, Park IK, Jeong YY. Magnetic iron oxide nanoparticles for multimodal imaging and therapy of cancer. Int J Mol Sci. 2013;14:15910-30.

[113] Pereira C, Pereira AM, Rocha M, Freire C, Geraldes CFGC. Architectured design of superparamagnetic $Fe_3 O_4$ nanoparticles for application as MRI contrast agents: mastering size and magnetism for enhanced relaxivity. J. Mater Chem B. 2015;3(30):6261–73.

[114] Murphy CJ, Gole AM, Hunyadi SE, Stone JW, Sisco PN. Alkilany A, Kinard BE, Hankins P. Chemical sensing and imaging with metallic nanorods, Chem. Commun., 2008, 544–557.

[115] Loo C, Lin A, Hirsch L, Lee M-H, Barton J, Halas N, West J, Drezek R. Nanoshell-enabled photonics-based imaging and therapy of cancer. Technol Cancer Res Treat. 2004;3(1):33–40.

[116] Yuan Z, Chen Y, Li H, Chang H, Chen Y. Fluorescent silver nanoclusters stabilized by DNA scaffolds. Chem Commun. 2014;50:9800–15.

[117] Gangula A, Podila R, Ramakrishna M, Karanam L, Janardhana C, Rao AM. Catalytic reduction of 4-nitrophenol using biogenic gold and silver nanoparticles derived from breynia rhamnoides. Langmuir. 2011;27:15268–74.

[118] Chen J, Zhang X, Cai S, Wu D, Chen M, Wang S, Zhang J. A fluorescent aptasensor based on DNA-scaffolded silver-nanocluster for ochratoxin A detection. Biosens Bioelectron. 2014;57:226–31.

[119] Farhadi K, Forough M, Molaei R, Hajizadeh S, Rafipour A. Highly selective Hg^{2+} colorimetric sensor using green synthesized and unmodified silver nanoparticles. Sens Actuator: B-Chem. 2012;161(1):880–5.

[120] Peng J, Ling J, Zhang X, Bai H, Zheng L, Cao Q, Ding Z. Sensitive detection of mercury and copper ions by fluorescent DNA/Ag nanoclusters in guanine-rich DNA hybridization. Spectrochim Acta A Mol Biomol Spectrosc. 2015;137:1250–7.

[121] Sondi I, Salopek-Sondi B. Silver nanoparticles as antimicrobial agent: a case study on *E. coli* as a model for Gram-negative bacteria. J Colloid Interface Sci. 2004;275:177–82.

[122] Hernández-Sierra JF, Ruiz F, Cruz Pena DC, Martínez-Gutiérrez F, Martínez AE, Guillén ADJP, Tapia-Pérez H, Castañón GM. The antimicrobial sensitivity of Streptococcus mutans to nanoparticles of silver, zinc oxide, and gold. Nanomedicine. 2008;4:237–40.

[123] Kaviyaa S, Santhanalakshmi J, Viswanathanb B, Muthumary J, Srinivasan K. Biosynthesis of silver nanoparticles using citrus sinensis peel extract and its antibacterial activity. Spectrochim Acta. 2011 (Part A); 79: 594–59.

[124] El-Sherif H, El-Masry M, Kansoh A. Hydrogels as template nanoreactors for silver nanoparticles formation and their antimicrobial activities. Macromol Res. 2011;19(11):1157–65.

[125] Velmurugan P, Cho M, Lee SM, Park JH, Bae S, Oh BT. Antimicrobial fabrication of cotton fabric and leather using green-synthesized nanosilver. Carbohydr Polym. 2014;106:319–25.

[126] Speshock JL, Murdock RC, Braydich-Stolle LK, Schrand AM, Hussain SM. Interaction of silver nanoparticles with Tacaribe virus. J Nanobiotechnology. 2010;8(19):1–9.

[127] Gusseme BD, Sintubin L, Baert L, Thibo E, Hennebel T, Vermeulen G, Uyttendaele M, Verstraeten W, Boon N. Biogenic silver for disinfection of water contaminated with viruses. Appl Environ Microbiol. 2010;76(4):1082–7.

[128] Bart De Gusseme BD, Hennebel T, Christiaens E, Saveyn H, Verbeken K, Fitts JP, Boon N, Verstraete W. Virus disinfection in water by biogenic silver immobilized in polyvinylidene fluoride membranes. Water Res. 2011;45: 1856-64.

[129] Elechiguerra JL, Burt JL, Morones JR, Camacho-Bragado A, Gao X, Lara HH, Yacaman MJ. Interaction of silver nanoparticles with HIV-1. J Nanobiotechnology. 2005;3(6):1–10.

[130] Lara HH, Ayala-Nuñez NV, Ixtepan-Turrent L, Rodriguez-Padilla C. Mode of antiviral action of silver nanoparticles against HIV-1. J Nanobiotechnology. 2010;8:1–10.

[131] Panáček A, Kolář M, Večeřová R, Prucek R, Soukupová J, Kryštof V, Hamal P, Zbořil R, Kvítek L. Antifungal activity of silver nanoparticles against Candida spp. Biomater. 2009;30(31):6333–40.

[132] Ponarulselvam S, Panneerselvam C, Murugan K, Aarthi N, Kalimuthu K, Thangamani S. Synthesis of silver nanoparticles using leaves of Catharanthus roseus Linn. G. Don and their antiplasmodial activities. Asian Pac J Trop Biomed. 2012:574–80.

[133] Gurunathan S, Lee KJ, Kalishwaralal K, Sheikpranbabu S, Vaidyanathan R, Eom SH. Antiangiogenic properties of silver nanoparticles. Biomater. 2009;30:6341–50.

[134] Tsuzuki T. Commercial scale production of inorganic nanoparticles, Int J Nanotechnol. 2009;6(5/6): 567–78.

Section 2

Application of Biopolymers

Chapter 3

Biopolymers – Application in Nanoscience and Nanotechnology

Sneha Mohan, Oluwatobi S. Oluwafemi, Nandakumar Kalarikkal, Sabu Thomas and Sandile P. Songca

Additional information is available at the end of the chapter

http://dx.doi.org/10.5772/62225

Abstract

In order to reduce the use of non-renewable resources and to minimize the environmental pollution caused by synthetic materials, the quest for utilizing biomaterials is on a rise. Biopolymers in nature are produced by a range of microorganisms and plants. Biopolymers produced by microorganisms require specific nutrients and controlled environmental conditions. This chapter discusses the recent developments and trends of biopolymers especially in the field of nanotechnology. A basic introduction regarding biopolymers is included at the beginning of the chapter. A detailed discussion on various characterization techniques used for characterizing biopolymers and various frequently used biopolymers is also included. Applications of biopolymers in various fields, especially in the field related to nanoscience and nanotechnology, is elaborated at the end of the chapter. Biopolymers together with nanotechnology have already found many applications in various fields including water treatment, biomedical application, energy sector, and food industry. This chapter is intended to give an overview on the importance of biopolymers in nanotechnology-based applications.

Keywords: Application, Characterization, Biodegradable, Biopolymers, Nanotechnology

1. Introduction

Biopolymers are polymeric biomolecules which contain monomeric units that are covalently bonded to form larger molecules. The prefix 'bio' means they are biodegradable materials produced by living organisms. A wide variety of materials usually derived from biological sources such as microorganisms, plants, or trees can be described using the term "biopolymer". Materials produced by synthetic chemistry from biological sources such as vegetable oils, sugars, fats, resins, proteins, amino acids, and so on can also be described as biopolymer [1].

As compared to synthetic polymers which have a simpler and more random structure, biopolymers are complex molecular assemblies that adopt precise and defined 3D shapes and structures. This is one important property which makes biopolymers active molecules *in vivo*. Their defined shape and structure are indeed keys to their function. For example, hemoglobin would not be able to carry oxygen in the blood if it was not folded in a quaternary structure.

The main property that distinguishes biopolymers from fossil-fuel-derived polymers is their sustainability, especially when combined with biodegradability. Biodegradable biopolymers from renewable resources have been synthesized to provide alternatives to fossil-fuel-based polymers. They are often synthesized from starch, sugar, natural fibers, or other organic biodegradable components in varying compositions. The biopolymers are degraded by exposure to bacteria in soil, compost, or marine sediment. Furthermore, subjecting biodegradable biopolymers to waste disposal by utilizing their characteristic of being degradable by the bacteria in the ground significantly reduces emission of CO_2 compared with conventional incineration. Therefore, attention is drawn to the use of biodegradable biopolymers from the viewpoint of global warming prevention. In recent years, with the critical situation of the global environment worsening due to global warming, the construction of systems with sustainable use of materials has been accelerated from the viewpoint of effectively using limited carbon resources and conserving limited energy resources. Furthermore, the cost of petroleum feedstocks has risen dramatically and there is a rising consumer interest in using "green" (or renewable resources) as the basis for consumer products.

One of the fastest-growing materials sectors in the past several years has been the production of polymers from renewable resources. Their development is fueled by the potential these polymers hold to replace fossil-fuel-based polymers. The main reasons for this drive can be summarized as follows: (1) inadequate fossil-fuel resources; (2) pricing instability of fossil fuel; (3) contribution of fossil fuel as a feedstock to climate change; (4) its occasional role as a political weapon; and (5) its association with the waste disposal problem created by the fossil-fuel-derived polymers.

The main objectives of this chapter are to give a basic introduction of biopolymers, its classification and different sources, and method of preparation. This chapter also describes various physical and morphological characterization techniques used for biopolymers. Some widely used biopolymers are discussed in detail followed by applications of biopolymers in various fields especially in the field of nanotechnology.

2. Different classes of biopolymers

Biopolymers are classified into different ways based on different scales. Based on their degradability, biopolymers can be divided into two broad groups, namely biodegradable and non-biodegradable, and alternatively, into bio-based and non-bio-based biopolymers (Fig. 1). On the basis of their polymer backbone, biopolymers can be classified roughly into the following groups: polyesters, polysaccharides, polycarbonates, polyamides, and vinyl polymers. These groups are again classified into several subgroups based on their origin.

Biopolymers can be classified, depending on the nature of the repeating unit they are made of, into three groups: (i) polysaccharides are made of sugars (e.g. cellulose found in plants), (ii) proteins are made of amino acids (e.g. myoglobin found in muscle tissues), and (iii) nucleic acids are made of nucleotides (DNA, genetic material of a given organism). Based on application, biopolymers can be classified as bioplastics, biosurfactant, biodetergent, bioadhesive, biofloculant, and so on.

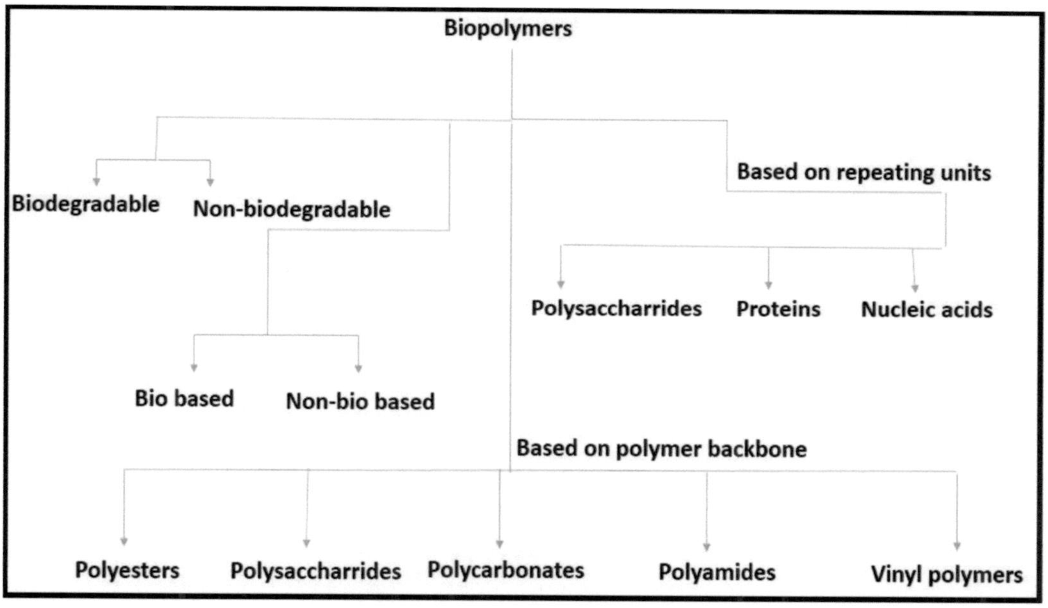

Figure 1. Different class of biopolymers

3. Sources and preparation of biopolymers

Polymeric biomaterials are synthetically derived or modified polymers designed for various applications. Engineering the production of novel biopolymers in plants provides a truly biorenewable avenue for their synthesis. Like all polymer industries, these polymers are also produced in bulk and then shaped for a specific end use. Microorganisms also play an important role in producing a huge variety of biopolymers, such as polysaccharides, polyesters, and polyamides which range from viscous solutions to plastics (Table 1). Their physical properties are dependent on the composition and molecular weight of the polymer [2]. The properties of various biopolymers synthesized by the aid of microorganisms can be tailored by the genetic manipulation of microorganisms, thereby making it suitable for high-value medical application such as tissue engineering and drug delivery.

Biopolymers which are produced with the help of microorganisms require specific nutrients and controlled environmental conditions. They are produced either directly via fermentation or by chemical polymerization of monomers, which are in turn produced through fermentation. Most biopolymers are biocompatible with no adverse effects on biological systems. The mechanism of synthesis of biopolymers from bacterial origin is believed to be either as a result of their defense mechanism or as storage material [3]. They can be degraded by natural processes, microorganisms, and enzymes so that it can be finally reabsorbed in the environment. Biopolymers or organic plastics are a form of plastics derived from renewable biomass sources such as vegetable oil, corn starch, pea starch, and so on (Fig.2). By shifting the focus more into the biopolymers, the conservation of fossil resources and reduction in CO_2 emissions can be achieved thereby promoting sustainable development [4]. Among the microorganisms, algae serve as an excellent feedstock for plastic production because of their high yield and the ability to grow in a range of environments. The use of algae opens up the possibility of utilizing carbon and neutralizing greenhouse gas emissions from factories or power plants. Algae-based plastics have been a recent trend in the era of bioplastics compared to traditional methods of utilizing feedstocks of corn and potatoes as plastics. While algae-based plastics are in their infancy, once they are into commercialization they are likely to find applications in a wide range of industries. Currently, microbial plastics are considered as an important source of polymeric material that have a great potential for commercialization. They can modify the flow characteristics of fluids, stabilize suspensions, flocculate particles, encapsulate materials, and produce emulsions [5].

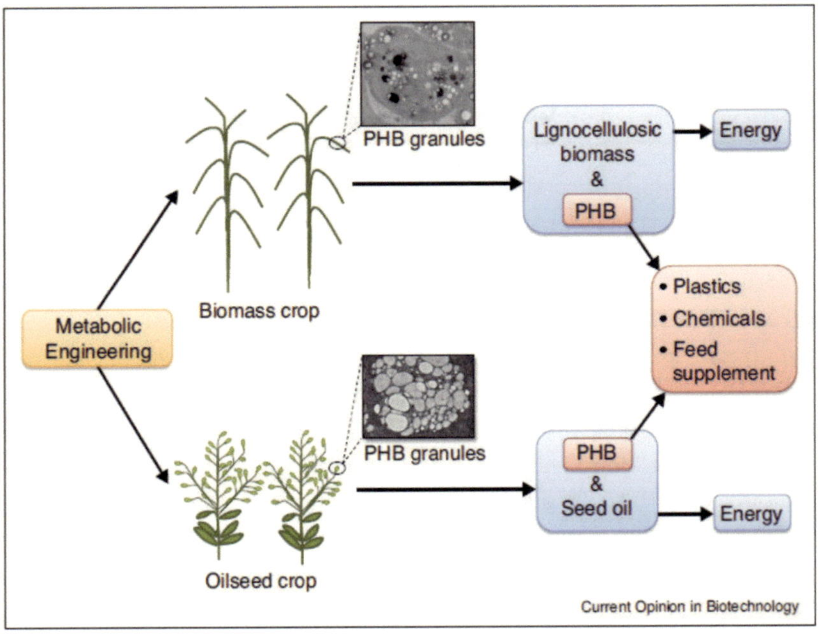

Figure 2. Schematic representation of the production of polyhydroxybutyrate (PHB) in both biomass and oilseed crops [6].

http://dx.doi.org/10.5772/62225

Polymer	Method/ Source	Example of bacteria used for synthesis
Hybrid plastics	Adding denatured algae biomass to petroleum-based plastics like polyurethane and polyethylene as fillers	Filamentous green algae, Cladophorales
Cellulose-based plastics	Biopolymer of glucose	30 % of the biomass produced after extraction of algal oil is known to contain cellulose
Poly-lactic acid (PLA)	Polymerization of lactic acid lactic acid	Bacterial fermentation of algal biomass
Bio-polyethylene	Ethylene produced from ethanol, by a chemical reaction called cracking. Ethanol derived from natural gas or petroleum	Bacterial fermentation of algal biomass
Poly esters	Biomass	Bacteria like Akaligenes eurrophus, E. coli, etc.

Table 1. Some biopolymers, their sources, and preparation

4. Structure of biopolymers

The structure of biopolymers has been a subject of investigation ever since they were discovered. So far, the most accurate methods for structure determination have been X-ray diffraction techniques. These techniques are, however, restricted to crystallizable molecules, and it has not been established if the molecular structure in a crystal is identical to that in solution. Many methods have been developed or adopted to probe biopolymer properties in solution [7]. The composition and sequence of the monomer units of biopolymers determine their so-called primary structure. Conformation, that is, one or another spatial form of the biopolymer molecules, is determined by their primary structure. Depending on the chemical structure and external conditions, the molecules may be in one or more preferred conformations.

Specific form and structure provide functionality to biological elements microscopically and macroscopically. The combination of cell and cell tissues results in the outward appearance of biological organisms. The causes for the shape of these building blocks of life are manifold. Biopolymers are important for the organization and functionality of cells. Prominent examples are the DNA as the carrier of the genetic code and the cytoskeletal filaments building the scaffold of the cell. The unique property of DNA depends on its equilibrium polymer conformations based on its internal characteristics like flexibility, topology and filament diameter and external constraints such as confinement. Any variation in these properties will alter the genetic characteristics of that living system which shows the significance of structure of biopolymers in living systems.

5. Different characterization techniques

Most of the techniques used for the characterization of polymers can be utilized for the characterization of biopolymers. Characterization of biopolymers has two purposes:

1. Development of parameters for processing;

2. Determination of end-use performance characteristics.

Through characterization, the most important properties of interest are molecular mass, polydispersity, size, degree of association, conformation, interaction, and so on. Some of the characterization techniques which are used for the analysis of biopolymers are listed in the following.

5.1. Fourier Transform Infrared Spectroscopy (FTIR)

FTIR can be used to study the various functional groups and interactions present in biopolymers. Interactions like hydrogen bonding, amide linkage, etc., can be easily detected by analyzing the spectra [8]. Among various techniques, attenuated total reflection (ATR) sampling device-based Fourier transform infrared (ATR-FTIR) spectroscopy is considered as a potential one. It requires minimal sample preparation, permits routine analysis at both laboratory and on-field environments, and is easy to operate. Visible and near-infrared (VIS–NIR) spectroscopy is a well-established alternative for measuring constituents of biological materials. The NIR spectrum is influenced by the different vibrational modes of the molecules, which are caused by their interaction with electromagnetic radiation absorbed at specific wavelengths. This technique can be used for the identification of several complex components such as proteins and carbohydrates in biological materials. Chemometrics allows the extraction of relevant information contained in the spectra for the development of calibration models that allow the classification and prediction of organic sample's composition. The thermotropic phase behavior of a suite of newly developed self-forming synthetic biopolymers has been investigated by variable-temperature Fourier transform infrared (FT–IR) absorption spectroscopy [9].

Various studies have extensively used the FTIR technique to explore the physical and mechanical properties of biopolymers. Two-dimensional (2D) correlation analysis was applied to characterize the ATR spectral intensity fluctuations of immature and mature cotton fibers [10]. FTIR analysis can also be used to verify the interactions between various biopolymers in its composites [11].

Figure 3 shows FTIR spectra of chitosan-starch films, keratin, and the composites synthesized with 20 %. Figure 3A (a) shows the band at 3400 cm^{-1} which is assigned to the hydrogen-bonded hydroxyl groups due to the presence of starch in chitosan-starch film. Fig. 3A (b) shows that the main vibrations attributed to the keratin structure were identified in the region around 3300 cm^{-1} and correspond to a range of amide bands. Furthermore, the peak at 2945 cm^{-1} is assigned to the asymmetric vibration of the CH of the methyl group. In Figure 3B (a) the main peaks related to chitosan-starch films are seen at 700–770 cm^{-1} which correspond to the saccharide structure, and the three-characteristic peaks between 995 and 1150 cm^{-1} are attributed to C–O bond stretching.

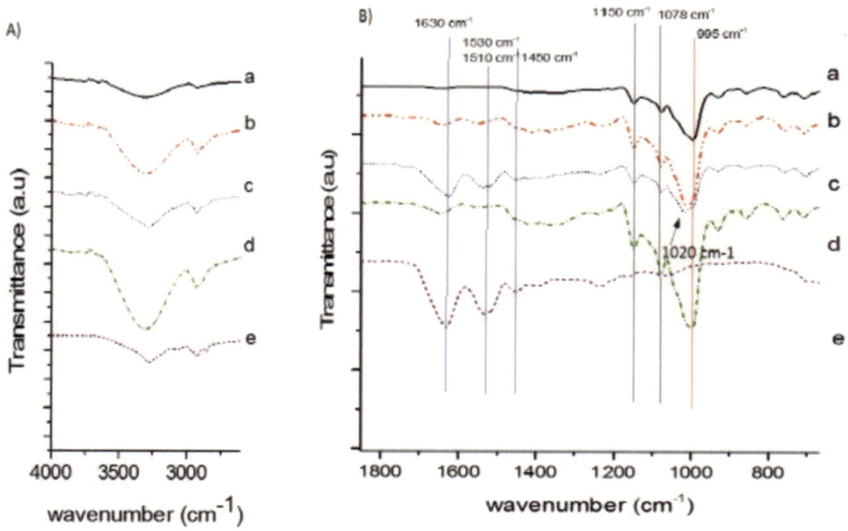

Figure 3. IR spectra of the composite films: A) from 4000–2500 cm^{-1}; B) from 1800–700 cm^{-1}. (a) chitosan-starch film; (b) chitosan-starch-short biofiber 20%; (c) chitosan-starch-long biofiber 20 %; (d) chitosan-starch-ground quill 20 %; (e) keratin [11].

In Fig 3A(e) the band at 1630 cm^{-1} is assigned to the C=O group and the peaks at 1530 cm^{-1} and 1510 cm^{-1} are attributed to the in plane bending of NH group, while the signal at 1,430 cm^{-1} is related to bending of the CH_3 group for keratin. FTIR results correlate to the thermomechanical properties and the morphological features of the composite. The spectrum indicates that a better storage modulus is obtained for composites with short and large fibers than the composite with ground quill. Also better distribution is obtained in these composites.

5.2. X-ray diffraction analysis

X-ray diffraction techniques can be successfully used to study the crystallinity of the biopolymers. Figure 4 represents the X-ray diffraction studies of native wheat starch, compressed starch foam, and compressed starch/fiber foam [12]. The native wheat starch had a semi-crystalline structure with peaks typical of A-type crystals. The gelatinization/compression processes during the composite preparation cause decrease of crystallinity resulting in diffractograms exhibiting a broad amorphous peak with no significant residual A-type crystallinity. The starch granule structure was disrupted by the processes of gelatinization and high compression which is clearly visible from the spectrum. Small peaks were found around 2Θ 12.0, 17.3, and 19.5, which is characteristic for crystallographic parameters of VH-type crystals for destructured starch prepared from solution. For the starch/fiber sample, the spectrum obtained indicates that the crystallinity of the fiber overlaps with the gelatinized starch.

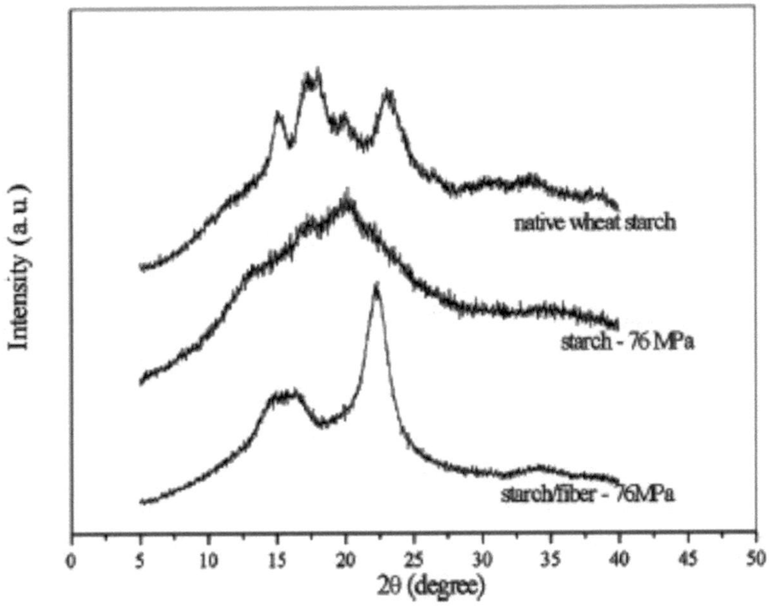

Figure 4. X-ray diffraction profile for native starch, starch (76 MPa) and starch/fiber (76 MPa) [11].

5.3. NMR spectroscopy

The molecular dynamics of water adsorbed on biopolymer surfaces and functional groups of biopolymers can easily be determined by using the NMR technique [13]. NMR spectroscopic technique relaxation times and their dependence on temperature and magnetic field have proved to be a useful source of information for molecular structure, phase change, conformational exchange, solubility, and diffusion of biopolymers. Many studies of spin–lattice relaxation time (T_1), spin–spin relaxation time (T_2), and spin–lattice relaxation time in the rotating frame (T_{1q}) of different biopolymers have been already reported [14]. The molecular origin of nuclear spin relaxation has been examined in cellulose and its derivatives [15]. In order to understand the behavior of main functional groups, the effect of temperature on nuclear-spin relaxation has been probed extensively in biopolymers in cell walls [16, 17]. The temperature dependence of relaxation times of starch chain has been studied in order to investigate structure–hydration relationships in the biopolymer [18].

5.4. Thermal analysis

Thermal analysis (TA) method has become an indispensable analytical technique for the characterization of polymers. The characterization technique provides relevant information about thermal and chemical stability, phase transition temperatures and kinetics, rheology and molecular relaxation times of the polymers. It provides information regarding the relation between the polymers chemical structure and their physical properties. The properties of

http://dx.doi.org/10.5772/62225

processed polymer depend on morphology and molecular chain orientation of the polymer backbone. In conventional thermal analysis techniques, the method usually involves the monitoring of material properties as a function of temperature. Differential scanning calorimetry (DSC) is one important method used to study the thermal behavior of biopolymers.

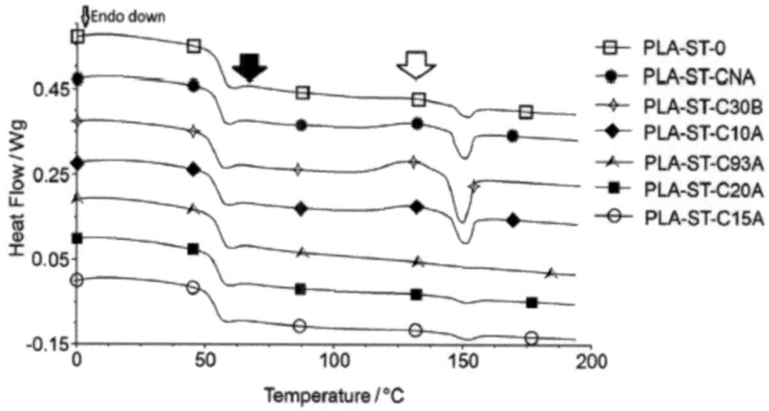

Figure 5. DSC thermograms of PLA/butylated starch blend and blend/MMT composites with MMTclays of different hydrophobicity. The samples had a clay and starch content of 4.5 wt % and 25.5 wt % of the total composite weight, respectively. The control sample PLA-ST-0 had no clay [19].

Figure 5 shows the standard/conventional DSC curves of the neat blend and blend/clay composites. All of the samples showed a glass transition with enthalpy relaxation at approximately 55 °C. There was some apparent cold crystallization for the samples that could not be clearly separated from the melting endotherm [19]. The melting endotherms appeared at approximately 150 °C, and the values could not be accurately determined because of the apparent influence of the cold crystallization exotherms. It was clear that the area under the cold crystallization and melting peaks (enthalpy of cold crystallization and melting, respectively) was higher for the control sample and the samples with less to moderate hydrophobic clays when compared with that of the composites with hydrophobic to extremely hydrophobic clays. This indicated the tendency of the more hydrophobic clays to hinder the re-arrangement of the PLA molecules into organized crystalline structures. Such a hindrance was highest in the case of the PLA-ST-C93A composite because of the high level of compatibility between two phases.

5.5. VIS–NIR spectroscopy

There is an increasing interest in the use of polysaccharides and proteins for the production of biodegradable films. VIS–NIR spectroscopy is a reliable analytical tool for objective analyses of biological sample attributes. It can be used to investigate the compositional characterization of biodegradable materials and correlation of their mechanical properties [20]. Partial least square regression can be used to investigate the correlation between spectral information and mechanical properties. Large-scale production of films requires fast and non-invasive analytical methods in order to assess film composition and mechanical properties. VIS–NIR spectroscopy is a well-established alternative as a non-destructive process analytical technology for measuring constituents of biological materials. The NIR spectrum is influenced by the different vibrational modes of the molecules, which are caused by their interaction with electromagnetic radiation absorbed at specific wavelengths. This technique has been used for the identification of several complex components such as proteins and carbohydrates in biological materials. Chemometrics allows the extraction of relevant information contained in the spectra for the development of calibration models that allow the classification and prediction of the composition of organic samples [21].

5.6. Scanning Kelvin Probe Microscopy (SKPM)

SKPM has become a very strong tool in material science [22], giving up a surface mapping of electrostatic potential [23]. Electrical properties have an important role in the rejection of charged species in membrane applications such as water softening and heavy metal removal. Increasing membrane charge may improve ion rejection for nanofiltration (NF) and reverse osmosis (RO) processes. If membrane is electrically charged, electrostatic interaction becomes significant for ion removal. Nanocomposites which presented the best flows were characterized by SKPM for getting information about their electrical potential.

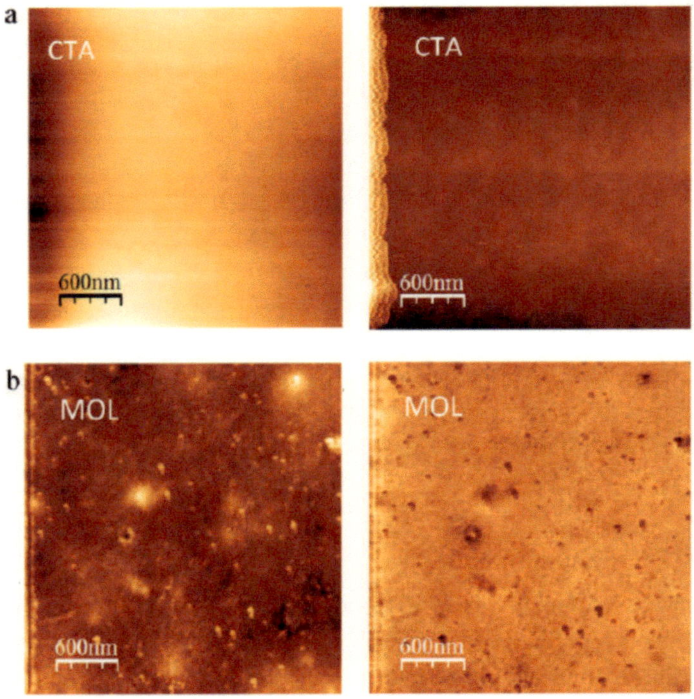

Figure 6. SKPM views of CTA (cellulose triacetate) and nanofiltration composites MOL (Lignin). Topography and surface potential mapping are shown at the left and at the right side of the figure, respectively [23].

Figure 6 shows SKPM image of cellulose triacetate and lignin. The dark zones in the images correspond to a potential difference between surface and AFM tip, whereas no difference corresponds to bright zones. For neat CTA (cellulose triacetate) membranes, no indication of electrical potential at the surface of the material was found. When lignin is incorporated into the polymer membrane, the resulting nanofiltration membranes showed a surface electrical potential.

6. Frequently studied biopolymers

6.1. Gelatin

Gelatin is a protein made from animal products. It is produced by the partial hydrolysis of collagen obtained from skin, bones, and tissues of animals. They are mostly derived from type I collagen containing no cysteine. There are mainly two types of gelatin based on their synthesis procedure and composition. Type A, which is obtained by acid hydrolysis of collagen, has 18.5 % of nitrogen and type B obtained by the alkaline hydrolysis of collagen has 18 % of nitrogen as they do not contain amide groups [24]. At elevated temperatures, it melts and exists as coils and on lowering the temperature, it undergoes coil to helix transformation. Collagen is resistant to proteases, whereas gelatin is susceptible to these enzymes. Due to the presence of many functional groups like $-NH_2$, –SH, and –COOH, gelatin can be modified with biomolecules and nanoparticles for various applications [25].

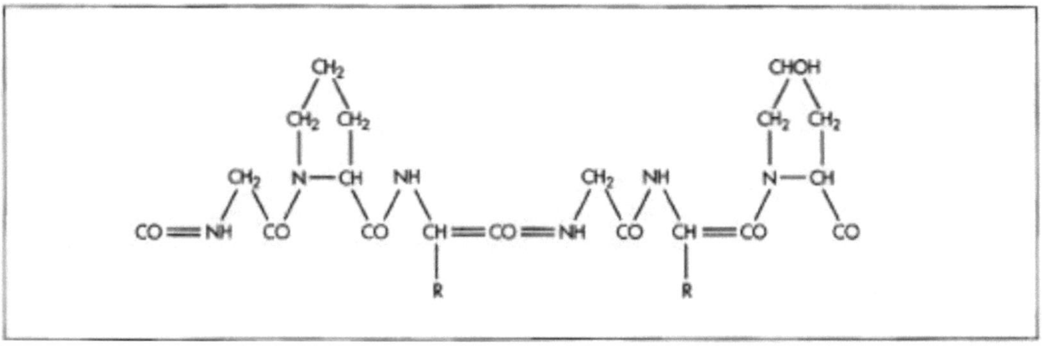

Figure 7. Structure of gelatin

It is hemostatic, pro-angiogenic, non-immunogenic, biodegradable, and biocompatible [26] and can be cross-linked to form hydrogels [27]. It is an ECM protein with wide number of applications in bone tissue engineering [28], wound dressing [29], drug delivery [30], and gene transfection [31]. Gelatin is used for weight loss and for treating osteoarthritis, rheumatoid arthritis, and brittle bones (osteoporosis). Gelatin is also used for improving hair quality and to shorten recovery after exercise and sports-related injury. In manufacturing, gelatin is used for preparation of foods, cosmetics, and medicines.

6.2. Starch

Starch is a semi-crystalline polymer composed of amylose and amylopectin molecules [32]. The primary carbohydrate used for energy storage in plants is starch, the equivalent of glycogen in animals. Both starch and glycogen are polymers of glucose, with different glycosidic linkages between glucose monomers and different degrees of branching. Amylose is a linear polymer composed of (1–4)-linked a-D-glucopyranosyl units while amylopectin is a very large molecule made up of (1–4)-linked a-D-glucopyranose chains with a-(1–6) branches. Starch is one of the most abundant natural polymers and it is cheap and degradable [33]. Properly processed starch can be made into various commercial products [34]. Due to their sensitivity to moisture and poor mechanical properties, they have only found use in various niche markets. Studies have shown that by incorporating fibers, the mechanical and barrier properties of starch products can be improved [35]. Nano and micro structured fibers when dispersed in the polymer matrix while preparing the composite, will improve its tensile strength and flexibility [36].

The significance of starch in plants and humans are undeniable. While plants consider it as a storage of energy, humans need it as a vital part of their diet and also for various commercial products. Starch and its derivatives have already been explored in various applications like in the manufacture of paper, textiles and adhesives. Their biodegradable and renewable nature make them an environmental friendly alternative to the use of synthetic additives in many other products, including plastics, detergents, pharmaceutical tablets, pesticides, cosmetics, and even oil-drilling fluids.

6.3. Cellulose

The monomer unit in cellulose is β-1, 4 linked glucan chains with hydrogen bonds formed between hydroxyl groups and oxygen atoms both within a single glucose chain and between neighboring chains. Hydrogen bonding and van der Waals forces aggregate glucan chains together side by side to form cellulose microfibrils. These cellulose microfibrils are then stacked together to form crystalline cellulose [37]. The polar –OH groups present in the cellulose chain form many hydrogen bonds with OH groups on adjacent chains, bundling the chains together. The stacking of chains is so regular that it forms hard, stable crystalline regions, and this gives the bundled chains more stability and strength. The length of the chain varies greatly from a few hundred sugar units in wood pulp to over 6000 for cotton. It is the most common natural organic polymer which is considered as an almost inexhaustible source of raw material for the increasing demand for environment friendly and biocompatible products. Due to the strong hydrogen bonding, it is insoluble in common solvents. As a result, it is chemically modified into ether, ester, and acetal derivatives.

Cellulose, the most abundant biological material on Earth, is also composed of glucose monomers but joined by beta glycosidic bonds, giving it a straighter shape that packs closely and provides mechanical strength in wood [38].

The remarkable strength of wood is due to cellulose which is a long chain of linked sugar molecules. Plant cell walls, which are the basic building block for textiles and paper, also have

http://dx.doi.org/10.5772/62225

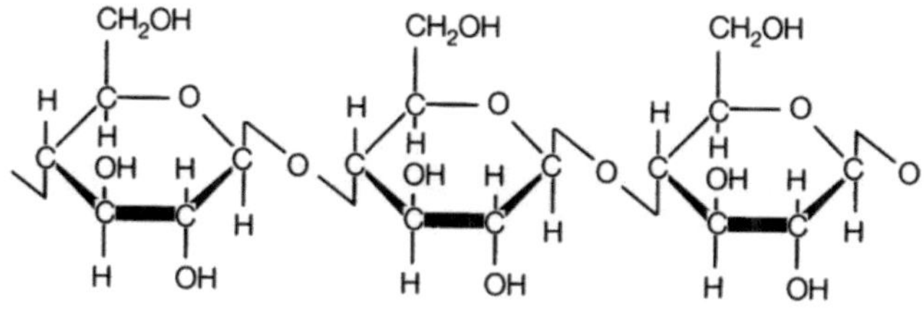

Figure 8. The structural representation of cellulose

cellulose as main component. The purest natural form of cellulose is cotton which is used in many textile applications.

Recently, cellulose has been excessively used in the form of nanofibrils. Cellulose nanofibrils (CNF) or nanocellulose is typically generated by mechanical grinding or high-pressure fluidization of cellulose to remove its lignin content. CNF consists of very thin (5–20 nm) and long (several μm) fibrils with high aspect ratio. At low concentrations, it forms a transparent gel-like material which can be used for producing biodegradable and environmentally safe, homogeneous, and dense films for various applications, especially in biomedical field. Extraction of CNF has been reported from various sources like coir, banana, sugar beet, hemp, softwood, and hardwood pulps. After using various plasticizers, various thermal, mechanical, barrier, and physical properties of the cellulose can be improved so that it can be used in various commercial applications. Plasticizing improves properties like grease proofness, and high barrier against oxygen transmission especially at dry conditions. All cellulose-synthesizing organisms including bacteria, algae, tunicates, and higher plants have cellulose synthase proteins, which catalyze the polymerization of glucan chains.

6.4. Chitin and chitosan

Chitin is a nitrogen-containing polysaccharide, related chemically to cellulose that forms a semitransparent horny substance, and is a principal constituent of the exoskeleton, or outer covering, of insects, crustaceans, and arachnids [39]. The structure of chitin is comparable to the polysaccharide cellulose, which consists of β-(1 → 4) linked nacetylglucosamine, forming crystalline nanofibrils or whiskers. Chitin is the second most abundant naturally occurring polysaccharide found in the outer shell of crustaceans and insect exoskeleton. Chitin is insoluble in most of the organic and inorganic solvents. Chitosan, the de-acetylated derivative of chitin, is a cationic polysaccharide that can form polyelectrolyte complexes with other polysaccharides and is the second most abundant natural biopolymer after cellulose (Fig. 9). It consists of β-(1 → 4)-linked D-glucosamine and N-acetyl-D-glucosamine. Due to easy applicability, water sorptivity, oxygen permeability, blood coagulating property, and cytokine induction (interleukin-8), which activate the fibroblast migration and proliferation, it is used as wound dressing material.

Chitosan, due to its good film forming property, has been widely used for casting free-standing films. It can be applied as a coating on various surfaces to improve the permeability of the

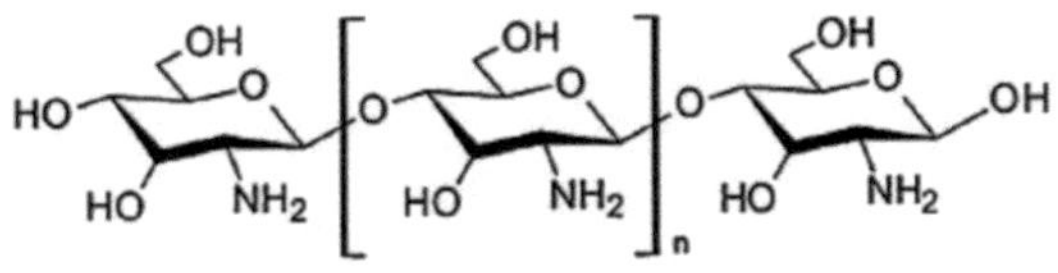

Figure 9. The structure of (a) chitosan biopolymer

surface against moisture and various gases. The water vapor permeability of the paper increased as a result of the chitosan coating. Chitosan coatings can be made on to NH_3- or CO_2-plasma activated polypropylene films, thereby making the film surfaces hydrophilic and prevent oxygen, carbon dioxide, and ethylene transmission. Barrier properties can be further improved by using nanoclays incorporated into chitosan.

6.5. Polylactic Acid (PLA)

Polylactic acid or polylactide (PLA, Poly) is a biodegradable thermoplastic aliphatic polyester derived from renewable resources, such as corn starch, tapioca roots, chips, starch, or sugarcane. PLA is obtainable primarily by the ionic polymerization of lactide, a ring closure of two lactic acid molecules. At temperatures between 140 and 180 °C and under the action of catalytic tin compounds (such as tin oxide), a ring-opening polymerization takes place. Lactide itself can be made through lactic acid fermentation from renewable resources such as starch by means of various bacteria. PLA can also be produced directly from lactic acid by polycondensation. However, this process yields low molecular weight polymers, and the disposal of the solvent is a problem in the industrial production. The properties of PLA primarily depend on the molecular mass, the degree of crystallinity, and possibly the proportion of co-monomers. A higher molecular mass raises T_g, as well as T_m, tensile strength, elastic modulus, and lowers the strain after fracture. Due to the CH_3 side group, the material has water-repellent or hydrophobic behavior. PLA is soluble in many organic solvents, such as dichloromethane or the like. PLA has higher transparency than other biodegradable polymers, and is superior in weather resistance and workability.

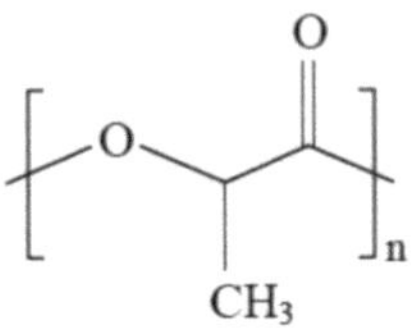

Figure 10. Structural formula of PLA

6.6. Poly(e-Caprolactone) (PCL)

Poly(e-caprolactone) (PCL) is a biodegradable polyester which is commonly used for preparing scaffolds for various tissue engineering applications. It has a low melting point of around

60 °C and a glass transition temperature of about −60 °C. It is used in the manufacture of specialty polyurethanes with good water, oil, solvent, and chlorine resistance. PCL can be used as an additive for resins to improve its impact resistance and their processing characteristics. PCL is compatible with a range of other materials which enable it to be mixed with other biodegradable polymers to lower its cost and increase biodegradability. It can also be used as a polymeric plasticizer to other polymers like PVC. Polycaprolactone is also used for splinting, modeling, and as a feedstock for prototyping systems such as a 3D printer. PCL is biocompatible, biodegradable polymer which has shown to be capable of supporting a wide variety of cell types. PCL is also a well-known FDA-approved biomaterial which is most widely used in biomedical field [40]. The PCL is intrinsic hydrophobic in nature. Its poor surface wetting and interaction with the biological fluids avoid cells adhesion and proliferation [41]. PCL is prepared by ring opening polymerization of ε-caprolactone in the presence of a catalyst. The significance of PCL in various tissue engineering applications increased because it is degraded by hydrolysis of its ester linkages in physiological conditions. This property makes it useful for various implantable biomaterials. Since its degradation rate is lower than that of polylactide, it is proposed to be used as long-term implantable devices.

6.7. Poly-(Vinyl Alcohol) (PVA)

Poly(vinyl alcohol) (PVOH or PVA) is a white, colorless and odorless, water-soluble synthetic polymer with molecular formula $[CH_2\ CH(OH)]_n$. It is commercially available as beads or as solutions in water. It is used as thickener in glues, paper-making, sizing agent in textiles, water-soluble films useful for packing, and a variety of coatings. Polyvinyl alcohol is used as an emulsion polymerization aid, as protective colloid, to make polyvinyl acetate dispersions. PVA is prepared by first polymerizing vinyl acetate, and the resulting polyvinyl acetate is converted to the PVA. PVA has a wide range of potential applications in optical, pharmaceutical, medical, and membrane fields. It is a water-soluble polymer and allows the development of environment-friendly material processes [42].

6.8. Polyvinyl Acetate (PVAc)

Polyvinyl acetate (PVAc) belongs to polyvinyl ester family and is a thermoplastic resin produced by the polymerization of vinyl acetate monomer. Its general formula is $[CH_3COOCHCH_2]_n$ with a solids content of 50–55 % in water [43]. Most polyvinyl acetate emulsions available in market contain various co-monomers such as n-butyl acrylate, 2-ethyl hexyl acrylate, ethylene, dibutyl maleate, and dibutyl fumarate. Polymerization of vinyl acetate with ethylene can also be used to produce solid vinyl acetate/ethylene copolymers with more than 50 % vinyl acetate content. Polyvinyl alcohol (PVOH) is produced by methanolysis or hydrolysis of polyvinyl acetates. The reaction can be controlled to produce any degree of replacement of acetate groups. Polyvinyl acetate has immense applications in various fields like biomedical, synthesis of metal nanoparticles, sensing activity, and so on [44].

6.9. Collagen

Collagen is the most abundant protein in human. It is found in the bones, muscles, skin, and tendons, where it forms a scaffold to provide strength and structure. Endogenously produced, collagen plays numerous important roles in health, with the breakdown and depletion of the body's natural collagen associated with a number of health problems [45]. Exogenous collagen is supplied to the body for medical and cosmetic purposes, including helping with healing and repairing of the body's tissues. Around 30 % of the protein in the human body is collagen and 80–90 % of the collagen in the body consists of types I, II, and III. Among this, type I collagen fibrils are stronger than steel hence is the substance that holds the whole body together. Collagen gives the skin its strength and structure, and also plays a role in the replacement of dead skin cells. Collagen in medical products can be derived from human, bovine, porcine, and ovine sources. Collagen dressings attract new skin cells to wound sites [46]. Controllable factors that damage the production of collagen include sunlight, smoking, and high sugar consumption.

6.10. Poly(3-hydroxybutyrate-co-3-hydroxyvalerate) PHBV

Poly(3-hydroxybutyrate-co-3-hydroxyvalerate) PHBV, is a polyhydroxyl- alkanoate- type thermoplastic linear aliphatic polyester. PHBV (Fig. 11) is synthesized by bacteria as an intercellular carbon and energy storage compounds under growth limiting conditions. It is biodegradable, nontoxic, and biocompatible plastic which can be produced from glucose and propionate by the recombinant *Escherichia coli* strains [47]. PHBV is a copolymer of 3-hydroxybutanoic acid and 3-hydroxypentanoic acid and can also be synthesized from butyrolacone and valerolactone in presence of oligomeric aluminoxane as catalyst. PHBV has many applications especially for the development of implanted medical devices for dental, orthopedic, hernioplastic, and skin surgery. Various potential medical devices like bioresorbable surgical sutures, biodegradable screws and plates for cartilage and bone fixation, biodegradable membranes for periodontal treatment, surgical meshes with PHBV coating for hernioplastic surgery, wound coverings, and the like have been developed using this biopolymer [48].

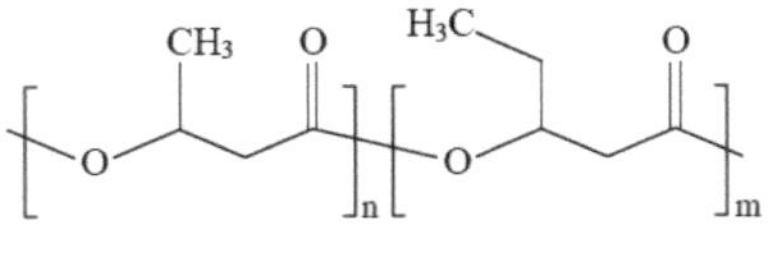

Figure 11. Structural formula of PHBV

7. Applications of biopolymers

Biopolymers, due to its biocompatible and biodegradable nature, can be used to improve the performances of other biologically active molecules in a product. They can also be modified to suite various potential applications which include the following.

7.1. Synthesis of nanomaterials

Nanotechnology is the science of nanomaterials which deals with its synthesis, characterization, and applications. Researchers are currently focusing on developing more eco-friendly processes for the synthesis of nanoparticles. The main focus for the synthesis protocol has shifted from physical and chemical processes towards "green" chemistry and bioprocesses. Metal nanoparticles, due to their quantum size effects, possess various novel properties. However, most of their synthesis protocol imposes a major threat to the environment [49]. In common synthetic methods, the reducing agents used which include organic solvents and toxic-reducing agents like hydrazine, N-dimethylformamide, and sodium borohydride are considered to be highly toxic for the environment [50]. All these chemicals are highly reactive and pose potential environmental and biological risks. With the increasing interest in minimization/elimination of waste and adoption of sustainable processes, the development of green chemistry approaches is desirable. Biopolymers have been extensively used as capping and reducing agent for the synthesis of various nanoparticles. Biopolymers like chitosan, heparin, soluble starch, cellulose, gelatin, PVA, PVP, and so on can be used to replace various toxic regents in synthesizing different nanoparticles [51–53].

7.2. Biomedical applications

In recent years, biopolymer materials have aroused great interest because of their biomedical applications, such as those in tissue engineering, pharmaceutical carriers, and medical devices [54, 55]. A common biopolymer, gelatin, was widely applied in medicine for dressing wounds, as an adhesive, and so on. Porous gelatin scaffolds and films were produced with the help of solvents or gases as simple porogens, which enable the scaffolds to hold drug or nutrients to be supplied to the wound for healing [56]. Electrospun PLGA-based scaffolds have been applied extensively in biomedical engineering, such as tissue engineering and drug-delivery system [57]. MWCNT-incorporated electrospun nanofibers with high surface area-to-volume ratio and porous characteristics have also shown potential applications in many aspects of tissue engineering [58, 59].

Biomaterials made from proteins, polysaccharides, and synthetic biopolymers are preferred but lack the mechanical properties and stability in aqueous environments necessary for medical applications. Cross-linking improves the properties of the biomaterials, but most cross-linkers either cause undesirable changes to the functionality of the biopolymers or result in cytotoxicity. Glutaraldehyde, the most widely used cross-linking agent, is difficult to handle and contradictory views have been presented on the cytotoxicity of glutaraldehyde-cross-linked materials [60].

7.3. Food industry

Replacing the oil-based packaging materials with biobased films and containers might give not only a competitive advantage due to more sustainable and greener image but also some improved technical properties. Biopolymers are currently used in food coatings, food packaging materials, and encapsulation matrices for functional foods. They provide unique solutions to enhance product shelf life while also reducing the overall carbon footprint related to food packaging [61]. Within food-related applications, these biobased materials are particularly useful in three main areas: food packaging, food coating, and edible films for food and encapsulation. The most commercially viable materials in food packaging are certain biodegradable polyesters and thermoplastics like starch, PLA, PHA, and so on, which can be processed by conventional equipment. These materials are already used in a number of monolayer and multilayer applications in the food-packaging field. Starch and PLA biopolymers are potentially the most attractive types of biodegradable material. This is due to the balance of their properties and the fact that they have become commercially available. PLA is of particular interest in food packaging, due to its excellent transparency and relatively good water resistance. The challenge for these specific biomaterials is to improve their barrier and thermal properties so that they perform like polyethylene terephthalate (PET). Other materials extracted from biomass resources, such as proteins (e.g., zein), polysaccharides (e.g., chitosan), and lipids (e.g., waxes), also have excellent potential as gas and aroma barriers. The inherent high rigidity and the difficulty of processing them in conventional equipment are the main drawbacks of these types of materials. The hydrophilic nature of most of the biopolymers affects their use as high-end products. The absorption of moisture causes plasticization of these materials thereby deteriorating the barrier properties of these materials.

Renewable polymers have also been used for encapsulation purposes. Encapsulation has previously been described as a technology to protect sensitive substances against the influences of adverse environments. The term "microencapsulation" refers to a defined method of wrapping solids, liquids, or gases in small capsules, which can release their contents under specific circumstances [62]. Such technologies are of significant interest to the pharmaceutical sector. The increasing interest in edible films and coatings using biopolymers is due to their ability to incorporate a variety of functional ingredients. Plasticizers, such as glycerol, acetylated monoglycerides and polyethylene glycol, which are used to modify the mechanical properties of the film or coating, make significant changes to the barrier properties of the film. However, the major advantage of coatings is that they can be used as a vehicle for incorporating natural or chemical active ingredients, such as antioxidants and antimicrobial agents, enzymes, or functional ingredients, like probiotics, minerals, and vitamins. These ingredients can be consumed with the food, thus enhancing safety, nutritional, and sensory attributes. Edible films can be used as flavor or aroma carriers in addition to providing a barrier to aroma loss [63].

Chitosan has shown great potential as an antimicrobial packaging agent to preserve food against a wide variety of microorganisms. Incorporating antimicrobial compounds into edible films or coatings provides a novel way to improve the safety and shelf life of ready-to-eat foods. Lysozyme is one of the most frequently used antimicrobial enzymes in packaging materials, since it is a naturally occurring enzyme. Biopolymers such as amylose, when mixed with plasticizers have excellent potential in forming thin films for various food and packaging applications [64]. Starch has high sensitivity to relative humidity (RH) due to its hydrophilic nature and this can be reduced by introducing plasticizers which enhances the flexibility of the matrix [65,66]. But this technique also has some limitations due to the complex interactions between the hydrophilic plasticizers and the starch. An "anti-plasticization" process takes place with increased stiffness of the matrix if the structure of the plasticizer molecule and the polymer matrix is not compatible [67].

7.4. Packaging applications

Currently, the most commercially viable materials in food packaging are certain biodegradable polyesters, which can be processed by conventional equipment. These materials are already used in a number of monolayer and multilayer applications in the food-packaging field. Among the most widely researched thermoplastics, the sustainable biopolymers used in monolayer packaging include starch, PHA, and PLA [68]. Starch and PLA biopolymers are potentially the most attractive types of biodegradable material. This is due to the balance of their properties and the fact that they have become commercially available. The challenge for these specific biomaterials is to improve their barrier and thermal properties so that they perform like polyethylene terephthalate (PET). Other materials extracted from biomass resources, such as proteins (e.g., zein), polysaccharides (e.g., chitosan), and lipids (e.g., waxes), also have excellent potential as gas and aroma barriers. The main drawbacks of these types of materials are their inherently high rigidity and the difficulty of processing them in conventional equipment.

For bio-based food-packaging applications, the most important parameter to be considered is its barrier properties. Hydrophilic polymers usually have poor moisture resistance which cause water vapor transmission through packaging and thus affect the quality of foods. This results in shorter shelf lives, increased costs, and eventually more waste. Another technique to improve the barrier properties of biopolymers is to add various nanofillers like nanoclays, metal oxide nanoparticles, and so on [69]. Among the bioplastics, polyglycolic acid (PGA) has excellent barrier properties thus it's one of the most promising new commercially available barrier polymers. Its precursor, glycolic acid, can now be produced via a natural metabolic route, the glyoxylate cycle.

7.5. Water purification

Safe drinking water is a significant, but simple indicator of development. Nanotechnology has shown promising developments in providing safe drinking water through effective purifying mechanisms. Several nanomaterials have already proved to have antibacterial and antifungal properties. Developing affordable materials which can constantly release these antibacterial materials like silver nanoparticles to water is an effective way of providing microbially safe drinking water for all. Developing various nanocomposites with functional materials which can scavenge various toxic metals like arsenic, lead, etc., from water together with the antibacterial agents can result in affordable water purifiers that can function without electricity. The main challenge in this technology is developing stable materials which can release nanoparticles continuously overcoming the scaling on nanomaterials caused by various complex species present inside water.

A relatively new biopolymer, chitosan, shows superior performance where many conventional polymers fail. It is a versatile polymer with applications in water treatment, biomedical and dietary supplement industries. Chitosan is used as a flocculant in water treatment processes and will biodegrade in the environment over periods of weeks or months rather than years. Compared to chitosan, many aggressive and cheap synthetic flocculants are available but they leave a residual impact on the environment. Chitosan removes metals from water by forming chelates.

Chelation is a process by which multiple binding sites along the polymer chain bind with the metal to form a metal cage like structure, to remove it from a solution. This property of chitosan together with its biodegradability make it an eminent candidate for treating difficult industrial storm water and waste water, where conventional methods failed to reduce the contaminant levels. Porous GO-biopolymer gels can efficiently remove cationic dyes and heavy metal ions from wastewater [70]. Nanocrystalline metal oxyhydroxide-chitosan granular composite materials prepared at near room temperature through an aqueous route was also efficient in water purification [71]. Nanofiber membranes can improve the water-filtration process without adversely affecting environment. Combining various nanomaterials together with biopolymers can effectively restrict the formation of biofilms on the polymer surface.

8. Conclusions

Increasing awareness towards the sustainable development has caused the researchers to think about natural and biodegradable polymers for replacing synthetic polymers for various applications. Nanotechnology has already emerged as the future technology which can make tremendous changes to the current existing technology. Various biopolymers are produced by the bacteria under limiting conditions to store carbon and energy resources. Incorporation of various nanofillers to these biopolymers will improve its mechanical and barrier properties which will improve its various applications. The biocompatibility together with the biodegradability of these materials encourages the use of these materials in day-to-day applications.

Author details

Sneha Mohan[1,2], Oluwatobi S. Oluwafemi[2,3*], Nandakumar Kalarikkal[1,4], Sabu Thomas[1,5] and Sandile P. Songca[6]

*Address all correspondence to: Oluwafemi.oluwatobi@gmail.com

1 International and Inter University Centre for Nanoscience and Nanotechnology, Mahatma Gandhi University, Kottayam, Kerala, India

2 Centre for Nanomaterials Science Research, University of Johannesburg, Johannesburg, South Africa

3 Department of Applied Chemistry, University of Johannesburg, Johannesburg, South Africa

4 School of Pure and Applied Physics, Mahatma Gandhi University, Kottayam, Kerala, India

5 School of Chemical Sciences, Mahatma Gandhi University, Kottayam, Kerala, India

6 Department of Chemistry, Walter Sisulu University, Mthatha, South Africa

References

[1] Hernandez N, Williams RC, Cochran, EW. The battle for the "green" polymer. Different approaches for biopolymer synthesis: bioadvantaged vs. bioreplacement. Org Biomol Chem. 2014; 12: 2834–2849. DOI: 10.1039/C3OB42339E

[2] Bozell JZ, Petersen GR. Technology development for the production of biobased products from biorefinery carbohydrates—The US Department of Energy's "Top 10" revisited. Green Chem. 2010; 12: 539–554. DOI: 10.1039/B922014C

[3] Sukan A, Roy I, Keshavarz T. Dual production of biopolymers from bacteria. Carbohydr Polym. 2015; 126: 47–51. DOI: 10.1016/j.carbpol.2015.03.001

[4] Koyama N, Doi Y. Continuous production of poly(3-hydroxybutymte-co-3-hydroxyvalerate) by Alcaligenes eutrophus. Biotechnol Lett. 1995; 17: 281–204. DOI: 10.1007/BF01190637

[5] Tirrell JG, Tirrell DA. Synthesis of biopolymers: proteins, polyesters, polysaccharides and polynucleotides. Curr Opin Solid State Mater Sci. 1996; I: 407–411. DOI: 10.1016/S1359-0286(96)80033-7

[6] Snell KD, Singh V, Brumbley SM. Production of novel biopolymers in plants: recent technological advances and future prospects. Curr Opin Biotechnol. 2015; 32: 68–75. DOI: 10.1016/j.copbio.2014.11.005

[7] Banachowicz E, Gapinski J, Patkowski A. Solution structure of biopolymers: A new method of constructing a bead model. Biophys J. 2000; 78: 70–78. DOI: 10.1016/S0006-3495(00)76573-8

[8] Theophile T, editor. Infrared spectroscopy - Anharmonicity of biomolecules, crosslinking of biopolymers, food quality and medical applications. InTech. 2015; 172. DOI: 10.5772/58483

[9] Bista RK, Brucha RF, Covingtona AM. Infrared spectroscopic study of thermotropic phase behavior of newly developed synthetic biopolymers. Spectrochim Acta Part A. 2011; 81: 583– 589. DOI: 10.1016/j.saa.2011.06.055

[10] Liu Y, Thibodeaux D, Gamble G. Characterization of attenuated total reflection infrared spectral intensity variations of immature and mature cotton fibers by two-dimensional correlation analysis. Appl Spectrosc. 2012; 66: 198–207. DOI: 10.1366/11-06440

[11] Flores-Hernández CG, Colin-Cruz A, Velasco-Santos C, Castano VM, Rivera-Armenta JL, Almendarez-Camarillo A, García-Casillas PE, Martinez-Hernandez AL. All green composites from fully renewable biopolymers: chitosan-starch reinforced with keratin from feathers. Polymers. 2014; 6: 686–705. DOI: 10.3390/polym6030686

[12] Teixeira EM, Campos A, Marconcini JM, Bondancia TJ, Wood D, Klamczynski A, Mattosoa LHC, Glenn GM. Starch/fiber/ poly(lactic acid) foam and compressed foam composites. RSC Adv. 2014; 4: 6616–6623. DOI: 10.1039/C3RA47395C

[13] Blicharska B, Peemoeller H, Witek M. Hydration water dynamics in biopolymers from NMR relaxation in the rotating frame. J Magn Reson. 2010; 207(2): 287–293. DOI: 10.1016/j.jmr.2010.09.012

[14] Rachocki A, Kowalczuk J, Tritt-Goc J. How we can interpret the T1 dispersion of MC, HPMC and HPC polymers above glass temperature?, Solid State Nucl Magn Reson. 2006; 30: 192–197. DOI: 10.1016/j.ssnmr.2006.10.001

[15] Rachocki A, Tritt-Goc J. The molecular origin of nuclear magnetic relaxation in methyl cellulose and hydroxypropylmethyl cellulose. J Polym Res. 2006; 13: 201–206. DOI: 10.1007/s10965-005-9026-6

[16] Tang HR, Wang YL, Belton PS. 13C CPMAS studies of plant cell wall materials and model systems using proton relaxation-induced spectral editing techniques. Solid State Nucl Magn Reson. 2000; 15(4): 239–248. DOI: 10.1016/S0926-2040(99)00064-8

[17] Tang HR, Belton PS, Ng A, Waldron KW, Ryden P. Solid state H-1 NMR studies of cell wall materials of potatoes. Spectrochim Acta Part A. 1999; 55: 883–894. DOI: 10.1016/S1386-1425(98)00235-2

[18] Calucci L, Galleschi L, Geppi M, Mollica G. Structure and dynamics of flour by solid state NMR: effects of hydration and wheat aging. Biomacromolecules. 2004; 5: 1536–1544. DOI: 10.1021/bm0499177

[19] Wokadala OC, Ray SS, Bandyopadhyay J, Wesley-Smith J, Emmambux NM. Morphology, thermal properties and crystallization kinetics of ternary blends of the polylactide and starch biopolymers and nanoclay: The role of nanoclay hydrophobicity. Polym. 2015; 71: 82–92. DOI: 10.1016/j.polymer.2015.06.058

[20] Barbin DF, Valous NA, Dias AP, Camisa J, Hirooka EY, Yamashita F. VIS–NIR spectroscopy as a process analytical technology for compositional characterization of film biopolymers and correlation with their mechanical properties. Mater Sci Eng C Mater Boil Appl. 2015; 56: 274–279. DOI: 10.1016/j.msec.2015.06.029

[21] Williams PC, Norris K, editors. Implementation of near-infrared technology in the agricultural and food industries, 2nd ed. American Association of Cereal Chemists, St. Paul, Minnesota, USA; 2001: 145–169.

[22] Kitamura S, Suzuki K, Iwatsuki M. Observation of silicon surfaces using ultrahigh vacuum noncontact atomic force microscope. Jpn J Appl Phys. 1998; 373: 765–3768. DOI: 10.1143/JJAP.37.3765

[23] Koley G, Spencer MG, Bhangale HR. Cantilever effects on the measurement of electrostatic potentials by scanning Kelvin probe microscopy. App Phys Lett. 2001; 79: 545–548. DOI: 10.1063/1.1384004

[24] Ikada, Y. Integrated Biomaterials Science, Barbucci, R., Ed. Springer US: 2002; p1–23.

[25] Lee, Y-H., Bhattarai, G., Aryal, S., Lee, N-H., Lee, M-H., Kim, T-G., Jhee, EC., Kim, H-Y., Yi, H-K. Modified titanium surface with gelatin nano gold composite increases os-

teoblast cell biocompatibility. Appl Surf Sci. 2010; 256(20): 5882–5887. DOI: 10.1016/j.apsusc.2010.03.069

[26] Yao HHI, Hong MKH, Drummond KJ. Haemostasis in neurosurgery: What is the evidence for gelatin-thrombin matrix sealant? J Clinl Neurosci. 2013; 20(3): 349–356. DOI: 10.1016/j.jocn.2012.09.005

[27] Dash R, Foston M, Ragauskas AJ. Improving the mechanical and thermal properties of gelatin hydrogels cross-linked by cellulose nanowhiskers. Carbohy Polym. 2013: 91(2): 638–645. DOI: 10.1016/j.carbpol.2012.08.080

[28] Miranda SCCC, Silva GAB, Hell RCR, Martins MD, Alves JB, Goes AM. Three-dimensional culture of rat BMMSCs in a porous chitosan-gelatin scaffold: A promising association for bone tissue engineering in oral reconstruction. Arch Oral Biol. 2011; 56(1): 1–15. DOI:10.1016/j.archoralbio.2010.08.018

[29] Wang T, Zhu XK, Xue XT, Wu DY, Wang T, Zhu XK, Xue XT, Wu DY. Hydrogel sheets of chitosan, honey and gelatin as burn wound dressings. Carbohy Polym. 2012; 88(1): 75–83. DOI:10.1016/j.carbpol.2011.11.069

[30] Curcio M, Spizzirri GU, Iemma, F, Puoci F, Cirillo G, Parisi OI, Picci N. Grafted thermo-responsive gelatin microspheres as delivery systems in triggered drug release. Eur J Pharm Biopharm. 2010: 76(1): 48–55. DOI:10.1016/j.ejpb.2010.05.008.

[31] Kido Y, Jo J, Tabata Y. A gene transfection for rat mesenchymal stromal cells in biodegradable gelatin scaffolds containing cationized polysaccharides. Biomaterials 2011; 32(3): 919–925. DOI: 10.1016/j.biomaterials.2010.09.056

[32] French ID. Starch: Chemistry and Technology. 2nd ed. Academic Press Inc., 1984p. 183–247.

[33] Glenn GM, Orts WJ, Nobes GAR. Starch, fiber and CaCO3 effects on the physical properties of foams made by a baking process. Ind. Crops Prod. 2001; 14: 201–212. DOI: 10.1016/S0926-6690(01)00085-1

[34] Orts WJ, Nobes JAR, Glenn GM, Gray GM, Imam S, Chiou B. Blends of starch with ethylene vinyl alcohol copolymers: effect of water, glycerol, and amino acids as plasticizers. Polym Adv Technol. 2007; 18: 629–635. DOI: 10.1002/pat.869

[35] Curvelo AAS, de Carvalho AJF, Agnelli JAM. Thermoplastic starch–cellulosic fibers composites: preliminary results. Carbohydr Polym. 2001; 45: 183–188. DOI: 10.1016/S0144-8617(00)00314-3

[36] Wollerdorfer M, Bader H, Wollerdorfer M, Bader H. Influence of natural fibres on the mechanical properties of biodegradable polymers. Ind Crops Prod. 1998; 8: 105–112. DOI: 10.1016/S0926-6690(97)10015-2

[37] Brett CT. Cellulose microfibrils in plants: biosynthesis, deposition, and integration into the cell wall. Intl Rev Cytol. 2000; 199: 161–199. DOI: 10.1016/S0074-7696(00)99004-1

[38] Rinaudo M. Chitin and chitosan: Properties and applications. Prog Polym Sci. 2006; 31: 603–632. DOI: 10.1016/j.progpolymsci.2006.06.001

[39] Ven TV, Godbout L, editors. Cellulose - Fundamental aspects. InTech; 2013: 376. DOI: 10.5772/2705

[40] Chon EJ, Phan TT, Lim IJ, Zhang YZ, Bay BH, Ramakrishna S, Lim CT. Evaluation of electrospun PCL/gelatin nanofibrous scaffold for wound healing and layered dermal reconstitution. Acta Biomater. 2007; 3: 321–330. DOI: 10.1016/j.actbio.2007.01.002

[41] Fabbri P, Bondioli F, Messori M, Bartoli C, Dinucci D, Chiellini F. Porous scaffolds of polycaprolactone reinforced with in situgenerated hydroxyapatite for bone tissue engineering. J Mater Sci: Mater Med. 2010; 21: 343–351. DOI: 10.1007/s10856-009-3839-5.

[42] Clemenson S, David L, Espuche E. Structure and morphology of nanocomposite films prepared from polyvinyl alcohol and silver nitrate: influence of thermal treatment. J Polym Sci Part A: Polym Chem. 2007; 45: 2657–2672. DOI: 10.1002/pola.22020

[43] Kim ST, Chob KB, Kang G. Preventive effects of polyurethane foam and polyvinyl acetate on bleeding and pain in young patients undergoing conchotomies. Int J Pediatric Otorhinolaryngology 2013; 77: 113–116. DOI: 10.1016/j.ijporl.2012.10.007

[44] Guzman A, Arroyo J, Verde L, Rengifo J. Synthesis and characterization of copper nanoparticles/polyvinyl chloride (Cu NPs/PVC) nanocomposites. Procedia Mater Sci. 2015; 9: 298–304.

[45] Jiang Y, Wang H, Deng M, Wang Z, Zhang J, Wang H, Zhang H. Effect of ultrasonication on the fibril-formation and gel properties of collagen from grass carp skin. Mater Sci Eng., C. 2016; 59: 1038–1046. DOI: 10.1016/j.msec.2015.11.007

[46] Lee CH, Singlaa A, Lee Y. Biomedical applications of collagen. Int J Pharm.; 2001; 221: 1–22. DOI: 10.1016/S0378-5173(01)00691-3

[47] Scheithauer EC, Li W, Ding Y, Harhaus L, Roether JA, Boccaccini AR. Preparation and characterization of electrosprayed daidzein–loaded PHBV microspheres. Mater Lett. 2015; 158: 66–69. DOI: 10.1016/j.matlet.2015.05.133

[48] Fedorov MB, Vikhoreva GA, Kildeeva NR, Maslikova AN, Bonartseva GA, Galbraikh LS. Modeling of surface modification of sewing thread. Fibre Chem. 2005; 37(6): 441–446.

[49] Alivisatos AP. Semiconductor clusters, nanocrystals, and quantum dots. Sci. 1996; 271(5271): 933–937. DOI: 10.1126/science.271.5251.933

[50] Hebeish AA, El-Rafie MH, Abdel-Mohdy FA, Abdel-Halim ES, Emam HE. Carboxymethyl cellulose for green synthesis and stabilization of silver nanoparticles. Carbohy Polym. 2010; 82: 933–941. DOI: 10.1016/j.carbpol.2010.06.020

[51] Panacek A, Kvitek L, Prucek R, Kolar M, Vecerova R, Pizurova N, Sharma VK, Nevena T, Zboril Z. Silver colloid nanoparticles: synthesis, characterization, and their antibacterial activity. J Phys Chem B. 2006; 110: 16248–16253. DOI: 10.1021/jp063826h

[52] Darroudi M, Ahmad MB, Abdullah AH, Ibrahim NA. Green synthesis and characterization of gelatin-based and sugar-reduced silver nanoparticles. Int J Nanomed. 2011; 6: 569–574. DOI: 10.2147/IJN.S16867

[53] Mohan S, Oluwafemi OS, George SC, Jayachandran VP, Lewu FB, Songca SP, Kalarikkal N, Thomas S. Completely green synthesis of dextrose reduced silver nanoparticles, its antimicrobial and sensing properties. Carbohyd Polym. 2014; 106: 469–474. DOI: 10.1016/j.carbpol.2014.01.008

[54] Hoffman AS. Hydrogels for biomedical applications. Adv Drug Delivery Rev. 2002; 43: 3–12. DOI: 10.1016/S0169-409X(01)00239-3

[55] Meinel AJ, Kubow KE, Klotzsch E, Fuentes M, Smith ML, Vogel V, Merkle H P, Meinel L. Optimization strategies for electrospun silk fibroin tissue engineering scaffolds. Biomater. 2009; 30: 3058–3067. DOI: 10.1016/j.biomaterials.2009.01.054

[56] Vlierberghe SV, Cnudde V, Dubruel P, Masschaele U, Cosijns A, Paepe ID, Jacobs PJS, Hoorebeke LV, Remon JP, Schacht E. Porous gelatin hydrogels: 1. Cryogenic formation and structure analysis. Biomacromolecules 2007; 8: 331–337. DOI: 10.1021/bm060684o

[57] Zhao W, Li J, Jin K, Liu W, Qiu X, Li C. Fabrication of functional PLGA-based electrospun scaffolds and their applications in biomedical engineering. Mater Sci Eng: C. 2016; 59: 1181–1194. DOI: 10.1016/j.msec.2015.11.026

[58] Liao H, Qi R, Shen M, Cao X, Guo R, Zhang Y, Shi X. Improved cellular response on multiwalled carbon nanotube-incorporated electrospun polyvinyl alcohol/chitosan nanofibrous scaffolds. Colloids Surf, B. 2011; 84: 528–535. DOI:10.1016/j.colsurfb.2011.02.010

[59] Reddy N, Reddy R, Jiang Q. Crosslinking biopolymers for biomedical applications. 2015; 33: 362–369. DOI: 10.1016/j.tibtech.2015.03.008

[60] Othman SH. Bio-nanocomposite materials for food packaging applications: Types of biopolymer and nano-sized filler. Agric Agric Sci Procedia. 2014; 2: 296–303. DOI: 10.1016/j.aaspro.2014.11.042

[61] Mohy Eldin MS, Soliman EA, Hashem AI, Tamer TM. Biopolymer Modifications for Biomedical Applications, In: Theophanides T, editor. Infrared Spectroscopy - Life and Biomedical Sciences. InTech; 2012. DOI: 10.5772/35489

[62] Champagne CP, Fustier P. Microencapsulation for the improved delivery of bioactive compounds into foods. Curr Opin Biotechnol. 2007; 18(2): 184–190. DOI: 10.1016/j.copbio.2007.03.001

[63] Shit SC, Shah PM. Edible polymers: challenges and opportunities. J Polym. 2014; 4: 1–13. DOI: 10.1155/2014/427259

[64] Chaudhary D, Dong Y, Kar KK. Hydrophilic plasticized biopolymers: Morphological influence on physical properties. Mater Lett. 2010; 64: 872–875. DOI:10.1016/j.matlet.2010.01.048

[65] Myllärinen P, Partanen R, Seppälä J, Forsella P. Effect of glycerol on behaviour of amylose and amylopectin films. Carb Polym. 2002; 50: 355–361. DOI: 10.1016/S0144-8617(02)00042-5

[66] Gaudin S, Lourdin D, Botlan DL, Ilari JL, Colonna P. Plasticisation and mobility in starch-sorbitol films. J Cereal Sci. 1999; 29: 273–284. DOI:10.1006/jcrs.1999.0236

[67] Chaudhary DS. Understanding amylose crystallinity in starch–clay nanocomposites. J Polym Sci: Polym Phys. 2008; 46: 979–988. DOI: 10.1002/polb.21437

[68] Siracus V, Blanco I, Romani S, Tylewicz U, Rocculi P, Rosa MD. Poly(lactic acid)-modified films for food packaging application. Physical, Mechanical, and Barrier Behavior. 2012; 125: E390–E401. DOI: 10.1002/app.36829

[69] Majeed K, Jawaid M, Hassan A, Abu Bakar A, Abdul Khalil HPS, Salema AA, Inuwa I. Potential materials for food packaging from nanoclay/natural fibres filled hybrid composites. Mater Des. 2013; 46: 391–410. DOI: 10.1016/j.matdes.2012.10.044

[70] Cheng C, Deng J, Lei B, He A, Zhang X, Ma L, Li S, Zhao C. Toward 3D graphene oxide gels based adsorbents for high-efficient water treatment via the promotion of biopolymers. J Hazard Mater. 2013; 263: 467–478. DOI: 10.1016/j.jhazmat.2013.09.065

[71] Sankar MU, Aigal S, Maliyekkal SM, Chaudhary A, Avula A, Kumar A, Chaudhari K, Pradeep T. Biopolymer-reinforced synthetic granular nanocomposites for affordable point-of-use water purification. PNAS. 2013; 110: 8459–8464. DOI: 10.1073/pnas.1220222110

Chapter 4

Biopolymer Thin Films Synthesized by Advanced Pulsed Laser Techniques

Emanuel Axente, Felix Sima, Carmen Ristoscu, Natalia Mihailescu and Ion N. Mihailescu

Additional information is available at the end of the chapter

http://dx.doi.org/10.5772/61734

Abstract

This chapter provides an overview of recent advances in the field of laser-based synthesis of biopolymer thin films for biomedical applications. The introduction addresses the importance of biopolymer thin films with respect to several applications like tissue engineering, cell instructive environments, and drug delivery systems. The next section is devoted to applications of the fabrication of organic and hybrid organic–inorganic coatings. Matrix-assisted pulsed laser evaporation (MAPLE) and Combinatorial-MAPLE are introduced and compared with other conventional methods of thin films assembling on solid substrates. Advantages and limitations of the methods are pointed out by focusing on the delicate transfer of bio-macromolecules, preservation of properties and on the prospect of combinatorial libraries' synthesis in a single-step process. The following section provides a brief description of fundamental processes involved in the molecular transfer of delicate materials by MAPLE. Then, the chapter focuses on the laser synthesis of two polysaccharide thin films, namely Dextran doped with iron oxide nanoparticles and Levan, followed by an overview on the MAPLE synthesis of other biopolymers. The chapter ends with summary and perspectives of this fast-expanding research field, and a rich bibliographic database.

Keywords: Biopolymers, MAPLE and Combinatorial-MAPLE, thin films, biomedical applications

1. Introduction

During the last decades, thin film layers have proved to be in the forefront for continuous developments in nanosciences and nanotechnologies [1]. Several research fields are nowadays related to thin films with a broad range of potential applications. Thin (<1 μm) and very thin films (<100 nm) have already proven as key prospectives for advances in semiconductor devices, wireless and telecommunications, integrated circuits, solar cells, light emitting diodes, liquid crystal displays, magneto-optic memories, audio and video systems, compact disks, electro-optic coatings, memories, multilayer capacitors, flat-panel displays, smart windows, computer chips, magneto-optic disks, lithography, micro-electro-mechanical systems, and multifunctional protective coatings, as well as other emerging cutting-edge technologies [1]. Among them, biopolymer thin films area hot topic for biomedical applications like tissue engineering and cell instructive environments, drug delivery systems, antimicrobial surfaces, or biosensors.

Both natural and synthetic biopolymer thin films are representatives for engineering of the cell/biomaterial interface in view of controlling cell behavior, with a high impact for the *design of instructive surfaces* for tissue repair or cell supports [2]. Several techniques for the synthesis of biopolymer thin films such as dip-coating, spin-coating, drop-casting, and Langmuir–Blodgett are currently used. Separately, each technique exhibits advantages and drawbacks, generally allowing the assembling of a limited class of compounds [3, 4].

Pulsed laser technologies have extensively confirmed to be versatile for the fabrication of high-quality biomaterial thin films as they ensure the stoichiometry preservation [5]. However, in case of biopolymers, Pulsed Laser Deposition (PLD) induces an irreversible damage of the organic materials' composition due to high laser intensities [6]. This limitation is avoided by the new technique called matrix-assisted pulsed laser evaporation (MAPLE), which allows transferring delicate, large molecular-mass organic compounds [7-9]. Indeed, after the first implementation in the late 1990s [10], MAPLE proved to attain its maturity with the synthesis of functional, high-quality organic thin films, as reported by Guo *et al.* [11]. Laser technologies have thus been shown to successfully apply to transfer polysaccharides [12-18], proteins [19-23], or even living organisms like bacteria [24, 25], fungi [26], or mammalian cells [27-29].

Among the three main classes of natural biopolymers (polynucleotides, polypeptides, and polysaccharides) [30], this chapter focuses on the advanced laser synthesis of two polysaccharide thin films, namely Dextran doped with iron oxide nanoparticles and Levan. A detailed literature survey overviews the MAPLE synthesis of other biopolymers. Moreover, the possibility to fabricate combinatorial libraries of biopolymers by advanced laser technique in a single-step process is mentioned.

The use of polysaccharides as biomaterials has evolved over the past several decades, covering biomimetic approaches. A simple classification divides the polysaccharides as derived from non-mammalian or mammalian sources. The first group includes Alginate, Chitin, and Dextran, among others, which possess similar saccharide structure although having different origins [31, 32]. The interest in these materials is due to their relatively easy extraction and purification of large quantities at low cost. Mammalian polysaccharides, such as the Glycosaminoglycans chondroitin sulfate, Hyaluronan, and Heparin, possess chemical similarities to the non-mammalian ones but their isolation is more complicated [33-35]. Their unique biological functionality, e.g., specific binding with multiple proteins, has raised increased interest for application in the field of biomaterials, as reviewed in Ref. [36].

Like other non-mammalian polysaccharides, Dextran is not present in human tissues, being expressed by bacteria such as either *Leuconostoc mesenteroides* or *Streptococcus mutans* [32]. It is a complex branched glucan (polysaccharide made of many glucose molecules) consisting of chains of different lengths (from 3 to 2000 kDa). Dextran is highly water-soluble and easily functionalized through its reactive hydroxyl groups [37]. As a biodegradable and biocompatible biopolymer, it was investigated as a blood plasma replacement in the early 1940s [38]. Dextran could be functionalized with nanoparticles [18, 39, 40] or conjugated with different polymers [41-44] for various biomedical applications as described in *Section 4.1.*

Among natural polysaccharides, Levan is a high molecular weight, water-soluble bacterial exopolysaccharide (β2,6-linked fructan) [45]. It is produced by microbial fermentation of *Halomonas smyrnensis* $AAD6^T$ batch cultures grown on pretreated sugar beet molasses [46]. Due to its anticancer [47], anti-inflammatory [48], anticytotoxic [46], and antitumoral [49-51] properties, Levan is a promising biopolymer with a huge potential for biomedical applications. Some other application fields are in textiles, cosmetics, wastewater treatment, food, and pharmaceutical industries [46, 50, 52]. Sima *et al.* [12] reported for the first time on pure and oxidized Levan nanostructured thin films assembling by MAPLE (*Section 4.2*). The authors evidenced a high potential for cell proliferation for both coatings (with certain predominance for oxidized Levan) by *in vitro* colorimetric assays.

The latest progress achieved in the development of new materials or innovative properties is often based on combinatorial processes. Usually, the fabrication of a multicomponent organic coating is performed by film casting procedures [53, 54]. Compositional and/or thickness gradient thin films of polymers are produced by premixing different polymer solutions, followed by applying a linear temperature gradient [55]. Subsequently, stem cells were exposed to molecule combinations arrays, looking for synergistic effects that could direct cell fate [56]. Other studies reported binary combinations of an adhesive molecule and a growth factor in view of parallel testing of several environmental media to control the evolution of neural stem cells [57]. This approach is based on previously tested well-defined concentrations of biomolecules to obtain the desired combinations.

This chapter introduces an innovative solution for the synthesis of combinatorial libraries of biopolymer thin films by Combinatorial-MAPLE (C-MAPLE) technique (*Section 4.3*). The compositional gradient occurs in this case naturally along the substrate by the simultaneous laser irradiation of two cryogenic targets and thin-film co-deposition. A representative example is for Levan (L) and oxidized Levan (OL) biopolymers, *in vitro* cell culture assays illustrating characteristic responses of cells to specific surface regions [13]. The versatility of the method for the synthesis of hybrid materials and further possible developments are summarized.

2. Laser-based techniques for the synthesis of biopolymer thin films

2.1. Complementary techniques for organic thin films' fabrication

Nowadays, there is a large body of experimental studies focused on the deposition of thin, uniform, and adherent films of numerous types of polymers, organic materials, and biomaterials (soft materials). The goal is the fabrication of controlled structures (or nanostructures) essential to be used, for instance, in diverse areas such as medical field, packaging, cosmetics, clothing fabrics, food additives, industrial plastics, water treatment chemicals absorbents, biosensors and detectors, or data storage elements [58]. For this purpose, one needs to choose an appropriate deposition method which should depend on the physical–chemical properties of the biomaterial, requirements for film quality, type of the substrate, and the production costs [25].

There exist techniques that can be used to deposit highly uniform biopolymers thin films and micro-patterns [59], like laser-induced forward transfer (LIFT) [60] or Matrix-assisted laser desorption/ionization (MALDI) [61]. Besides, sol-gel, layer by layer (LbL), aerosol spraying, dip coating, and spin coating are techniques that entail liquid solutions of the material in a volatile solvent [7, 62, 63]. A common method to obtain surfaces with a single biomolecular layer is Langmuir–Blodgett (LB) dip coating, using self-assembled monolayers [7, 64, 65].

In spite of the rich list reported in the literature of biomaterials deposited thin films, these techniques have their own merits but also disadvantages (e.g., manufacturing of a limited class of biomaterials). In order to obtain a better quality of thin films, these techniques should allow the control of several parameters during and after deposition. In short, in the case of LB method, key parameters have been identified to be crucial for obtaining high-quality thin films, e.g., deposition speed or transfer surface pressure [66]. These parameters directly affect the adhesion strength between film and substrate, causing delamination of the layer or generating discontinuities in the structure, which worsen the homogeneity of the films [67]. Also, this technique is limited to very thin layers [68, 69].

Sol-gel is a largely used method, successfully applied to obtain organic/inorganic hybrid coatings [70, 71] through precipitation by chemical reactions in liquid medium. In spite of its advantages, sol-gel technique will never reach its full potential due to some limitations, e.g., poor coating adhesion, low wear-resistance, involvement of liquid media that could impede the multilayer assembling (affect interfaces), high permeability, limit of the maximum coating thickness (~0.5 μm) [72], and difficulties in controlling porosity. On the other hand, a thick organic coating often results in failure during thermal process. Moreover, sol-gel is a substrate-dependent technique, and the thermal expansion mismatch hampers the wide application.

Other drawbacks that restrict the application of these methods are related to the choice of the solvent or liquid media issues during multilayer assembling, difficulties in obtaining large-area uniform thin coatings, or that the methods are time- and material-consuming [73-77].

It is widely accepted that the surface topography has a significant influence in a wide range of organic materials applications. In addition to surface topography and chemistry, thin film adhesion to the substrate also plays an important role. According to the literature, the current coating techniques [78, 79] provide inadequate coating adhesion to the substrate.

Nowadays, it is considered that laser-based technologies are among the main, most powerful tools for fabrication of micro- and nano-arrays of a wide range of different biomaterials with controlled thickness (with the precision of 1 Å), good adhesion to the substrate, and specific surface properties. Moreover, these methods permit the relatively uniform spreading of material over rather large areas, control of substrate temperature, low material consumption, and stoichiometry conservation of the growing film.

Nevertheless, when using UV lasers such as excimers operating at 193- or 248-nm or frequency tripled Nd:YAG lasers at 355 nm (6.4–3.5 eV/photon) to obtain thin films of very complex delicate biomolecules, irreversible damage of the chemical bonds and consequent compositional modification are induced. Consequently, PLD technique is not a viable option for fabrication of complex biomolecules such as polymer thin films [6].

2.2. Matrix-Assisted Pulsed Laser Evaporation (MAPLE)

Discovered at the end of the 1990s at the Naval Research Laboratory, MAPLE has become nowadays an active area of research [9, 80]. Developed as a complementary method to PLD, MAPLE has introduced new advancements in laser methods deposition of thin films. In short, MAPLE provides gentler pulsed laser evaporation, a less damaging approach for transferring many different compounds including small or large molecular weight species, such as organic and polymeric molecules [81].

MAPLE technique has been successfully applied to obtain thin films of sensitive materials avoiding thermal decomposition and irreversible degradation under the action of electric or magnetic fields. Applications targeted development of biosensing, chemical sensing, and biochemical analysis, as well as drug delivery systems and the developing of a new generation of implants [19, 82-86]. In MAPLE, the laser-induced material ejection is generated backward from a solid cryogenic target. The expulsed substance is assembled on substrates, where it forms a thin film with a thickness from a few to several hundreds of *nm*. The incident laser pulse used for MAPLE initiates two photo-thermal processes in the matrix: evaporation of the frozen composite target and transfer of the material onto the substrate. The schematic of MAPLE setup is presented in Figure 1.

Typically, the target consists of base material (less than 10% wt.) dissolved/suspended into a laser wavelength absorbing solvent when in frozen state. The organic material molecules reach sufficient kinetic energy by collective collisions with the evaporating solvent molecules, ensuring a controlled transfer on the substrate, in gas phase. Since the receiving substrate is usually kept at room temperature and the sticking coefficient of the solvent is nearly zero, the evaporated solvent is efficiently pumped away by the vacuum system. The solvent and solute concentration should be selected so that: the solute can be dissolved to form a dilute, particulate-free solution; most of the laser energy has to be absorbed by the solvent molecules rather than the solute ones; and no photochemical reaction is produced between solvent and solute. By optimization of the MAPLE deposition parameters (laser wavelength, repetition rate, laser fluence, solvent type, solute concentration, substrate temperature, background gas and pressure), the process can proceed without significant material decomposition [6, 25, 81]. It was demonstrated that MAPLE could provide more crystalline layers as compared to PLD method from the same materials [87]. Recent comprehensive reviews on MAPLE deposition of organic, biological and nanoparticle thin films illustrated large potential (drug delivery, biosensors, etc.) applications of thin coatings obtained by this method [9, 88].

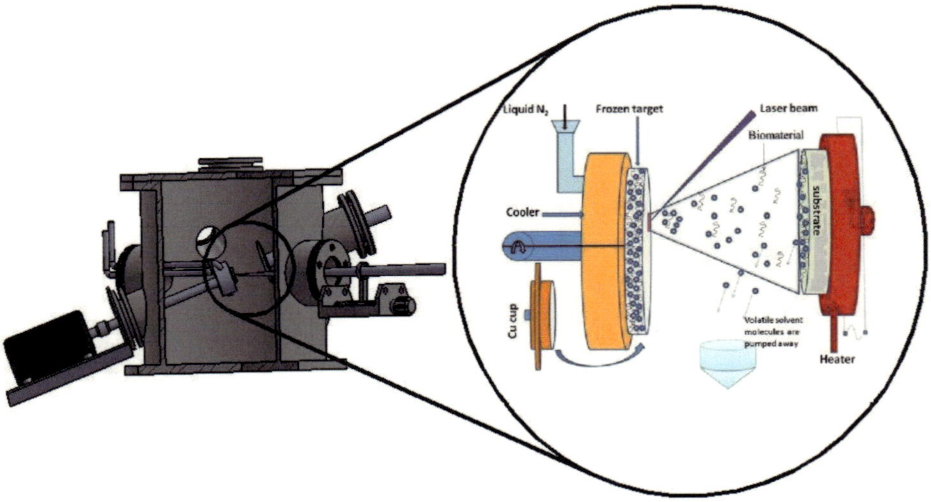

Figure 1. The experimental setup of MAPLE method.

2.3. Compositional library thin films fabrication by Combinatorial-MAPLE

Recently, improvements have been made on MAPLE technique based on developing a new concept for the synthesis of functionalized biomaterials surfaces. Combinatorial-MAPLE technique was introduced as a new approach for the fabrication of gradient organic/inorganic thin films, for the identification of best bioactive surfaces able to modulate and control cell behavior [13]. In C-MAPLE experiments, two targets are simultaneously evaporated by laser beams. Different lasers could be used, having different characteristics (pulse duration, wavelength, repetition rate), or the beam of one laser is divided into two beams (Figure 2) by an optical splitter.

The two beams are independently focused onto the surface of each target, containing the frozen solutions to be irradiated. To grow uniform thin films, the targets are continuously rotated (from 1 to 80 rpm). This arrangement is preferred to avoid drilling and allow for the expulsion of materials in an accurate mode. To deposit high-quality films with controlled thickness, one should choose for each compound the appropriate fluence and number of laser pulses to reasonably balance the deposition rate along the length of the substrate. This particular combinatorial setup allows for the smooth and isotropic interpenetration and mixing of the two substance fluxes evaporated form the targets, resulting in the deposition of a continuous and uniform composition gradient. A gradient of composition from 100% A material to 100% B material is thus obtained on a substrate, as schematically depicted in Figure 3.

This method opens the possibility to both combine and immobilize two or more materials, dissolved in different solvents, using diverse wavelengths. Further, by investigating the obtained structures by physical, chemical, and biological methods, one could select the best compositions that can be synthesized from the two components. The advent of Combinatorial-MAPLE could open new research frontiers in identification of the best dosage between several organic and/or inorganic materials with great prospective for many applications.

http://dx.doi.org/10.5772/61734

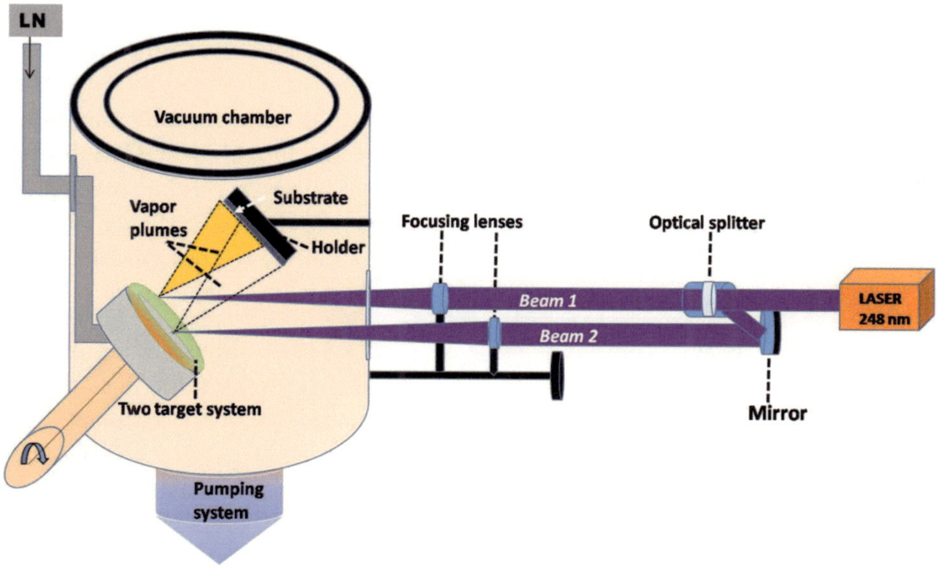

Figure 2. The experimental setup of C-MAPLE method.

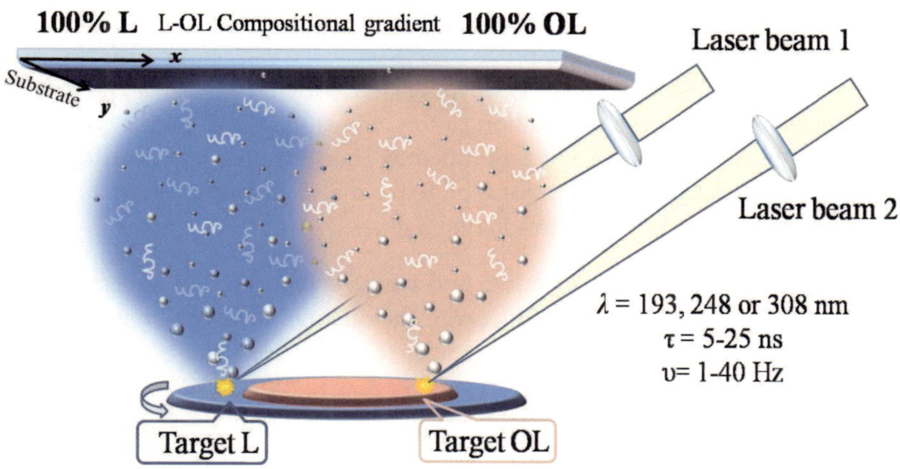

Figure 3. The design of the experimental setup.

3. Theoretical aspects of laser–matter interactions: Insights for MAPLE

The understanding of the fundamental processes responsible for the molecular transfer (physical mechanisms governing the ejection) of delicate materials when suspended in a volatile matrix in form of an icy target should be investigated in order to explain the experimental results. More concretely, the large surface roughness specific to MAPLE depositions could not be accounted for in terms of the first explanations advanced in Ref. [6], where it is hypothesized that the matrix absorbs laser energy and converts it into thermal energy which will vaporize the solvent molecules. Then, the active material molecules are displaced onto the collector surface after collisions with solvent molecules. Moreover, the studies have proved that it is mandatory to make an appropriate choice of transfer parameters (laser wavelength, incident fluence and pulse duration, type of solvent, substrate nature and temperature, and nature and pressure of the background gas). When these conditions are met, MAPLE ensures the "soft" ejection and transfer of delicate material molecules preserving their chemical structure and very likely their functionality and biologic activity. Accordingly, the proper choice of solvent and deposition conditions is essential for getting the best possible compromise between films bioactivity and morphology.

Before 2007, no theoretical or computational works for a better understanding of MAPLE process have been conducted. Leveugle and Zhigilei [89] developed for the first time a computational model (a coarse-grained molecular dynamic (MD) model) to explain the basic mechanisms related to laser–material interaction and non-equilibrium processes and the resulting film characteristics, especially morphology. The authors demonstrated that even at low concentration (0.1–5% wt.) of active material in the matrix, the active molecules can influence the molecular ejection and subsequently the morphology of the films. The MD simulations were conducted for a laser wavelength of 337 nm with pulse duration of 50 ps and incident laser fluences in the range 3–9 mJ/cm^2. It was revealed that below the ablation threshold of the matrix (3.5 mJ/cm^2), only an evaporative process occurs. In this regime, the matrix molecules solely get vaporized and no active molecules are ejected. For laser fluences superior to this threshold, an explosive process takes place accompanied by ejection of clusters and liquid particles from the MAPLE target. This results in the deposition of thin coatings with a high roughness morphology. In addition to this, the simulations predicted that also the composition of the target surface can be modified by the ablation process induced by matrix evaporation. It is expected that an increase of active material concentration in the target is produced, especially in a multipulse irradiation regime.

The ejection of clusters composed of solute and matrix molecules and the rough surface of the MAPLE coatings seem inevitable. However, the quality of the growing films can be, at least partially, controlled by the temperature of the substrate and possible post-deposition treatments.

Based on this model, it is possible to avoid or minimize the deposition of molecular clusters in MAPLE and achieve a molecule-by-molecule deposition of ultrathin films without significant roughness by selecting the appropriate set of transfer parameters.

Water, "the universal solvent," is perhaps the most versatile matrix for biopolymers. A frozen aqueous solution is an attractive medium since such an icy matrix has turned out recently to yield promising results for biomolecule transfer from targets to selected substrates [21, 90, 91]. However, the laser light at 248 nm is not very efficiently absorbed by the ice matrix, but is on the other hand less harmful to the bonds in the polymer than light at shorter wavelengths [92].

When using a water matrix, the ablation process could be related to local overheating of absorbing areas constituted by biomolecules in the outmost surface layer, heating the solvent in their vicinity [93]. In vacuum conditions, the water solvent starts boiling at room temperature, the vapors transporting the biomolecules toward the substrate surface. The material ejection is consequently produced at lower temperature than the degradation threshold.

Another mechanism based on *nonhomogeneous absorption* could be applied for low laser fluence, when the mass ejection is produced by surface evaporation, but also for higher fluences when the expulsion is governed by the hydrodynamic ablation [94]. A frozen MAPLE target contains not only the molecules of active materials and of the matrix but also different phases such as ice cracks, air bubbles, or other defects. These phases were suggested to be involved in light absorption or scattering processes during laser irradiation of the heterogeneous frozen target. Accordingly, the absorption was found to be higher in ice as compared to water. The laser absorption can be increased by the addition of other compounds in the solution, which introduce local modifications of material properties [9].

These MD simulation and models allow achieving two main objectives of MAPLE, which are as follows: (*i*) to avoid photo-chemical and photo-thermal molecular fragmentation (also called "bond scission") characteristic to PLD, and (*ii*) to achieve deposition of highly uniform thin films that cannot be obtained by solvent-based coating methods.

4. Biopolymer thin films for biomedical applications

4.1. Hybrid Dextran–iron oxide thin films

The development of hybrid biomaterials, in particular in the form of thin films, has received a growing interest in the last decades mainly due to their biomedical applications. It is generally accepted that both synthetic and natural biopolymers could be used in biomaterials research, because of their unique structures that allows for a specific functionalization for desired applications [36]. Moreover, embedding metal and/or metal oxide nanoparticles (NPs) into an organic and/or inorganic matrix could lead to the fabrication of a novel generation of *smart biomaterials*, with optimized properties [95].

As known, Dextran is a natural biopolymer that can be synthesized from fermentation of sucrose-containing media [31, 37]. Its structure consists of linear (1 → 6)-α-D-glucose, with branches extending mainly from (1 → 3) and occasionally from (1 → 4) or (1 → 2) positions accounting for a 5% degree of branching [36]. Due to its specific properties (neutral and water-soluble, easy to functionalize through its reactive hydroxyl groups, biodegradable, biocompatible, long-term stability), it is intensively used in several biomedical applications like an antithrombotic (antiplatelet) to reduce blood viscosity, and as a volume expander in hypovolemia [96]. Moreover, Dextran-based coatings were proven to develop well-defined surface modifications that could induce specific cell interactions and enhanced performances in long-term biomaterial implants [97].

In the last decades, biocompatible iron oxide NPs have attracted increased consideration due to promising properties for the biomedical field. Applications reported in the literature are related to: contrast agents in magnetic resonance imaging (MRI) [98, 99], *in vitro* cell separation

[100], *in vivo* diagnosis of cancer [101], targeted destruction of tumor tissues by hyperthermia [102, 103], and targeted drug delivery systems since allows activation by applying an external magnetic field [104, 105]. However, for such applications the NPs must combine several different properties like high magnetic saturation, biocompatibility, and interactive functions at the surface. Accordingly, in most of the cases, a further modification of their surfaces is mandatory by applying thin organic and/or inorganic coatings [106] in view of binding to drugs, proteins, enzymes, antibodies, or nucleotides [107]. The most studied iron oxide NPs are maghemite γ-Fe_2O_3, and magnetite Fe_3O_4 with single domains of about 5–20 nm in diameter. Details about their chemical synthesis, surface engineering, and effectiveness for biomedical applications were reviewed by Gupta and Gupta [107].

Although emerging applications envisioned, it was only recently reported that the nanosized feature of particles could be associated to cytotoxicity [108, 109], at least when large amounts of NPs have to be used. Only NP concentrations below 100 μg/ml are considered safe [110]. The growth of hybrid thin films consisting of Dextran and maghemite γ-Fe_2O_3 NPs using MAPLE technique was reported by Predoi and coworkers [18]. The authors investigated the biocompatibility, an essential requirement for the introduction of iron oxide into the human body, but also the influence of the NP concentration on the biomimetic properties of the synthesized coatings.

The chemical synthesis of iron oxide NPs was performed following a classical co-precipitation procedure, according to Bee *et al.* [111]. For investigations, the obtained particles were dispersed in deionized water, the pH being adjusted to 7 using aqueous ammonia. The total iron concentration of the suspensions determined by redox-titration was 0.38 mol L^{-1}. Well-crystallized NPs, having an average size of 8.3±0.3 nm were obtained, as visualized in high resolution Transmission Electron Microscopy (HRTEM) images, the corresponding Selected Area Electron Diffraction (SAED) pattern indicating the reflection of the cubic maghemite γ-Fe_2O_3 phase [18].

Dextran and Dextran–iron oxide composite thin films were deposited by MAPLE. Different solutions consisting of 25,000 Da molecular weight Dextran (10 % wt.), iron oxide NPs (0–5 % wt.), and distilled water as matrix solvent were used for target preparation. Before each deposition, 5 ml of the obtained solution was dropped in a copper holder of 3 cm diameter and 5 mm height and immersed in liquid nitrogen (77 K) to freeze a solid target. The pure and hybrid coatings were grown on SiO_2 glass substrates by applying 25×10^3 subsequent laser pulses. After optimization trials, the incident laser fluence on the target surface was set at 0.5 J/cm^2.

The structure of hybrid Dextran–iron oxide thin films obtained from the composite targets was first analyzed by X-Ray Diffraction (XRD). The diffraction patterns revealed the presence of the peaks assigned to the cubic maghemite γ-Fe_2O_3 structure. This observation is in good agreement with SAED analysis indicating that the NPs' crystalline phase was preserved during laser processing. Moreover, the average size <*d*> of the nanocrystalites computed using Scherrer's formula [112] evidenced values around 7.7 nm, in accordance with the average size of the NPs determined from HRTEM micrographs [18].

Figure 4 illustrates the structural characteristics of the coatings inferred by Fourier transform infrared spectroscopy (FTIR). The graphs illustrate the spectra of iron oxide NPs (Figure 4.A) and Dextran used for the preparation of the MAPLE targets (Figure 4.B), as well as the spectra of the Dextran–iron oxide thin films synthesized by MAPLE from the composite targets containing 5% wt. (Figure 4.C) and 1% wt. (Figure 4.D) maghemite γ-Fe_2O_3, respectively. All spectra exhibit the bands assigned to OH stretching (ν OH) and HOH bending (δ OH) vibrational bands at 3480 cm^{-1} and 1700 cm^{-1} due to adsorbed water molecules [113]. The bands observed at 620 cm^{-1} and 580 cm^{-1} in the spectrum of the iron oxide NPs correspond to the Fe–O vibration modes of γ-Fe_2O_3 [114, 115]. In the FTIR spectrum of Dextran-NPs, the characteristic absorption bands of the polysaccharide can be observed [114, 116-118]. They are summarized in Table 1. One could notice that the spectra recorded in case of the hybrid coatings are very similar to the spectra of the starting materials. Furthermore, the intensity of the bands corresponding to the maghemite γ-Fe_2O_3 phase increases with increasing NPs concentration in the composite targets used in MAPLE experiments.

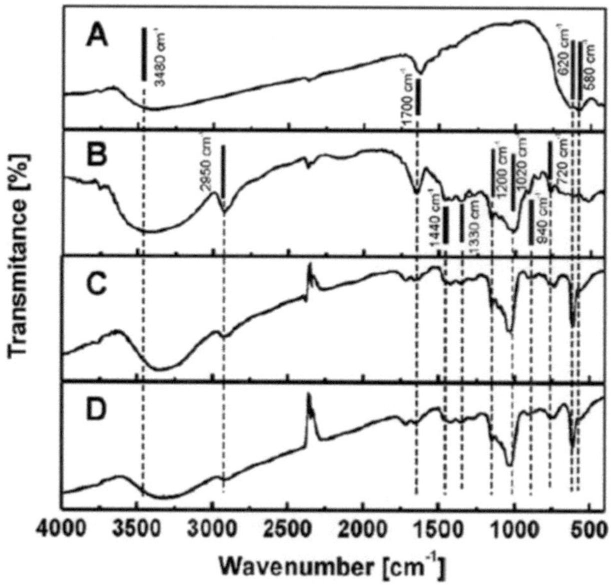

Figure 4. FTIR spectra of iron oxide NPs (A), pure Dextran (B), as well as Dextran–iron oxide thin films containing 10% wt. Dextran, 5% wt. (C) and 1% wt. (D) iron oxide NPs. (Reproduced with permission from [18])

The typical surface morphology of thin films deposited by MAPLE technique is characterized by an aggregated structure, consisting of micrometer-sized particles [17, 18, 22, 39, 119]. It is worth noting that a larger specific surface area was proven to induce an enhanced bioactivity, able to promote osteoblast differentiation, as reported in case of hybrid organic–inorganic thin films deposited by MAPLE [120, 121].

The biocompatibility of the Dextran–iron oxide thin films was demonstrated by 3-(4.5 dimethylthiazol-2yl)-2.5-diphenyltetrazolium bromide-based colorimetric assay, using human liver hepatocellular carcinoma (Hep G2) cell line [18]. In Figure 5 are visible the inverted light microscope images of Hep G2 cells cultivated on pure Dextran (Figures 5.A, 5.B) and Dextran–iron oxide thin films obtained from composite targets (Figures 5.C, 5.D) after 24 (A, C) and 48 h (B, D) incubation time. The cultured Hep G2 cells form polygonal multicellular aggregates [122] as could be observed from figures, this morphology being preserved even after 48 h incubation time. At both iron oxide concentrations, the aggregates' size increased with the incubation time, but is still close to those grown on uncoated plastic slides.

ν (cm^{-1})	*Characteristic modes*
3480	OH stretching (ν OH)
1700	HOH bending (δ OH)
2950	ν (C\H) and δ (C\H)
1440	vibrational modes
1200	ν (C\O) vibrations
940	α Glucopyranose ring
720	deformation modes

Table 1. Characteristic absorption bands of Dextran biopolymers [18].

The results of the viability tests (MTT) of Hep G2 cells on pure Dextran drop-casted solution, Dextran and Dextran–iron oxide composite thin films obtained by MAPLE are presented in Figure 5.E, and compared to the cells cultivated on control samples (considered as having a viability of 100%). A small decrease of viability (~8%) was observed for Dextran–iron oxide thin films after 24 h incubation time. When increasing the incubation time, this drop increased but still remained below 12% for the Dextran thin film containing 5% wt. iron oxide, pointing to good biocompatibility [18]. Moreover, in a similar study, Ciobanu *et al.* [39] showed that Hep G2 cells adhered very well to thin films of Dextran-doped maghemite and exhibited a normal actin cytoskeleton, proving that these cells underwent normal cell cycle progression. As a result, the authors consider that hepatocytes adhered to hybrid thin films could be used as biosensors for different xenobiotics.

In summary, due to its properties, relatively low cost and availability, Dextran and its conjugates have increased utilization in the field of biomaterials. Non-laser-based techniques are intensively used to fabricate thin films and hydrogels as well [123-125]. The influence of Dextran and albumin-derived iron oxide nanoparticles on fibroblasts *in vitro* was studied by Berry *et al.* [126]. Magnetic composite thin films of Fe_xO_y nanoparticles and photo-cross-linked Dextran hydrogels are promising candidates for a broad field of applications from medicine to mechanical engineering [40]. Dextran-conjugated materials have been successfully investigated as controlled release delivery vehicles of indomethacin (a low molecular weight

http://dx.doi.org/10.5772/61734

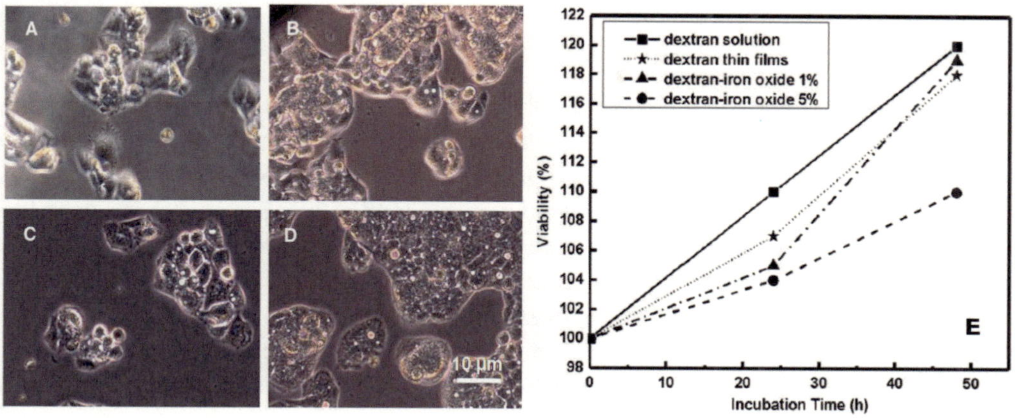

Figure 5. Inverted light microscopic images of Hep G2 cells grown on pure Dextran thin films (A, B), and Dextran–iron oxide thin films obtained from composite targets (C, D) after 24 (A, C) and 48 h (B, D) incubation time. Viability of Hep G2 cells grown on Dextran, and on composite thin films deposited by MAPLE technique (E). (Reproduced with permission from [18])

hydrophobic anti-inflammatory drug) [42], bovine serum albumin [43, 44], lysozyme, and immunoglobulin G [41, 44].

4.2. Nanostructured Levan thin films

High-purity biopolymers are now obtained by microbial fermentation. Levan is a natural polysaccharide produced from fructose by many microorganisms [127]. It is composed of d-fructofuranosyl monomers linked by β(2 → 6) units and β(2 → 1) branches. The carbohydrate structure of Levan synthesized by different microorganisms is rather similar, while small differences appear in degree of polymerization and branching unit [128]. Its specificity is related to furanose form of carbohydrate conformation with an important role in molecule dissolution [129]. Levan is less studied than other known polysaccharides such as Dextran or Pullulan, mainly due to the lack of information about its biocompatibility [130]. Because of unique combined properties like solubility in both oil and water, high molecular weight, and strong adhesion, Levan can be considered as a novel functional polymer with huge potential in industry, from foods, cosmetics, pharmaceutics to chemistry [131]. Indeed, its use as drug delivery matrix, antitumor agent, or protective coatings, stabilizers, or emulsifiers represent only few raised applications of Levans [132, 133]. The important issue is that for most applications or mass production, thin films and coatings are required. These structures compensate the cost by limiting the interaction of the environment with the product to the surface only. Films of polymers are currently produced by solvent casting or thermal processing [134, 135]. A concrete example is the case of a drug dose slow release, when the tablet coating consists of biopolymer with plasticizers for stabilization. It was demonstrated that by extrusion and molding, thick films of Levan can be produced by adding glycerol for cohesion [136]. There is however a high risk of poor adhesion, cracking, and peeling due to rather bulky aspect and consequently dissolution of the film. Thin films of dry and pure Levan are brittle, while biopolymer nanocomposites exhibit improved mechanical properties in respect to the corresponding pure compound [137, 138].

MAPLE process was successfully applied to fabricate organized and nanostructured pure thin films of Levan (L) and oxidized Levan (OL) in vacuum. In order to produce functional aldehyde groups, the oxidation of Levan was carried out before laser transfer in dark oven at 50 °C for six days. MAPLE proved to be the only technique able to transfer nanostructured Levan thin coatings on solid substrates, which exhibited biocompatible properties *in vitro* [12]. The nanostructured aspect of the film that increases the specific surface area can consequently boost the material properties. Indeed, the behavior of Levan in aqueous solutions is difficult to predict and chemical methods fail in producing homogenous and uniform films. On the other hand, during MAPLE process the polymer molecules are transported by Dimethyl sulfoxide (DMSO) solvent molecules from which they separate during the transfer from the target to the solid substrate. Even though some solvent molecules accompany the polymer on the facing substrate, these proved beneficial in the film assembling. Most of the DMSO molecules are vaporized and removed from the reaction chamber by pumping system, whereas Levan molecules were transferred onto the substrates without degradation. In experiments, an excimer laser (KrF*, λ = 248 nm, τ = 25 ns) was used. Deposition process parameters such as fluence, pulse repetition rate, substrate temperature, or distance between the target and collector were optimized in order to grow biopolymer films on Si and glass substrates. DMSO was chosen as solvent as it does not chemically affect L or OL; it is highly volatile; and absorbs at 248 nm laser wavelength.

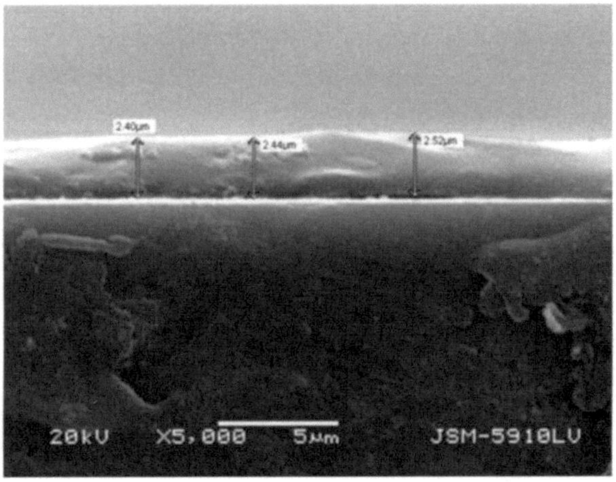

Figure 6. Typical XSEM of L thin films on glass obtained by MAPLE (Reproduced with permission from [12])

Levan films were compact in volume, exhibiting a good adhesion to substrate. As observed from cross-section SEM (XSEM) image in Figure 6, the film was rather compact, while the variation in height across films was low over a relatively large area, supporting the uniformity of the layer. The surface was smooth over large areas and homogenous. A growth film ratio of 0.012 nm/pulse was estimated.

Uncommon two-dimensional ordered array was evidenced at film surface due the most probably to a controlled aggregation during the growing of the film (Figure 7). The nano-structured assembling appears when the solvent DMSO molecules evaporate from the heated substrate.

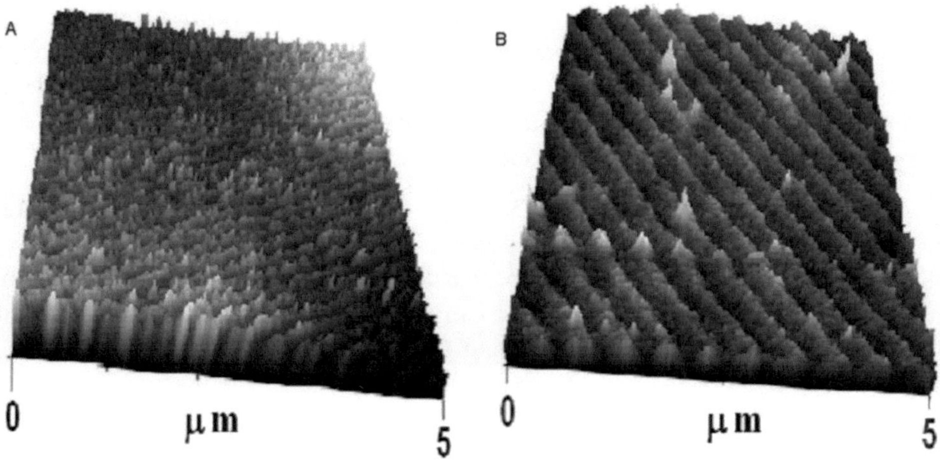

Figure 7. Typical AFM images of sample surfaces for (A) L and (B) OL coatings by MAPLE on Si (Reproduced with permission from [12])

The dynamics of polymers at surface is substantially altered especially when some solvent molecules induce rearrangements. Totally different to rigid ceramic or metal materials, the composition of the polymer varies also with the depth [139]. The morphology is quite similar for L and OL thin films obtained by MAPLE exhibiting a spatial orientation due to a collective influence of evaporation-induced assembly with the specific linkages of the linear structures of polysaccharides. These assembling morphologies were also found for nano-hydroxyapatite (nHA) – chitosan composites [140]. It is considered that the nanostructured assembling, which induces a larger specific surface area, boosts the surface properties of the biopolymer.

The effect of Levan films grown by MAPLE on cell viability and proliferation was investigated by interaction with bone cells. Their proliferation on Levan and control samples was found to be similar. The OL coatings induced an increased cell activity revealed by enhanced cell proliferation as compared with the simple L coatings. This is in accordance with the higher hydrophilicity of OL surfaces due to the acidic aldehyde–hydrogen bonds forming after oxidation [12].

4.3. Compositional libraries thin films by C-MAPLE

Combinatorial processes are required for the synthesis of new organic multicomponent thin coatings [54, 141]. In case of polysaccharides, co-electrodeposition is applied after materials are dissolved or suspended in aqueous solution and integrated into thin films [142]. To produce thin film compositional libraries, premixing of biopolymer solutions followed by temperature gradient over the coating [55], casting processes [143], or flow-coating methods [144] have been used.

A combinatorial technology based on MAPLE for the blending of novel organic compounds was introduced. The new processing method called Combinatorial-MAPLE (C-MAPLE) was proposed to biopolymer compounds [13, 14, 22]. The composition gradient between two materials is achieved by laser co-evaporation of two distinct cryogenic targets and thin-film co-deposition process on solid substrate as described in *Section 3.2*. Two similar compounds such as L and OL but with different physical–chemical and biological properties were chosen in order to generate a compositional discrete library of the two organic compounds.

In experiments, an excimer laser source (KrF*, λ = 248 nm, τ = 25 ns) was used for target evaporation. The Si substrates or glass slides were placed at 4 cm far and parallel to the targets and slightly heated during laser deposition. In a configuration with a distance between the plasmas' centers of 2 cm (see Figure 2), one can obtain a 4 cm long deposition with edges consisting of only L and OL, respectively, and in-between discrete areas of L–OL blended compositions. The soft mixing of the two compounds evaporated from the two distinct targets results in the deposition of a continuous and uniform film with compositional gradient. A gradient of composition from 100% L at left corner to 100% OL at right corner (Figure 8.B) was thus obtained.

The compositional gradient of the film was followed by fluorescence microscopy, as a change in fluorescence emission between L and OL occurs. Levan contains fructose, which is highly fluorescent under green excitation (488 nm). OL loses the fluorescence because fructose is oxidized to aldehyde groups [145]. In Figure 8.A are presented optical and fluorescent pair images in which one can observe the increasing of fluorescence intensity from OL to L along the deposited sample (Figure 8.A and 8.C). This confirms the compositional gradient in the structure [13].

Cellular adhesion, spreading, and proliferation are processes dependent on surface composition and roughness. An optimal cell response to the surface characteristics is of great significance for tissue engineering and nanomedicine. The biocompatibility and cellular behavior to gradient films on glass and silicon substrates was thus evaluated. Initial cell–substrate interaction is shown by cell attachment, followed by adhesion and proliferation. The cell attachment efficiency and morphology is indicative of material biocompatibility. To clearly discriminate between the cell responses, the samples were cut in four equivalent pieces (OL, OL-L, L-OL, and L, respectively). L regions should consist of Levan only, OL of oxidized Levan only, while in OL–L intermediary areas one can expect to contain more OL than L and in L–OL areas more L than OL. The density and actin morphology of cells were evaluated on the four regions (Figure 9.A) at 40 min after cell seeding, as the proof of primary attachment. Similarities on all four zones and standard microscopy cover slips were indicative of biocompatibility and of the dynamic interaction of L/OL gradient coatings with bone cells. The quantification (Figure 9.B) showed that the cells preferred OL as compared to L areas. The

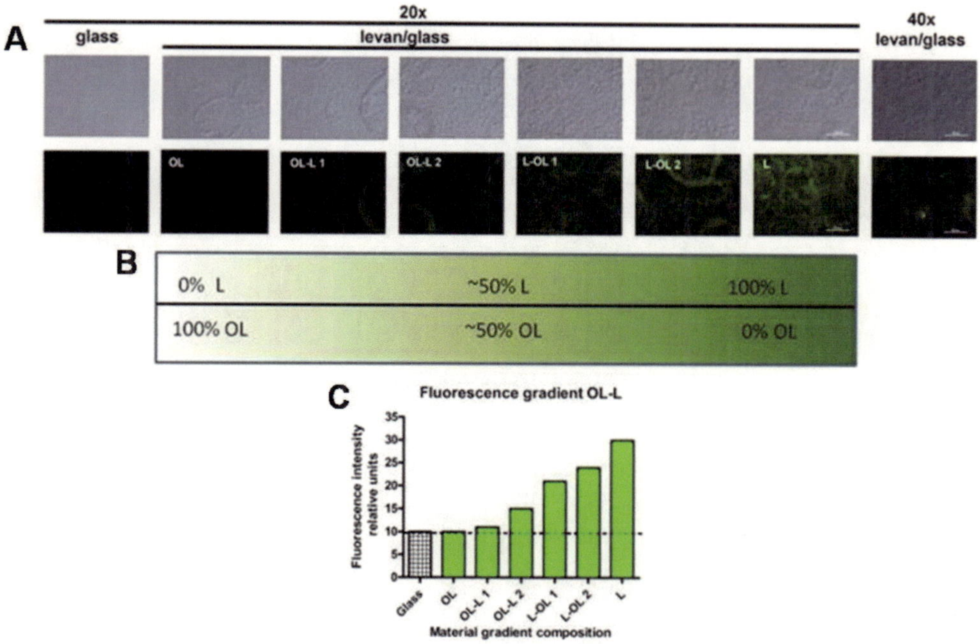

Figure 8. (A) Differential interference contrast and fluorescence microscopy images of OL (left) – L (right) gradient film obtained by C-MAPLE along the glass slide. Positive (Levan/glass) and negative (glass) controls are presented. Bar = 100 μm (20x) and 50 μm (40x). (B) Diagram of expected composition gradient obtained by C-MAPLE from OL and L targets. (C) Quantification of gradient regions fluorescence emission intensity using ImageJ histogram function. Glass background is set as threshold and depicted as dotted line (Reproduced with permission from [14]).

main result was the evident increased cell accumulation on L–OL film blends. This effect was explained by surface wettability associated with the presence of the appropriate amount of OL within L zones. Indeed, the degree of oxidation combined with surface hydrophilicity and roughness stay at the origin of improved bone cell proliferation on L–OL and OL zones. Interestingly, the cell density on OL areas was superior to standard cover slips. It was suggested that such compositional gradients could be used to screen specific nanostructured surface cues for tailoring cell proliferation or to modulate intracellular signaling pathways for specific biomedical applications [13, 14].

The new combinatorial laser technology opens the prospect to simultaneously combine and immobilize *in situ* and in well-defined manner two or more organic materials on a solid substrate by laser evaporation.

4.4. Overview of recent contributions in the field of thin films synthesis by MAPLE

An exhaustive list of other thin films grown by advanced laser techniques, as well as their physical- and biochemical characterizations for biomedical applications could be found in the literature [6, 81]. Several book chapters and review articles available to readers, spanning a broad coverage of both fundamental and applicative aspects, published in the last five years, are summarized in Table 2.

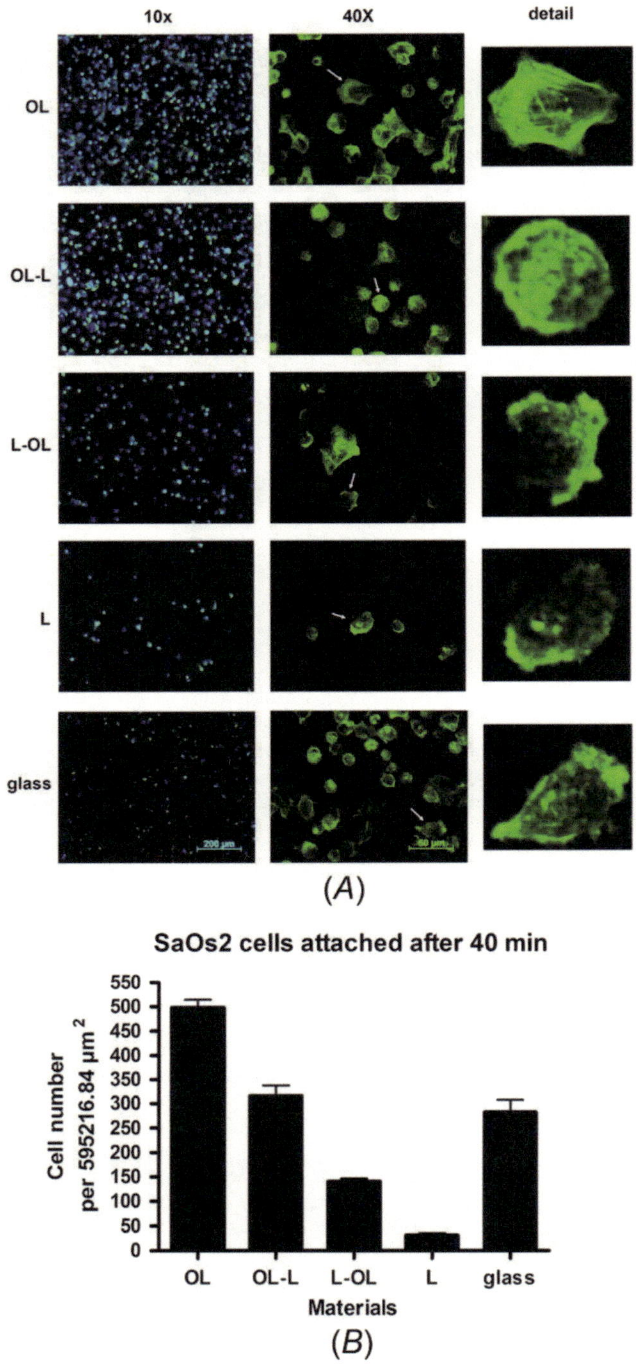

Figure 9. (A) Fluorescence microscopy of bone cells on combinatorial and control materials after 40 mins seeding. Different magnifications (10X and 40X) of cells labeled with Alexa Fluor 488-conjugated phalloidin (actin – green) and DAPI (nuclei – blue) are presented with details on cell morphology. (B) Quantification of cells by ImageJ nuclei counting function. Mean ± SEM is depicted on graph.

Materials	Applications	Title	Authors/Reference
Living cells enzymes, proteins and bioceramics.	Tissue engineering, stem cell and cancer research.	Topical Review: "*Laser-based direct-write techniques for cell printing*"	N.R. Schiele et al. [146]
Polymers (SXFA, POOPT (poly [3-(4-octyloxyphenyl) thiophene]), poly(9,9-dioctylfluorene) (PF8), Ge-corrole derivative (Ge(TPC)OCH3)), Proteins (horseradish peroxidase (HRP), insulin, bovine serum albumin (BSA)), Nanoparticles (TiO_2, SnO_2).	Biomaterials, gas sensing.	Chapter 9: "*Fundamentals and Applications of MAPLE*"	A. Luches and A. P. Caricato [7]
Hybrid organic–inorganic bionanocomposites [HA–sodium maleate (HA–NaM) copolymer, alendronate–HA].	Advanced biomimetic Implants.	Chapter 10: "*Advanced Biomimetic Implants Based on Nanostructured Coatings Synthesized by Pulsed Laser Technologies*"	I. N. Mihailescu et al. [8]
Polymers and biological molecules, biomaterials, nanoparticle films.	Drug delivery, tissue engineering, for gas and vapor detection, for light emitting devices, etc.	Review: "*Applications of the matrix-assisted pulsed laser evaporation method for the deposition of organic, biological and nanoparticle thin films: a review*"	A.P. Caricato, A. Luches [80]
Living mammalian cells and pluripotent stem cells (e.g., human dermal fibroblasts, rat neural stem cells, mouse embryonic stem cells).	In vitro cellular microenvironment, tissue engineering, regenerative medicine.	Review: "*Matrix-assisted pulsed laser methods for biofabrication*"	B.C. Riggs et al. [147]
Polymer and other soft matter thin films (Horseradish peroxidase, Ribonuclease A, Poly(ethylene glycol), Poly(3-hexyl thiophene), MEH-PPV).	Organic electronics, medical implants, drug delivery systems, and sensors.	Trends in Polymer Science: *MAPLE Deposition of Macromolecules*	Shepard, K. B. and Priestley, R. D. [148]
Biocompatible and biodegradable polymers ((PEG), (PLGA), mixtures	Biomimetic applications in drug delivery systems,	Chapter 5: "*Biomimetic Assemblies by Matrix-Assisted Pulsed*	F. Sima and I.N. Mihailescu [9]

Materials	Applications	Title	Authors/Reference
PEG-PLGA, poly(D,L-lactide), Levan); Extracellular matrix proteins (fibronectin, vitronectin); organic – inorganic composites.	biosensors and advanced implant coatings.	*Laser Evaporation"*	
Polysaccarides (Levan), Composite alendronate-HA, Enzyme ribonuclease A.	Biomimetic coating of medical implants, drug delivery systems, biosensing.	Chapter 11: *"Biomaterial Thin Films by Soft Pulsed Laser Technologies for Biomedical Applications"*	I. N. Mihailescu et al. [88]
Enzyme immobilization (Laccase).	Bio-Sensors.	Chapter 9: *"Deposition and Characterization of Laccase Thin Films Obtained by Matrix Assisted Pulsed Laser Evaporation"*	N. Cicco et al. [149]

Table 2. Overview of recent reviews and book chapters published in the field of thin films synthesis by MAPLE

5. Summary and perspectives

MAPLE synthesis of biopolymer thin film was applied to fabricate organized and nanostructured pure and hybrid polysaccharide layers. It was demonstrated that laser-based techniques allow for transferring complex, large molecular-mass organic compounds, avoiding their photo-thermal decomposition and/or irreversible damage. The functionality preservation was secured for Dextran and Levan coatings and derivatives, as revealed by cells' viability and proliferation *in vitro* tests. Combinatorial-MAPLE evidenced the possibility to generate compositional gradient thin films of two organic and/or inorganic compounds in a single-step process. Engineering the cell/biomaterial interface to control cell behavior has implications for the fabrication of instructive environments for tissue repair or cell supports.

The flexibility of the C-MAPLE method allows for the synthesis of new hybrid materials by correlating laser irradiation settings with the thermo-physical and chemical properties of the raw materials. This approach opens a great potential to the discovery of new drugs for the pharmaceutical industry and for drug release applications from biodegradable polymeric coatings. Intelligent materials synthesized on discrete areas exhibiting desired properties such as controlled rate of coating dissolution stands for a future challenge. An expansion of combinatorial organic domain can be stimulated by laser technologies. The strong advantages are the control of the preferred density of functional groups at the surface, among which we mention chemical composition and/or physical properties on nanometric areas and the fabrication of multicomponent gradient layers.

Acknowledgements

The authors acknowledge the support of UEFISCDI under the contracts 19_RO-FR/2014 and PNII-RU-TE-2014-4-1790, -1273 and -0396.

Author details

Emanuel Axente, Felix Sima, Carmen Ristoscu, Natalia Mihailescu and Ion N. Mihailescu*

*Address all correspondence to:

Lasers Department, National Institute for Lasers, Plasma and Radiation Physics, Magurele, Ilfov, Romania

References

[1] Nalwa HS. *Handbook of Thin Film Materials*. Academic Press; 2002;1 p. 1-698.

[2] Stevens MM, George JH. Exploring and engineering the cell surface interface. *Science*. 2005;310(5751):1135-8.

[3] Decher G, Schlenoff JB. *Multilayer Thin Films: Sequential Assembly of Nanocomposite Materials, 2nd Edition*. John Wiley & Sons; 2012. p. 1-1112.

[4] Schottner G. Hybrid sol-gel-derived polymers: applications of multifunctional materials. *Chem Mater*. 2001;13(10):3422-35.

[5] Nelea V, Mihailescu IN, Jelinek M. *Biomaterials: new issues and breakthroughs for biomedical applications. Pulsed Laser Deposition of Thin Films: Applications-Led Growth of Functional Materials*. John Wiley & Sons;2007.p. 421-60.

[6] Chrisey D, Pique A, McGill R, Horwitz J, Ringeisen B, Bubb D, et al. Laser deposition of polymer and biomaterial films. *Chem Rev*. 2003;103(2):553-76.

[7] Luches A, Caricato AP. *Fundamentals and Applications of MAPLE. Laser-Surface Interactions for New Materials Production*. Springer; 2010. p. 203-33.

[8] Mihailescu IN, Ristoscu C, Bigi A, Mayer I. Advanced biomimetic implants based on nanostructured coatings synthesized by pulsed laser technologies. *Laser-Surface Interactions for New Materials Production*. Springer; 2010. p. 235-60.

[9] Sima F, Mihailescu IN. Biomimetic assemblies by matrix-assisted pulsed laser evaporation. *Laser Technology in Biomimetics*. Springer; 2013. p. 111-41.

[10] McGill RA, Chrisey DB. Method of producing a film coating by matrix assisted pulsed laser deposition. USPatent;US6025036 A, 2000.

[11] Guo Y, Morozov A, Schneider D, Chung JW, Zhang C, Waldmann M, et al. Ultrastable nanostructured polymer glasses. *Natur Mater*. 2012;11(4):337-43.

[12] Sima F, Mutlu EC, Eroglu MS, Sima LE, Serban N, Ristoscu C, et al. Levan nanostructured thin films by MAPLE assembling. *Biomacromolecules*. 2011;12(6):2251-6.

[13] Sima F, Axente E, Sima L, Tuyel U, Eroglu M, Serban N, et al. Combinatorial matrix-assisted pulsed laser evaporation: Single-step synthesis of biopolymer compositional gradient thin film assemblies. *Appl Physics Lett*. 2012;101(23):233705.

[14] Axente E, Sima F, Sima LE, Erginer M, Eroglu MS, Serban N, et al. Combinatorial MAPLE gradient thin film assemblies signalling to human osteoblasts. *Biofabrication*. 2014;6(3):035010.

[15] Cristescu R, Dorcioman G, Ristoscu C, Axente E, Grigorescu S, Moldovan A, et al. Matrix assisted pulsed laser evaporation processing of triacetate-pullulan polysaccharide thin films for drug delivery systems. *Appl Surf Sci*. 2006;252(13):4647-51.

[16] Cristescu R, Jelinek M, Kocourek T, Axente E, Grigorescu S, Moldovan A, et al., Matrix assisted pulsed laser evaporation of pullulan tailor-made biomaterial thin films for controlled drug delivery systems. *J Physics*: Conference Series; 2007;59:144-49.

[17] Jelinek M, Cristescu R, Axente E, Kocourek T, Dybal J, Remsa J, et al. Matrix assisted pulsed laser evaporation of cinnamate-pullulan and tosylate-pullulan polysaccharide derivative thin films for pharmaceutical applications. *Appl Surf Sci*. 2007;253(19): 7755-60.

[18] Predoi D, Ciobanu CS, Radu M, Costache M, Dinischiotu A, Popescu C, et al. Hybrid dextran-iron oxide thin films deposited by laser techniques for biomedical applications. *Mater Sci Engin*: C. 2012;32(2):296-302.

[19] Sima F, Axente E, Ristoscu C, Mihailescu IN, Kononenko TV, Nagovitsin IA, et al. Tailoring immobilization of immunoglobulin by excimer laser for biosensor applications. *J Biomed Mater Res* Part A. 2011;96(2):384-94.

[20] Sima F, Davidson P, Pauthe E, Sima L, Gallet O, Mihailescu I, et al. Fibronectin layers by matrix-assisted pulsed laser evaporation from saline buffer-based cryogenic targets. *Acta Biomaterialia*. 2011;7(10):3780-8.

[21] Motoc M, Axente E, Popescu C, Sima L, Petrescu S, Mihailescu I, et al. Active protein and calcium hydroxyapatite bilayers grown by laser techniques for therapeutic applications. *J Biomed Mater Res* Part A. 2013;101(9):2706-11.

[22] Sima F, Axente E, Iordache I, Luculescu C, Gallet O, Anselme K, et al. Combinatorial matrix assisted pulsed laser evaporation of a biodegradable polymer and fibronectin for protein immobilization and controlled release. *Appl Surf Sci*. 2014;306:75-9.

[23] Sima F, Davidson PM, Dentzer J, Gadiou R, Pauthe E, Gallet O, et al. Inorganic–organic thin implant coatings deposited by lasers. *ACS Appl Mater Interf.* 2014;7(1): 911-20.

[24] Ringeisen B, Chrisey D, Pique A, Young H, Modi R, Bucaro M, et al. Generation of mesoscopic patterns of viable Escherichia coli by ambient laser transfer. *Biomaterials.* 2002;23(1):161-6.

[25] Cristescu R, Mihailescu I, Jelinek M, Chrisey D. Functionalized thin films and structures obtained by novel laser processing issues. *Functional Properties of Nanostructured Materials*. Springer; 2006. p. 211-26.

[26] Hopp B, Smausz T, Antal Z, Kresz N, Bor Z, Chrisey D. Absorbing film assisted laser induced forward transfer of fungi (Trichoderma conidia). *J Appl Physics.* 2004;96:3478-81.

[27] Barron JA, Ringeisen BR, Kim H, Spargo BJ, Chrisey DB. Application of laser printing to mammalian cells. *Thin Solid Films*. 2004;453:383-7.

[28] Hopp B, Smausz T, Kresz N, Barna N, Bor Z, Kolozsvári L, et al. Survival and proliferative ability of various living cell types after laser-induced forward transfer. *Tissue Engin*. 2005;11(11-12):1817-23.

[29] Guillemot F, Souquet A, Catros S, Guillotin B. Laser-assisted cell printing: principle, physical parameters versus cell fate and perspectives in tissue engineering. *Nanomedicine*. 2010;5(3):507-15.

[30] Sabu Thomas, Neethu Ninan, Sneha Mohan, Francis E. *Natural Polymers, Biopolymers, Biomaterials, and Their Composites, Blends, and IPNs*. Sabu Thomas, Mathew Sebastian, Anne George, Weimin Y, editors. Apple Academic Press 2012. 370 p.

[31] Dumitriu S. *Polysaccharides in Medicinal Applications*. CRC Press; 1996. p. 1-816.

[32] Phillips GO, Williams PA. *Handbook of Hydrocolloids, 2nd Edition*.Woodhead Publishing Limited; 2009. p. 592-614.

[33] Chong BF, Blank LM, Mclaughlin R, Nielsen LK. Microbial hyaluronic acid production. *Appl Microbiol Biotechnol*. 2005;66(4):341-51.

[34] Volpi N. *Chondroitin Sulfate: Structure, Role and Pharmacological Activity*. Academic Press; 2006. p. 1-600.

[35] Horton D. *Advances in Carbohydrate Chemistry and Biochemistry*. Academic Press; 2014. p. 1-453.

[36] Baldwin AD, Kiick KL. Polysaccharide-modified synthetic polymeric biomaterials. *Peptide Sci*. 2010;94(1):128-40.

[37] Naessens M, Cerdobbel A, Soetaert W, Vandamme EJ. Leuconostoc dextransucrase and dextran: production, properties and applications. *J Chem Technol Biotechnol.* 2005;80(8):845-60.

[38] Harrison JH. Dextran as a plasma substitute with plasma volume and excretion studies on control patients. *Annals Surg.* 1954;139(2):137.

[39] Ciobanu CS, Iconaru SL, Gyorgy E, Radu M, Costache M, Dinischiotu A, et al. Biomedical properties and preparation of iron oxide-dextran nanostructures by MAPLE. *Chemistry Central Journal.*2012;6:17.

[40] Brunsen A, Utech S, Maskos M, Knoll W, Jonas U. Magnetic Composite Thin Films of Fe x O y Nanoparticles and Photocrosslinked Dextran Hydrogels. *J Magnet Magnetic Mater.* 2012;324(8):1488-97.

[41] De Jong S, Van Eerdenbrugh B, van Nostrum Cv, Kettenes-Van Den Bosch J, Hennink W. Physically crosslinked dextran hydrogels by stereocomplex formation of lactic acid oligomers: degradation and protein release behavior. J Controlled Release. 2001;71(3):261-75.

[42] Zhang Y, Chu CC. Biodegradable dextran–polylactide hydrogel networks: Their swelling, morphology and the controlled release of indomethacin. *J Biomed Mater Res.* 2002;59(2):318-28.

[43] Sun G, Chu C-C. Synthesis, characterization of biodegradable dextran–allyl isocyanate–ethylamine/polyethylene glycol–diacrylate hydrogels and their in vitro release of albumin. *Carbohy polym.* 2006;65(3):273-87.

[44] Hiemstra C, Zhong Z, van Steenbergen MJ, Hennink WE, Feijen J. Release of model proteins and basic fibroblast growth factor from in situ forming degradable dextran hydrogels. *J Controlled Release.* 2007;122(1):71-8.

[45] Bostan MS, Mutlu EC, Kazak H, Keskin SS, Oner ET, Eroglu MS. Comprehensive characterization of chitosan/PEO/levan ternary blend films. *Carbohy Polym.* 2014;102:993-1000.

[46] Poli A, Kazak H, Gürleyendağ B, Tommonaro G, Pieretti G, Öner ET, et al. High level synthesis of levan by a novel Halomonas species growing on defined media. *Carbohy Polym.* 2009;78(4):651-7.

[47] Sarilmiser HK, Oner ET. Investigation of anti-cancer activity of linear and aldehyde-activated levan from Halomonas smyrnensis AAD6 T. *Biochem Engin J.* 2014;92:28-34.

[48] Kim KH, Chung CB, Kim YH, Kim KS, Han CS, Kim CH. Cosmeceutical properties of levan produced by Zymomonas mobilis. *J Cosmetic Sci.* 2004;56(6):395-406.

[49] Yoo S-H, Yoon EJ, Cha J, Lee HG. Antitumor activity of levan polysaccharides from selected microorganisms. *Int J Biologic Macromol.* 2004;34(1):37-41.

[50] Calazans GcM, Lima RC, de França FP, Lopes CE. Molecular weight and antitumour activity of Zymomonas mobilis levans. *Int J Biologic Macromol*. 2000;27(4):245-7.

[51] Leibovici J, Stark Y. Direct antitumor effect of the polysaccharide levan in mice: effects of drug concentration and time and temperature of incubation. *J Natl Canc Instit*. 1984;72(6):1417-20.

[52] Iyer A, Mody K, Jha B. Emulsifying properties of a marine bacterial exopolysaccharide. *Enzy Microb Technol*. 2006;38(1):220-2.

[53] Meier MA, Hoogenboom R, Schubert US. Combinatorial methods, automated synthesis and high-throughput screening in polymer research: The evolution continues. *Macromol Rapid Commun*. 2004;25(1):21-33.

[54] Steele TW, Huang CL, Kumar S, Irvine S, Boey FYC, Loo JS, et al. Novel gradient casting method provides high-throughput assessment of blended polyester poly (lactic-co-glycolic acid) thin films for parameter optimization. *Acta Biomaterialia*. 2012;8(6):2263-70.

[55] Meredith JC, Karim A, Amis EJ. High-throughput measurement of polymer blend phase behavior. *Macromol*. 2000;33(16):5760-2.

[56] Fisher OZ, Khademhosseini A, Langer R, Peppas NA. Bioinspired materials for controlling stem cell fate. *Acc Chemic Res*. 2009;43(3):419-28.

[57] Nakajima M, Ishimuro T, Kato K, Ko I-K, Hirata I, Arima Y, et al. Combinatorial protein display for the cell-based screening of biomaterials that direct neural stem cell differentiation. *Biomaterials*. 2007;28(6):1048-60.

[58] U.S. Congress, (1993), Office of Technology Assessment, Biopolymers: Making Materials Nature's Way-Background Paper, OTA-BP-E-102 (Washington, DC: U.S. Government Printing Office, ISBN 0-16042098-9

[59] Schmidt V, Belegratis MR. *Laser Technology in Biomimetics, Basics and Applications*. Berlin Heidelberg: Springer-Verlag; 2013. p. 1-262.

[60] Grigorescu S, Hindié M, Axente E, Carreiras F, Anselme K, Werckmann J, et al. Fabrication of functional fibronectin patterns by nanosecond excimer laser direct write for tissue engineering applications. *J Mater Sci: Mater Med*. 2013;24(7):1809-21.

[61] Zaima N, Goto-Inoue N,Hayasaka T.MALDI Mass Spectrometry Imaging of Biological Structures. *Encyclopedia of Analytical Chemistry*. John Wiley & Sons, Ltd; 2015. p. 1-20.

[62] Scriven L, editor. Physics and applications of dip coating and spin coating. *MRS proceedings*.1988;121:717.

[63] Raybaudi-Massilia RM, Mosqueda-Melgar J. *Polysaccharides as Carriers and Protectors of Additives and Bioactive Compounds in Foods*. INTECH Open Access Publisher; 2012. p. 429-54.

[64] Karthik C, Manjuladevi V, Gupta R, Kumar S. Pattern formation in Langmuir–Blodgett films of tricycloquinazoline based discotic liquid crystal molecules. *J Mol Struct*. 2014;1070:52-7.

[65] Soares CG, Caseli L, Bertuzzi DL, Santos FS, Garcia JR, Péres LO. Ultrathin films of poly (2, 5-dicyano-p-phenylene-vinylene)-co-(p-phenylene-vinylene) DCN-PPV/PPV: A Langmuir and Langmuir-Blodgett films study. *Coll SurfA: Physicochem Engin Asp*. 2015;467:201-6.

[66] Gupta RK, Manjuladevi V. *Molecular Interactions at Interfaces*. INTECH Open Access Publisher; 2012. p. 81-104.

[67] Girard-Egrot AP, Blum LJ. Langmuir-Blodgett technique for synthesis of biomimetic lipid membranes. *Nanobiotechnology of Biomimetic Membranes*. Springer; 2007. p. 23-74.

[68] Fujiwara I, Ohnishi M, Seto J. Atomic force microscopy study of protein-incorporating Langmuir-Blodgett films. *Langmuir*. 1992;8(9):2219-22.

[69] Chen X, Gu Y, Lee J, Lee W, Wang H. Multifunctional surfaces with biomimetic nanofibres and drug-eluting micro-patterns for infection control and bone tissue formation. *Eur Cells Mater*. 2012;24:237-48.

[70] Pandey S, Mishra SB. Sol–gel derived organic–inorganic hybrid materials: synthesis, characterizations and applications. *J Sol-Gel Sci Technol*. 2011;59(1):73-94.

[71] Wojciechowska P. The Effect of concentration and type of plasticizer on the mechanical properties of cellulose acetate butyrate organic-inorganic hybrids: INTECH Open Access Publisher; 2012. p. 141-164.

[72] Olding T, Sayer M, Barrow D. Ceramic sol–gel composite coatings for electrical insulation. *Thin Solid Films*. 2001;398:581-6.

[73] Wu P, Ringeisen B, Callahan J, Brooks M, Bubb D, Wu H, et al. The deposition, structure, pattern deposition, and activity of biomaterial thin-films by matrix-assisted pulsed-laser evaporation (MAPLE) and MAPLE direct write. *Thin Solid Films*. 2001;398:607-14.

[74] Colina M, Serra P, Fernández-Pradas JM, Sevilla L, Morenza JL. DNA deposition through laser induced forward transfer. *Biosens Bioelectron*. 2005;20(8):1638-42.

[75] Dinca V, Ranella A, Farsari M, Kafetzopoulos D, Dinescu M, Popescu A, et al. Quantification of the activity of biomolecules in microarrays obtained by direct laser transfer. *Biomed Microdevices*. 2008;10(5):719-25.

[76] Thibault MM, Hoemann CD, Buschmann MD. Fibronectin, vitronectin, and collagen I induce chemotaxis and haptotaxis of human and rabbit mesenchymal stem cells in a standardized transmembrane assay. *Stem Cells Dev*. 2007;16(3):489-502.

[77] Karagkiozaki V, Vavoulidis E, Karagiannidis PG, Gioti M, Fatouros DG, Vizirianakis IS, et al. Development of a nanoporous and multilayer drug-delivery platform for medical implants. *Int J Nanomed.* 2012;7:5327.

[78] Gray J, Luan B. Protective coatings on magnesium and its alloys—a critical review. *J Alloys Comp.* 2002;336(1):88-113.

[79] Hornberger H, Virtanen S, Boccaccini A. Biomedical coatings on magnesium alloys–a review. *Acta Biomaterialia.* 2012;8(7):2442-55.

[80] Caricato AP, Luches A. Applications of the matrix-assisted pulsed laser evaporation method for the deposition of organic, biological and nanoparticle thin films: a review. *Appl Physics* A. 2011;105(3):565-82.

[81] Piqué A. Deposition of polymers and biomaterials using the matrix-assisted pulsed laser evaporation (MAPLE) process. Pulsed Laser Deposition of Thin Films: Applications-Led Growth of Functional Materials.John Wiley & Sons; 2007. p. 63-84.

[82] Gorustovich AA, Roether JA, Boccaccini AR. Effect of bioactive glasses on angiogenesis: a review of in vitro and in vivo evidences. *Tissue Engin Part B: Rev.* 2009;16(2): 199-207.

[83] Zergioti I, Karaiskou A, Papazoglou D, Fotakis C, Kapsetaki M, Kafetzopoulos D. Femtosecond laser microprinting of biomaterials. *Appl Physics Lett.* 2005;86(16): 163902.

[84] Patz T, Doraiswamy A, Narayan R, He W, Zhong Y, Bellamkonda R, et al. Three-dimensional direct writing of B35 neuronal cells. *J Biomed Mater Res Part B: Appl Biomater.* 2006;78(1):124-30.

[85] Piqué A, Auyeung R, Stepnowski J, Weir D, Arnold C, McGill R, et al. Laser processing of polymer thin films for chemical sensor applications. *Surf Coat Technol.* 2003;163:293-9.

[86] Cristescu R, Doraiswamy A, Patz T, Socol G, Grigorescu S, Axente E, et al. Matrix assisted pulsed laser evaporation of poly (d, l-lactide) thin films for controlled-release drug systems. *Appl Surf Sci.* 2007;253(19):7702-6.

[87] Boanini E, Torricelli P, Fini M, Sima F, Serban N, Mihailescu IN, et al. Magnesium and strontium doped octacalcium phosphate thin films by matrix assisted pulsed laser evaporation. *J Inorganic Biochem.* 2012;107(1):65-72.

[88] Mihailescu IN, Bigi A, Gyorgy E, Ristoscu C, Sima F, Oner ET. Biomaterial thin films by soft pulsed laser technologies for biomedical applications. *Lasers in Materials Science.* Springer; 2014. p. 271-94.

[89] Leveugle E, Zhigilei LV. Molecular dynamics simulation study of the ejection and transport of polymer molecules in matrix-assisted pulsed laser evaporation. *J Appl Physics.* 2007;102(7):074914.

[90] Smausz T, Megyeri G, Kékesi R, Vass C, György E, Sima F, et al. Comparative study on pulsed laser deposition and matrix assisted pulsed laser evaporation of urease thin films. *Thin Solid Films*. 2009;517(15):4299-302.

[91] György E, Sima F, Mihailescu I, Smausz T, Hopp B, Predoi D, et al. Biomolecular urease thin films grown by laser techniques for blood diagnostic applications. *Mater Sci Engin: C*. 2010;30(4):537-41.

[92] Lippert T, Dickinson JT. Chemical and spectroscopic aspects of polymer ablation: special features and novel directions. *Chem Rev*. 2003;103(2):453-86.

[93] Ahn HS, Kim MJ, Seol HJ, Lee JH, Kim HI, Kwon YH. Effect of pH and temperature on orthodontic NiTi wires immersed in acidic fluoride solution. *J Biomed Mater Res Part B, Appl Biomater*. 2006;79(1):7-15. Epub 2006/02/14.

[94] György E, Pérez del Pino A, Sauthier G, Figueras A. Biomolecular papain thin films grown by matrix assisted and conventional pulsed laser deposition: A comparative study. *J Appl Physics*. 2009;106(11):114702.

[95] Hench LL, Polak JM. Third-generation biomedical materials. *Science*. 2002;295(5557): 1014-7.

[96] Lewis SL, Dirksen SR, Heitkemper MM, Bucher L. *Medical-Surgical Nursing: Assessment and Management of Clinical Problems*. Single Volume: Elsevier Health Sciences; 2014. p. 1-1824.

[97] Massia SP, Stark J, Letbetter DS. Surface-immobilized dextran limits cell adhesion and spreading. *Biomaterials*. 2000;21(22):2253-61.

[98] Johnson WK, Stoupis C, Torres GM, Rosenberg EB, Ros PR. Superparamagnetic iron oxide (SPIO) as an oral contrast agent in gastrointestinal (GI) magnetic resonance imaging (MRI): comparison with state-of-the-art computed tomography (CT). *Magnetic Resonance Imaging*. 1996;14(1):43-9.

[99] Qiao R, Yang C, Gao M. Superparamagnetic iron oxide nanoparticles: from preparations to in vivo MRI applications. *J Mater Chem*. 2009;19(35):6274-93.

[100] Jordan A, Scholz R, Wust P, Schirra H, Schiestel T, Schmidt H, et al. Endocytosis of dextran and silan-coated magnetite nanoparticles and the effect of intracellular hyperthermia on human mammary carcinoma cells in vitro. *J MagnetMagnetic Mater*. 1999;194(1):185-96.

[101] Huh Y-M, Jun Y-w, Song H-T, Kim S, Choi J-s, Lee J-H, et al. In vivo magnetic resonance detection of cancer by using multifunctional magnetic nanocrystals. *J Am Chem Soc*. 2005;127(35):12387-91.

[102] Lee J-H, Jang J-t, Choi J-s, Moon SH, Noh S-h, Kim J-w, et al. Exchange-coupled magnetic nanoparticles for efficient heat induction. *Natur Nanotechnol*. 2011;6(7):418-22.

[103] Ito A, Kuga Y, Honda H, Kikkawa H, Horiuchi A, Watanabe Y, et al. Magnetite nanoparticle-loaded anti-HER2 immunoliposomes for combination of antibody therapy with hyperthermia. *Canc Lett.* 2004;212(2):167-75.

[104] Jain TK, Morales MA, Sahoo SK, Leslie-Pelecky DL, Labhasetwar V. Iron oxide nanoparticles for sustained delivery of anticancer agents. *Mol Pharmaceut.* 2005;2(3): 194-205.

[105] Yu MK, Jeong YY, Park J, Park S, Kim JW, Min JJ, et al. Drug-loaded superparamagnetic iron oxide nanoparticles for combined cancer imaging and therapy in vivo. *Angewandte Chemie Int Ed.* 2008;47(29):5362-5.

[106] Berry CC, Curtis AS. Functionalisation of magnetic nanoparticles for applications in biomedicine. *J Physics D: Appl Physics.* 2003;36(13):R198.

[107] Gupta AK, Gupta M. Synthesis and surface engineering of iron oxide nanoparticles for biomedical applications. *Biomaterials.* 2005;26(18):3995-4021.

[108] Jiang W, Kim BY, Rutka JT, Chan WC. Nanoparticle-mediated cellular response is size-dependent. *Natur Nanotechnol.* 2008;3(3):145-50.

[109] Lewinski N, Colvin V, Drezek R. Cytotoxicity of nanoparticles. *Small.* 2008;4(1):26-49.

[110] Ankamwar B, Lai T, Huang J, Liu R, Hsiao M, Chen C, et al. Biocompatibility of Fe3O4 nanoparticles evaluated by in vitro cytotoxicity assays using normal, glia and breast cancer cells. *Nanotechnology.* 2010;21(7):075102.

[111] Bee A, Massart R, Neveu S. Synthesis of very fine maghemite particles. *J Magnet Magnetic Mater.* 1995;149(1):6-9.

[112] Schwertmann U, Cornell RM. *Iron Oxides in the Laboratory.* John Wiley & Sons; 2008. p. 1-204.

[113] Hair ML. Hydroxyl groups on silica surface. *J Non-Crystalline Solids.* 1975;19:299-309.

[114] Bautista MC, Bomati-Miguel O, del Puerto Morales M, Serna CJ, Veintemillas-Verdaguer S. Surface characterisation of dextran-coated iron oxide nanoparticles prepared by laser pyrolysis and coprecipitation. *J Magnet Magnetic Mater.* 2005;293(1):20-7.

[115] De Vidales JM, López-Delgado A, Vila E, Lopez F. The effect of the starting solution on the physico-chemical properties of zinc ferrite synthesized at low temperature. *J Alloys Comp.* 1999;287(1):276-83.

[116] Hradil J, Pisarev A, Babič M, Horák D. Dextran-modified iron oxide nanoparticles. *China Particuol.* 2007;5(1):162-8.

[117] Wang H-R, Chen K-M. Preparation and surface active properties of biodegradable dextrin derivative surfactants. *Coll Surf A: Physicochem Engin Asp.* 2006;281(1):190-3.

[118] Gil EC, Colarte AI, Sampedro JLL, Bataille B. Subcoating with Kollidon VA 64 as water barrier in a new combined native dextran/HPMC–cetyl alcohol controlled release tablet. *Eur J Pharmaceut Biopharmaceut*. 2008;69(1):303-11.

[119] Floroian L, Popescu A, Serban N, Mihailescu IN. Polymer-bioglass composite coatings: A promising alternative for advanced biomedical implants. In: Cuppoletti J, editor. *Metal, Ceramic and Polymeric Composites for Various Uses*. INTECH Open Access Publisher; 2011. p. 393-420.

[120] Bigi A, Boanini E, Capuccini C, Fini M, Mihailescu IN, Ristoscu C, et al. Biofunctional alendronate–hydroxyapatite thin films deposited by matrix assisted pulsed laser evaporation. *Biomaterials*. 2009;30(31):6168-77.

[121] Boanini E, Torricelli P, Sima F, Axente E, Fini M, Mihailescu IN, et al. Strontium and zoledronate hydroxyapatites graded composite coatings for bone prostheses. *J Coll Interf Sci*. 2015;448:1-7.

[122] Powers MJ, Rodriguez RE, Griffith LG. Cell-substratum adhesion strength as a determinant of hepatocyte aggregate morphology. *Biotechnol Bioengin*. 1997;53(4):415-26.

[123] Zhang X, Wu D, Chu C-C. Synthesis and characterization of partially biodegradable, temperature and pH sensitive Dex–MA/PNIPAAm hydrogels. *Biomaterials*. 2004;25(19):4719-30.

[124] Huang X, Zhang Y, Donahue HJ, Lowe TL. Porous thermoresponsive-co-biodegradable hydrogels as tissue-engineering scaffolds for 3-dimensional in vitro culture of chondrocytes. *Tissue Engin*. 2007;13(11):2645-52.

[125] Hiemstra C, van der Aa LJ, Zhong Z, Dijkstra PJ, Feijen J. Rapidly in situ-forming degradable hydrogels from dextran thiols through Michael addition. *Biomacromolecules*. 2007;8(5):1548-56.

[126] Berry CC, Wells S, Charles S, Curtis AS. Dextran and albumin derivatised iron oxide nanoparticles: influence on fibroblasts in vitro. *Biomaterials*. 2003;24(25):4551-7.

[127] Suzuki M, Chatterton NJ. *Science and Technology of Fructans*. CRC Press; 1993. p. 1-384.

[128] Han YW. Microbial levan. *Adv Appl Microbiol*. 1990;35(171194):2.

[129] French AD. Accessible conformations of the β-d-(2 → 1)-and-(2 → 6)-linked d-fructans inulin and levan. *Carbohy Res*. 1988;176(1):17-29.

[130] Mano J, Silva G, Azevedo HS, Malafaya P, Sousa R, Silva S, et al. Natural origin biodegradable systems in tissue engineering and regenerative medicine: present status and some moving trends. *J Royal Soc Interf*. 2007;4(17):999-1030.

[131] Rehm BHA.*Microbial Production of Biopolymers and Polymer Precursors: Applications and Perspectives*.Caister Academic Press; 2009.p. +294.

[132] Freitas F, Alves VD, Reis MA. Advances in bacterial exopolysaccharides: from production to biotechnological applications. *Trends Biotechnol.* 2011;29(8):388-98.

[133] Sezer AD, Kazak H, Öner ET, Akbuğa J. Levan-based nanocarrier system for peptide and protein drug delivery: optimization and influence of experimental parameters on the nanoparticle characteristics. *Carbohy Polym.* 2011;84(1):358-63.

[134] Frank C, Rao V, Despotopoulou M, Pease R, Hinsberg W, Miller R, et al. Structure in thin and ultrathin spin-cast polymer films. *Science.* 1996;273(5277):912-5.

[135] Rhim JW, Mohanty AK, Singh SP, Ng PK. Effect of the processing methods on the performance of polylactide films: thermocompression versus solvent casting. *J Appl Polym Sci.* 2006;101(6):3736-42.

[136] Barone JR, Medynets M. Thermally processed levan polymers. *Carbohy Polym.* 2007;69(3):554-61.

[137] Ray SS, Okamoto M. Polymer/layered silicate nanocomposites: a review from preparation to processing. *Progr Polym Sci.* 2003;28(11):1539-641.

[138] Johansson C, Bras J, Mondragon I, Nechita P, Plackett D, Simon P, et al. Renewable fibers and bio-based materials for packaging applications–a review of recent developments. *BioResources.* 2012;7(2):2506-52.

[139] Eroglu MS. Surface free energy analysis of poly (HEMA)-poly (perfluoroacrylate) copolymer networks. *J Appl Polym Sci.* 2006;101(5):3343-7.

[140] Kumar R, Prakash K, Cheang P, Gower L, Khor K. Chitosan-mediated crystallization and assembly of hydroxyapatite nanoparticles into hybrid nanostructured films. *J Royal Soc Interf.* 2008;5(21):427-39.

[141] Pilcher G. Meeting the challenge of radical change. *J Coat Technol.* 2001;73(921): 135-43.

[142] Sun F, Sask K, Brash J, Zhitomirsky I. Surface modifications of Nitinol for biomedical applications. *Coll Surf B: Biointerfaces.* 2008;67(1):132-9.

[143] Hoogenboom R, Meier MA, Schubert US. Combinatorialmethods, automated synthesis and high-throughput screening in polymer research: past and present. *Macromol Rapid Commun.* 2003;24(1):15-32.

[144] Meredith JC, Smith AP, Karim A, Amis EJ. Combinatorial materials science for polymer thin-film dewetting. *Macromolecules.* 2000;33(26):9747-56.

[145] Gordon E, Gallop M, Patel D. Strategy and tactics in combinatorial organic synthesis. Applications to drug discovery. *Acc Chem Res.* 1996;29(3):144-54.

[146] Schiele, N.R., et al., *Laser-based direct-write techniques for cell printing*. Biofabrication, 2010. 2: 032001 (14pp).

[147] Riggs, B.C., et al., *Matrix-assisted pulsed laser methods for biofabrication*. MRS BULLETIN, 2011. 36.

[148] Shepard, K. B. and Priestley, R. D., MAPLE *Deposition of Macromolecules*. Macromol. Chem. Phys. 2013. 214: p. 862–872.

[149] Cicco, N., et al., *Deposition and Characterization of Laccase Thin Films Obtained by Matrix Assisted Pulsed Laser Evaporation*. in Sensors, Lecture Notes in Electrical Engineering. 2015. 319. p. 47-51.

Chapter 5

Hydrogels for Regenerative Medicine

Divya Bhatnagar, Marcia Simon and Miriam H. Rafailovich

Additional information is available at the end of the chapter

http://dx.doi.org/10.5772/62044

Abstract

Regenerative medicine requires materials that are biodegradable, biocompatible, structurally and chemically stable, and that can mimic the properties of the native extracellular matrix (ECM). Hydrogels are hydrophilic three-dimensional networks that have long received attention in the field of regenerative medicine due to their unique properties. Hydrogels have a potential to be the future of regenerative medicine due to their desirable mechanical and chemical properties, ease of their synthesis, and their multiple applicability as drug delivery vehicles, scaffolds, and constructs for cell culture. In this chapter, we have described hydrogels in terms of their cross-linking and then discussed the most recent developments in the use of hydrogels for peripheral nerve regeneration, tooth regeneration, and 3D bioprinting.

Keywords: Hydrogels, nerve regeneration, 3-D printing, tooth regeneration

1. Introduction

Two-dimensional (2D) substrates such as tissue culture polystyrene (TCPS), thin films, and other flat surfaces have traditionally been used to culture mammalian cells *in vitro*. These experiments with the 2D cell constructs have not only provided the basis for understanding complex biological processes but have also led the way in exploring the stem cell differentiation, cell–material interactions, and cell–cell interactions [1]. Three-dimensional (3D) scaffolds were designed to mimic the important physiochemical features of the native cellular microenvironment for *in vitro* cell culture. Among these 3D scaffolds, hydrogels are defined as the cross-linked polymer networks with high water content. Hydrogels are viscoelastic in nature, encompassing both the viscous and the elastic properties. They swell strongly in aqueous media and are typically composed of a hydrophilic organic polymer component that is cross-linked into a network by either covalent or noncovalent interactions [2, 3]. Cross-linking provides structural stability, and the high water content provides fluid-like transport properties [4]. Variation in the cross-linking of these hydrogels also allows for tunable mechanical properties which can be used to evaluate the structure–function relationship at the cell/biomaterial surface. Currently, hydrogels used for mammalian cell culture are synthesized from natural and synthetic materials. Bioactive hydrogel constructs are extensively being used to repair, regenerate, or engineer tissues by being able to promote cell adhesion, migration, proliferation, and stem cell differentiation appropriate to particular tissues [5]. To fully understand the cell functional responses in the context of a particular tissue, recently, many researchers have tried to develop physiologically relevant, biocompatible, biodegradable hydrogel constructs that resemble native tissue and very closely mimic the actual *in vivo* conditions [5–7].

In this chapter, we will review the classification of natural and synthetic polymer-based hydrogels in terms of their cross-linking. Recent advances in the application of novel hydrogels for regenerative medicine areas such as their use in peripheral nerve regeneration, tooth regeneration, and 3-D printed scaffolds would also be addressed.

2. Classification of hydrogels

One way of classifying the hydrogels is through the type of cross-linking [8]. Cross-linking maintains the hydrogel network structure and prevents dissolution of the hydrophilic chains.

2.1. Physically cross-linked hydrogels

Physically cross-linked gels, also known as reversible gels, are networks that are held together by attractive noncovalent forces between the polymer chains (Figure 1). These hydrogels have a tendency of going through a transition from a three-dimensional stable state to eventually degrade and dissolve as a polymer solution. These forces that hold these polymer networks together to form a hydrogel, which includes hydrophobic interactions, hydrogen bonding, or ionic interactions [9, 10].

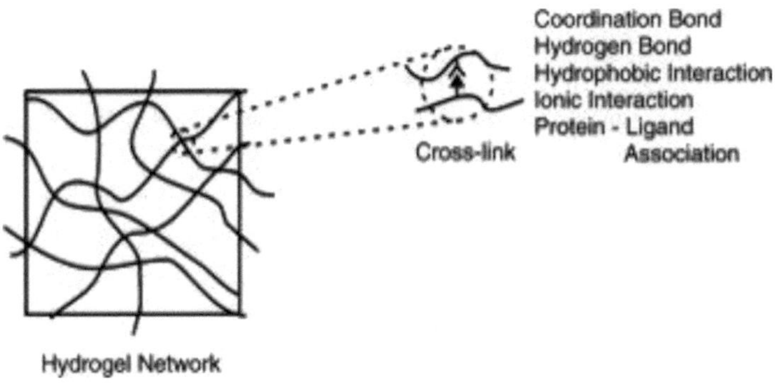

Figure 1. Physical cross-linking in hydrogels, in which the cross-links are formed via noncovalent interaction. Reproduced from ref. [11] © John Wiley and Sons.

http://dx.doi.org/10.5772/62044

Physically cross-linked hydrogels have found their use as matrices for cells/drug encapsulation and release, as scaffolds for cell growth, proliferation, and adhesion. Collagen, gelatin, hyaluronic acid (HA), and alginate are the most commonly used natural polymers, which form physical hydrogels. However, these physically cross-linked hydrogels are prone to premature degradation by proteolytic enzymes such as gelatinase for gelatin, collagenases for collagen, and hyaluronidase for HA [12]. On the other hand, physically cross-linked gels such as pure non-modified HA exhibits poor biomechanical properties [13] and gelatin dissolves into a solution at higher temperatures. Many researchers have therefore tried to formulate physically cross-linked hydrogels with improved mechanical properties and better cell adhesion properties. For example, a composite hydrogel of HA and gelatin was formulated by intercalating the polymer chains into laponite clay by ion exchange. The resulting hydrogel had improved mechanical properties and cell-adhesive surface [7, 14]. Another example of cross-linking by ionic interactions is that of dextran, which forms a hydrogel in the presence of potassium ions [15]. Alginate, a polysaccharide, can also be cross-linked with divalent calcium ions to form a hydrogel [8].

Synthetic polymers such as the triblock copolymer poly(ethylene oxide)99–poly(propylene oxide)67–poly(ethylene oxide)99 (PEO99-PPO67-PEO99, Pluronic F127) can also form a physical hydrogel via hydrogen bonding. Pluronic F127 is unique for its hydrophobic interactions between triblock copolymer chains. At low temperatures, both PPO and PEO chains are soluble in water. Above the critical solution temperature (CST) at which gelation occurs, the polymers dissolve due to the breaking of hydrogen bonds between water molecules and the chains, and PPO becomes hydrophobic PPO core and PEO corona, forming a face-centered cubic nanostructured hydrogel. At even higher temperatures, the micelles aggregate together into hexagonally packed cylinders [16, 17]. Blends and interpenetrating networks of two dissimilar polymers can also form physical hydrogels through noncovalent cross-links. The pure F-127 hydrogel has reduced mechanical properties and, therefore, it has been blended with HA and gelatin to improve its mechanical properties [6]. Other synthetic polymers such as poly(acrylic acid) and poly(methacrylic acid) form physical hydrogels by forming hydrogen bonds with poly(ethylene glycol). This kind of hydrogel formation is pH-dependent since the hydrogen bonds are formed only when the acid groups are protonated [18, 19].

2.2. Chemically cross-linked hydrogels

Chemically cross-linked hydrogels, also known as "permanent" gels, were cross-linked networks formed due to covalent bonds. These gels are usually more stable than the physically cross-linked hydrogels and have a permanent structure [8, 20, 21]. Polymerizing monomers in the presence of cross-linking agents typically forms chemically cross-linked gels. Poly(2-hydroxyethyl methacrylate) is a well-known hydrogel-forming polymer which is generally synthesized by radical polymerization of HEMA in the presence of a suitable cross-linking agent (e.g., ethylene glycol dimethacrylate) [8]. Figure 2 shows a schematic example of the formation of a chemically cross-linked hydrogel via radical polymerization. Hydrogels can also be formed by cross-linking of the various functional groups present in the polymer backbone. Polymers containing hydroxy, amine, or hydrazide groups can be cross-linked by using glutaraldehyde, which forms covalent bonds with each of these functionalities [4]. The swelling, mechanical strength, elastic modulus, diffusional, and other physical properties of these chemical hydrogels are mainly dependent upon their degree of cross-linking, method of preparation, polymer volume fraction, temperature, and swelling agent [22].

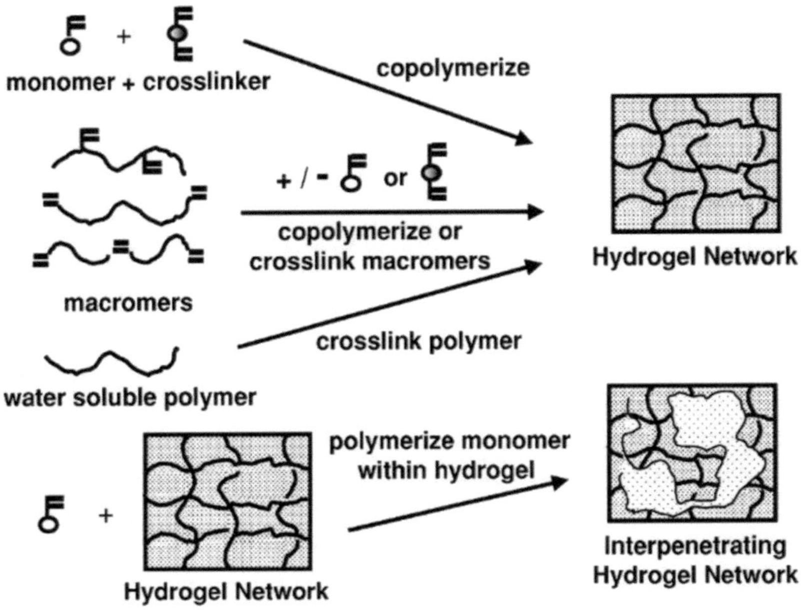

Figure 2. Schematic of methods for formation of cross-linked hydrogels by free radical reactions, including a variety of polymerizations and cross-linking of water-soluble polymers. Examples include cross-linked PHEMA and PEG hydrogels. Reproduced with permission from ref. [2] © Elsevier.

Covalently cross-linked hydrogels can also be formed via enzymatic cross-linking. For example, gelatin, which is chemically cross-linked using glutaraldehyde and formaldehyde to form a stable hydrogel, can also be cross-linked with microbial transglutaminase (mTG) to form an enzymatically cross-linked system. Transglutaminases are a class of natural enzymes that catalyze the acyl-transfer reaction between the ε-amino group of lysine and the γ-carboxyamide group of glutamine in proteins [23, 24]. Microbial transglutaminase (mTG) catalyzes the formation of N-ε-(γ-glutamyl) lysine amide bonds between individual gelatin strands to form a permanent network of cross-linked gelatin [25]. This permanent network of gelatin offers multiple focal adhesion sites for cell attachment, proliferation, and migration.

Another class of hydrogels is the stimuli-responsive hydrogels. These hydrogels can show significant changes in their swelling behavior owing to subtle changes in the pH, temperature, electric–magnetic field, and light [21]. The behavior of these stimuli-sensitive hydrogels depends on the type of the polymer used in making the gel and/or any post-polymerization modifications that are made [26, 27]. pH-responsive hydrogels are swollen ionic networks containing either acidic or basic pendant groups which in an aqueous environment of appropriate pH, ionize developing fixed charges on the gel and thus increasing the swelling forces

http://dx.doi.org/10.5772/62044

[22]. The use of stimuli-sensitive polymers in fabricating hydrogels has led to many interesting applications. Poly(N-isopropylacrylamide) (pNIPAm) is the most widely studied stimuli-responsive polymer. It is formed from the monomer N-isopropylacrylamide ($H_2C{=}CHCONHCH(CH_3)_2$) that exhibits temperature-sensitive swelling behavior over a temperature range of interest. pNIPAm has a lower critical solution temperature (LCST), below which the polymer is soluble. This is attributed to its coil-to-globule transition [28, 29]. Researchers have shown that it is possible to form a strong, thermally responsive nanocomposite hydrogel within a physiological temperature range by initiating free radical polymerization of NIPA from the clay surface [30–32]. Unique properties of cross-linked nanocomposite PNIPA hydrogels has enabled its use as drug delivery systems, rapid release cell culture substrates (Figure 3), and as wound healing dressings.

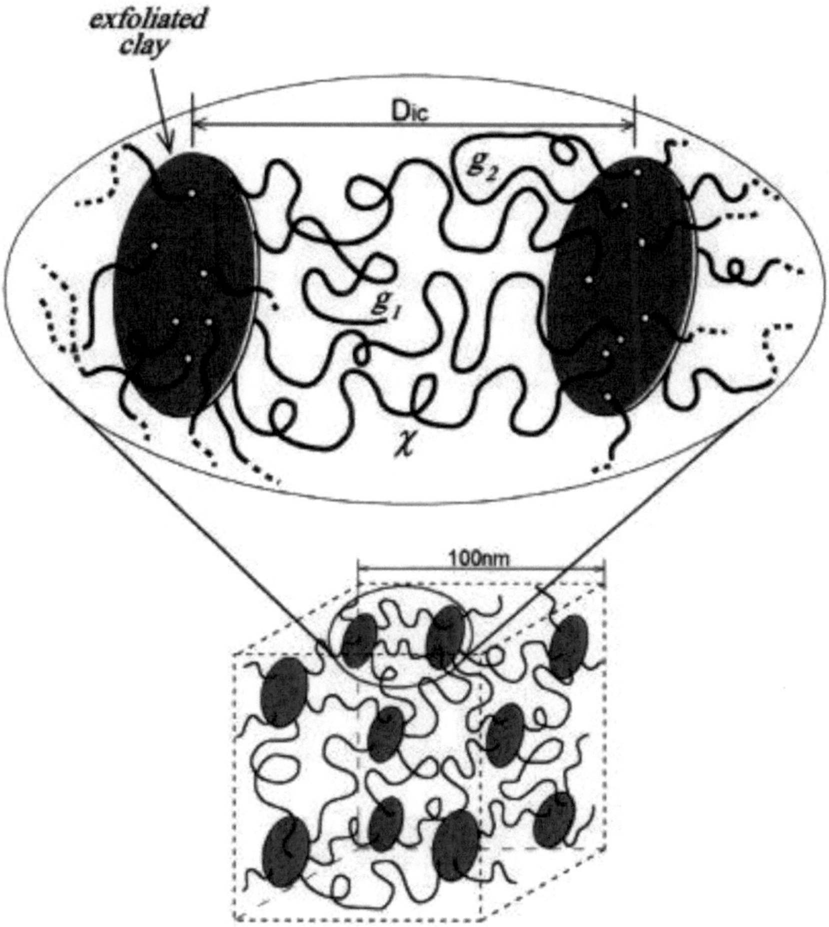

Figure 3. Schematic representation of the structural model with organic/inorganic networks in the NC gel. D_{ic} is an interparticle distance of exfoliated clay sheets. *ø*, *g*1, and *g*2 represent cross-linked chain, grafted chain, and looped chain. In the model, only a small number of polymer chains are depicted for simplicity. Reprinted (adapted) with permission from ref. [33] Copyright (2003) American Chemical Society

3. Application of hydrogels in regenerative field

Field of regenerative medicine works with a common goal of repairing and regenerating damaged tissues and organs. The regenerative process encompasses isolating living cells from patients, expanding them *in vitro* using polymeric scaffolds, and then to re-implanting the tissue-like constructs into the patient [34]. Because of their versatile properties, hydrogels have found several applications in the field of regenerative medicine as scaffolds for cell culture and delivery vehicles for cells and genes [35]. These hydrogels can be made biocompatible with tunable mechanical and degradation properties. They can be equipped with biological cues to guide adhesion, migration, and proliferation of cells and binding sites for growth factors, peptides, or cytokines. This allows for the formation of biomimetic hydrogels that can mimic the extracellular matrix (ECM) environment.

3.1. Hydrogels for peripheral nerve regeneration

Peripheral nervous system (PNS) can repair itself after an injury, but this process has its limitations beyond the critical size gap. Nerve grafts are an alternative to repairing severe peripheral nerve injuries. Nerve autograft and allografts are often used for nerve injuries that cannot be repaired by direct coaptation. However, nerve autografts have several limitations including donor site morbidity, limited availability of the donor tissue, and limited functional recovery. On the other hand, allografts require the use of immunosuppressants for over 18 months and hence, have a significant drawback in their applicability [36]. Nerve guidance tubes (NGTs) fabricated from natural or synthetic biomaterials, for this reason, have become an attractive alternative to repairing critical size nerve defects. NGTs act as a connecting bridge between the proximal and distal ends of the severed nerve, where the nerve stumps are inserted into the ends of the tube and sutured together. A protein-rich fluid containing growth-promoting substances is released into the NGTs. Within days, a fibrin cable is formed that supports the migration of Schwann cells (SCs) and facilitates axonal regeneration from the proximal to the distal stump (Figure 4) [37].

Hydrogels as conduit material: Collagen is an important extracellular matrix (ECM) component that has been studied quite extensively in peripheral nerve regeneration. Collagen hydrogels have been used successfully for the in-vitro culture of many neuronal cell types. Many researchers have developed collagen-based nerve conduits to repair short nerve gaps [38]. Few examples of commercially available, FDA-approved collagen-based tubes that have been clinically used are NeuroGen, NeuroFlex, NeuroMatrix, NeuroWrap, and NeuroMend [39–43]. A nerve tube fabricated from highly purified type I + III collagen derived from porcine skin, Revolnev, has also been used to repair 1 cm rat peroneal nerve with satisfactory functional recovery [44]. Aligned collagen conduits developed by Phillips et al. were shown to orient SCs *in vitro* and their implantation *in vivo* resulted in higher axonal regeneration in a rat sciatic nerve injury model [45]. Hyaluronic acid, a naturally occurring polysaccharide, is another ECM component that has been used to fabricate nerve conduits in a modified form. Sakai et al. made an HA-based nerve conduit that facilitated cellular and axonal ingrowth during peripheral nerve regeneration by identifying viability of disseminated Schwann cells and neuron cells

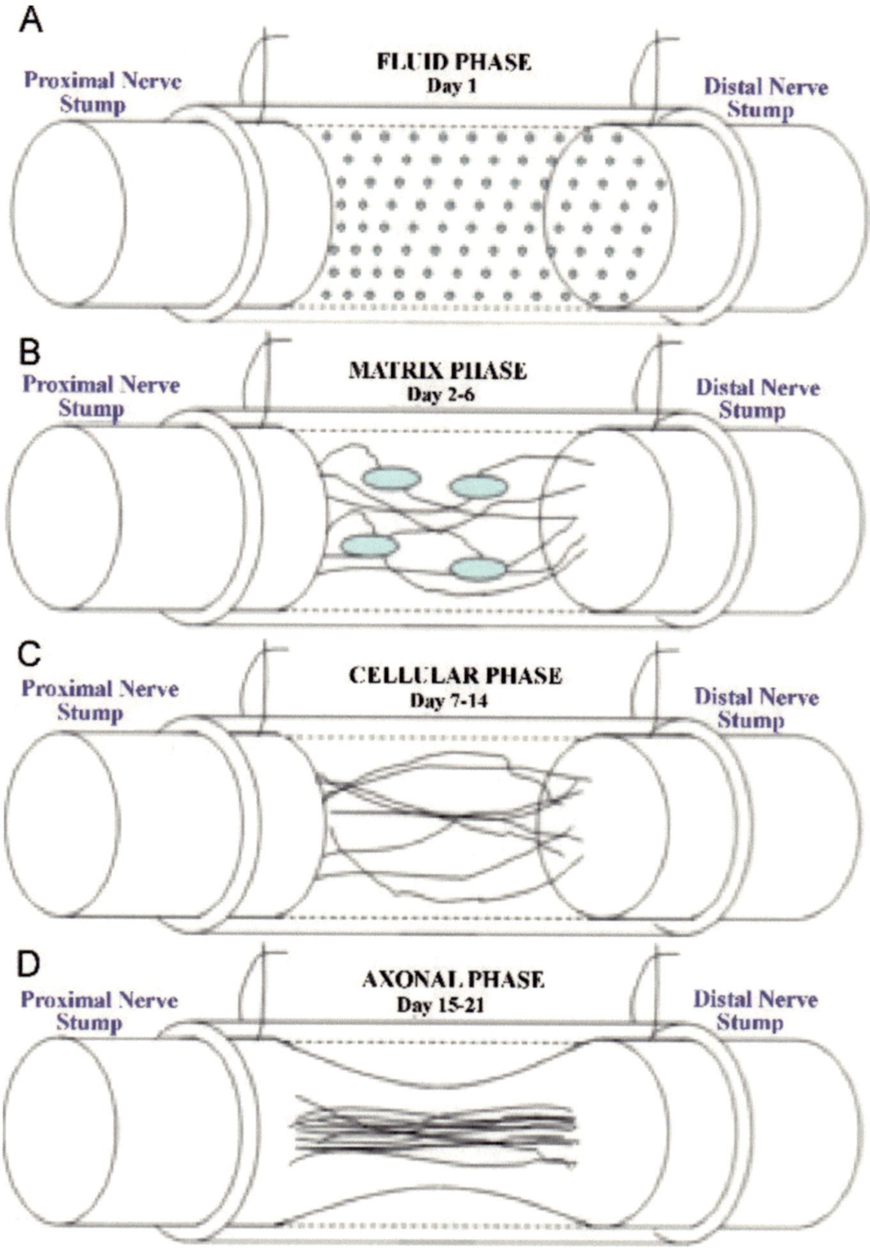

Figure 4. Principle of nerve entubulization and the sequence of events leading to the growth of a new nerve cable: (A) chamber walls with a protein-rich fluid (containing neurotrophic factors); (B) generation of a fibrin-rich scaffold; (C) cell migration (perineural, endothelial, and Schwann cells); (D) axonal cables elongation. Reproduced with permission from ref. [37] © Elsevier.

into HA conduits *in vitro* [46]. Jansen et al. prepared nerve conduits from an esterified hyaluronan derivative (Hyaff) by individual knitting of the strands and strengthening it by coating a thin layer of the same polymer [47]. Fibrin glue/gel is an FDA-approved sealant that contains fibrinogen and thrombin. Fibrin is a protein that is involved in normal blood clotting, while fibrin gel has been extensively used in peripheral nerve regeneration as a sealant for coaptation of nerves [48]. Fibrin gels have also been used as nerve conduit to promote nerve regeneration [49–52]. Pettersson et al. [52] showed that hollow fibrin conduits supported muscle recovery and axonal growth in short nerve gaps. However, in longer nerve gaps, the hollow tubes failed in comparison to autografts. In larger nerve defects, the fibrin cable that facilitates SCs migration is not formed in the hollow NGTs. Therefore, NGTs are limited to nerve lesions < 4 cm and result in poor functional recovery at longer gaps.

Hydrogels as luminal fillers: The empty lumen of a nerve conduit lacks the necessary support structure for the ingrowth and migration of SCs and axons, thus making it an undesirable environment for axonal repair. Hence, in the longer nerve gaps, there is a need for a substrate inside the lumen of the NGTs that can provide necessary mechanical and biological cues for SCs migration and enhance nerve regeneration [36, 53]. Hydrogels are owing to their injectable behavior and their drug encapsulation and delivery capability have found their use as luminal fillers inside the conduits. Figure 5 shows how NGTs filled with hydrogels provide mechanical support in addition to serving as a carrier of bioactive molecules needed for proper functional recovery. This property is absent in the hollow tubes where bioactive cues are not present to direct proper axonal regeneration [54].

Various hydrogels alone or supplemented with small molecules, growth factors, neurotrophic factors, and cellular components have been used as luminal fillers for nerve conduits. For example, agarose hydrogels containing gradients of laminin-1 and nerve growth factor (NGF) molecules have been used in polysulfone (PSU) tubes [55]. Various researchers have also investigated the role of collagen as a luminal filler [53]. However, just the mere presence of these hydrogels sometimes is insufficient to achieve enhanced functional recovery. Therefore, luminal collagen fillers have been supplemented with laminin, NGF, fibroblast growth factor (FGF), etc. to promote better nerve regeneration. Similarly, fibrin gel has also been used to enhance SC migration, myelination, and rate of regeneration inside silicone tubes in a 1 cm rat sciatic nerve model [53].

Seckel et al. used hyaluronic acid gel in the conduits to produce better conduction velocity, higher axon counts, and myelination [56, 57]. They postulated that HA improves fibrin matrix formation and decreases scarring that might interfere with nerve regeneration. Mohammad et al. [58] showed that when HA was used with NGF, there was a 45% increase in the myelinated axon count. Most recently, keratin-based hydrogels that were used to fill commercial nerve tubes showed an improved axonal area and myelination compared to the empty tube. Electrophysiological analysis such as conduction delay and impulse amplitude were also better than the hollow tube and comparable to the autografts [59, 60]. Luminal fillers in nerve conduits supplemented with essential growth factors are promising ways to achieve nerve regeneration at par with autologous grafts. With an appropriate nerve conduit designed for long nerve gap, bioactive luminal fillers can aid in enhanced functional recovery.

Hydrogels have an added advantage in the field of peripheral nerve regeneration, as they can serve as a support system inside the conduit and also as a mode of delivery of various growth factors necessary for nerve regeneration. However, if the mechanical properties of the

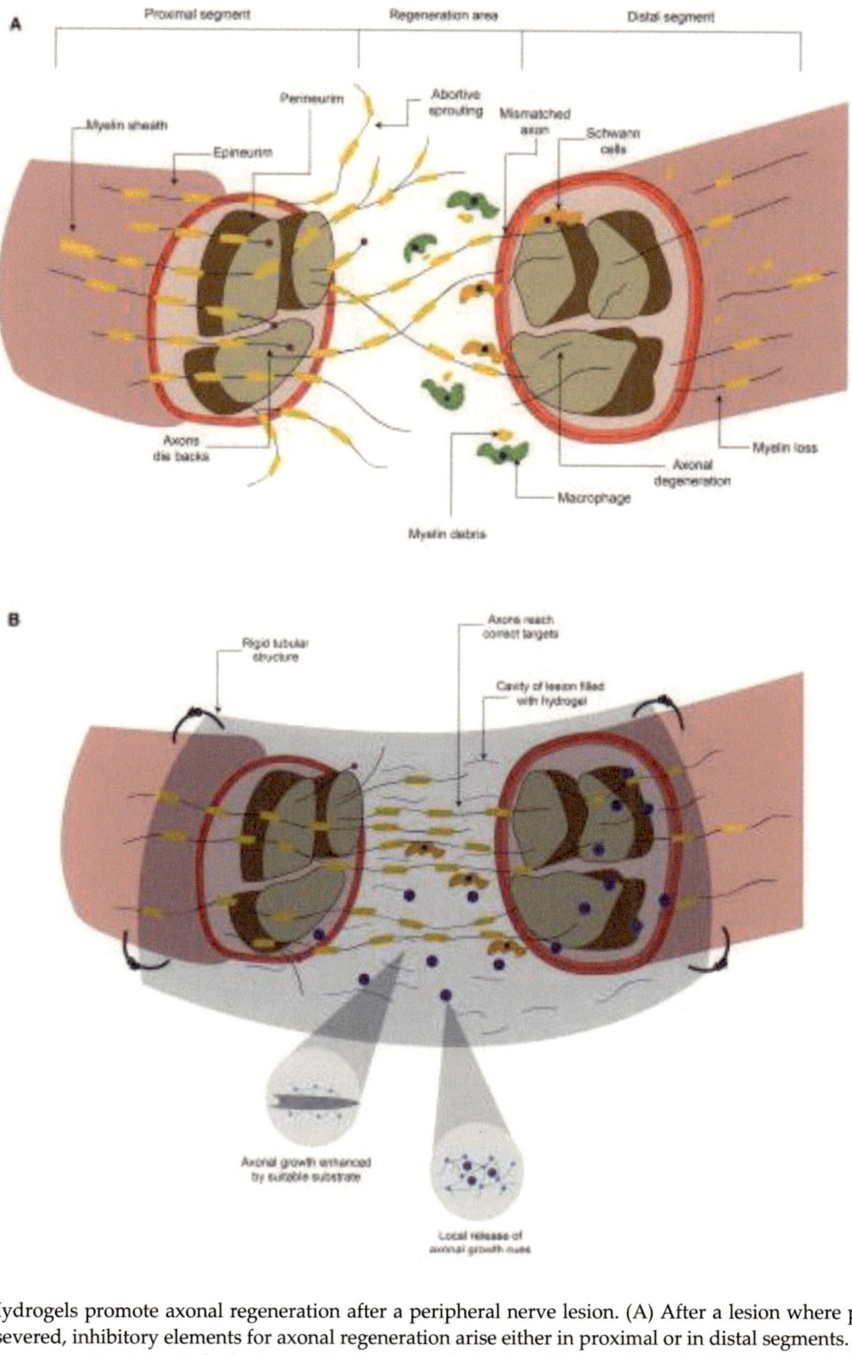

Figure 5. Hydrogels promote axonal regeneration after a peripheral nerve lesion. (A) After a lesion where peripheral nerves are severed, inhibitory elements for axonal regeneration arise either in proximal or in distal segments. Although there can be regeneration to unite both stumps, it is common that mismatches are formed. (B) When the lesion area is connected with a rigid tubular structure, and this is filled with a hydrogel, there is a mechanical support and a suitable substrate for axonal growth. In addition, the hydrogel can serve as a carrier of molecules that promote axonal regeneration and ultimately functional recovery. Reproduced from ref. [54] © Carballo-Molina and Velasco under the terms of Creative Commons Attribution License (CC BY).

hydrogels are not adjusted appropriately, they can hinder the nerve regeneration. Therefore, the limitation of using hydrogels as luminal fillers is primarily their cross-linking. Highly cross-linked viscous gels can be disadvantageous for nerve regeneration. At the same time, the rate of degradation of hydrogels plays an important role if they are used as conduit materials. Hence, it is essential to tune the mechanical and chemical properties of the hydrogels for their best use in peripheral nerve regeneration.

3.2. Hydrogels for tooth regeneration

Tooth regeneration similar to the construction of other tissues also requires an appropriate cell source, a biodegradable scaffold that can mimic the natural extracellular matrix (ECM) and bioactive molecules. Tooth organ is composed of enamel, dentin, cementum, and dental pulp. Cells such as ameloblasts form the enamel, odontoblasts form the dentin, cementoblasts form the cementum, and mesenchymal, fibroblastic, vascular, and neural cells form the dental pulp [61]. Scaffold materials play a critical role in determining how cells proliferate and differentiate. Those that mimic the characteristics of natural ECM can best promote appropriate cell and tissue maturation. The tooth scaffolds should be such that they provide chemical and mechanical integrity, are biocompatible, are able to restore the normal functioning of the tooth, and are able to integrate with the surrounding tissues [25]. For dentin-pulp tissue engineering, in particular, hydrogels come across as a favorable choice because they are injectable and have a 3D morphology that helps in the encapsulation of cells and growth factors. Hydrogel scaffolds made from natural biopolymers such as collagen, chitosan, hyaluronic acid, gelatin, fibrin, and alginate have been used quite extensively since they are readily cross-linkable and can be easily combined with various bioactive molecules [62]. Kim et al. [63] loaded collagen gels with a series of growth factors and injected them into pulp chambers and root canals of endodontically treated human teeth. They found that on *in vivo* implantation of endodontically treated human teeth in mouse dorsum for the tested 3 or 6 weeks, there was a recellularized and revascularized connective tissue that integrated to the native dentinal wall in root canals.

Collagen gels have also been used to deliver dental pulp stem cells (DPSCs) and dentin matrix protein-1 (DMP-1) *in vivo* where it led to the ectopic formation of dental pulp-like tissue [64]. Although, collagen is a major ECM component used to fabricate hydrogel scaffolds in tissue engineering, its moderate mechanical strength is a limitation [61]. For this reason, Bhatnagar et al. [25] used cross-linked gelatin scaffolds for dentin regeneration without any external chemical stimulus. These gelatin hydrogel scaffolds that were cross-linked with microbial transglutaminase (mTG) provided variable mechanical properties and were capable of differentiating DPSCs *in vitro* toward odontogenesis irrespective of their mechanical stimulus and external chemical inducer (Figure 6). Alginate hydrogels have also been used for dentin-pulp regeneration. When loaded with exogenous transforming growth factor TGFβ1, these hydrogels were shown to promote dentin matrix secretion and odontoblast-like cell differentiation [65]. Among synthetic polymers, PLGA hydrogels with recombinant human growth differentiation factor-5 (rhGDF-5) have been used in periodontal defects in a dog model. Periodontal pockets (3 × 6mm, width × depth) were surgically created over the buccal roots of the second and fourth mandibular premolars in mongrel dogs and progressive alveolar bone maturation was seen at 6 weeks on rhGDF-5/PLGA delivery [66].

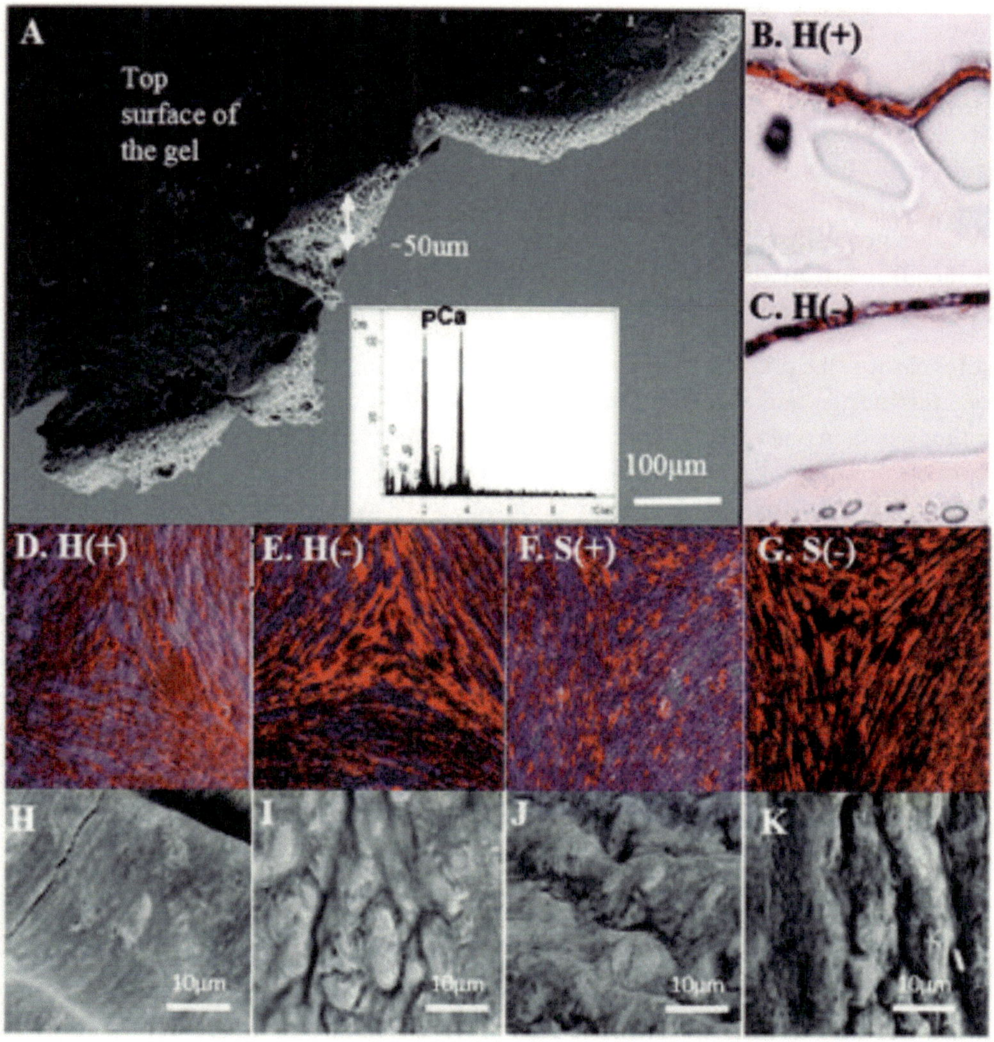

Figure 6. (A) Cross-section of a non-induced hard gel (H (−)) after 35 days of DPSCs culture showing a self-supporting sheet of biomineralized deposits present inside the gel. EDX spectra (inset in (A)) confirm the hydroxyapatite mineral. A cross-sectional view of the alizarin red-stained calcified biomineralized deposits in the (B) hard (+) and (C) hard (−) gel. Top view of the alizarin red-stained calcified deposits and their corresponding SEM images after 35 days of DPSCs cultured on: (D, H) hard (+); (E, I) hard (−); (F, J) soft (+); (G, K) soft (−) gels. The calcified deposits laid by the cells are stained dark red and have a defined pattern. Reproduced from ref. [25] © by permission of The Royal Society of Chemistry.

Hydrogels have shown their potential in regenerating dentin-pulp tissue. Researchers have demonstrated the successful use of hydrogel scaffolds for dentin-pulp matrix regeneration. However, hydrogels have a limitation when it comes to regenerating the whole tooth organ.

Not much research has been done in the field of using hydrogel scaffolds for regenerating the whole tooth structure.

3.3. Hydrogels for 3D printing

3D printing is emerging as a potential tool in regenerative medicine for building complex 3D structures across length scales ranging from micrometers to millimeters. 3D printing represents a way to pattern and assemble the cells with materials in a controlled and functional 3D architecture. The only limitation that arises is due to the materials being printed and necessitates a need for new inks to expand the utility of 3D printed structures [67]. 3D printing techniques generally comprises of: (a) extrusion-based printing that requires a material to be extruded through an orifice, (b) ink-jet based printing that requires a material to be ejected as droplets onto a substrate, and (c) laser based printing where a material is cured using a laser [67].

Hydrogels for 3D printing should be printable, biocompatible, have desired mechanical properties, shape, and structure (Table 2) [68]. Collagen has been extensively used for 3D printing where in one case, sodium hydrogen carbonate ($NaHCO_3$) vapor was applied to gel the printed collagen layer and in another instance, $NaHCO_3$ was mixed with collagen and cells and then printed using laser-assisted bioprinting [68]. Several researchers have utilized the temperature-responsive hydrogels, particularly pluronic F127 that gels in the temperature range of 10 to 40°C. Pluronics have been combined with collagen and cross-linked gelatin methacrylate (GelMa) to form bioinks. Kolesky et al. printed pluronic F127 as a sacrificial vascular network embedded in GelMa matrix that mimic natural fine capillaries [69].

Ideal bioprinting hydrogel properties	
Printability	Viscosity
	Shear-thinning property
	Response and transition time
	Sol–gel transition stimulus
Biocompatibility	Degradability
	Cell-binding motifs
	Non-toxic
	Non-immunogenic
Mechanical Properties	Stiffness
	Elasticity
	Strength
Shape and structure	Pore size
	Micro/Nanostructure

Table 1. Ideal bioprinting hydrogel properties. Reproduced from ref. [68] © Wang et al., under the terms of Creative Commons Attribution-Non-commercial 4.0 International License.

Photocross-linking property of the hydrogels has been utilized to bioprint tough and rigid hydrogel constructs with cells. For example, partially photocross-linking gelatin methacry-

http://dx.doi.org/10.5772/62044

late (GelMA) was combined with hyaluronic acid methacrylate (HAMA) to form a gel-like fluid which was then printed with a defined pattern. This printed layer was further irradiated to obtain a tubular tissue construct [70]. Hong et al. [71] combined sodium alginate and poly (ethylene glycol) (PEG) to constitute an interpenetrating network. Laponite clay was used to form a nanogel. Poly(ethylene glycol) diacrylate (PEGDA) and alginate mixture were combined with laponite clay to form a pre-gel solution. To cross-link PEGDA and alginate, a photoinitiator and calcium sulfate solution were added to the pre-gel solution. The PEGDA–alginate–nanoclay pre-gel solution was 3D printed via extrusion-based printing (Figure 7). The resulting hydrogels were tough and had the potential to encapsulate cells for tissue regeneration.

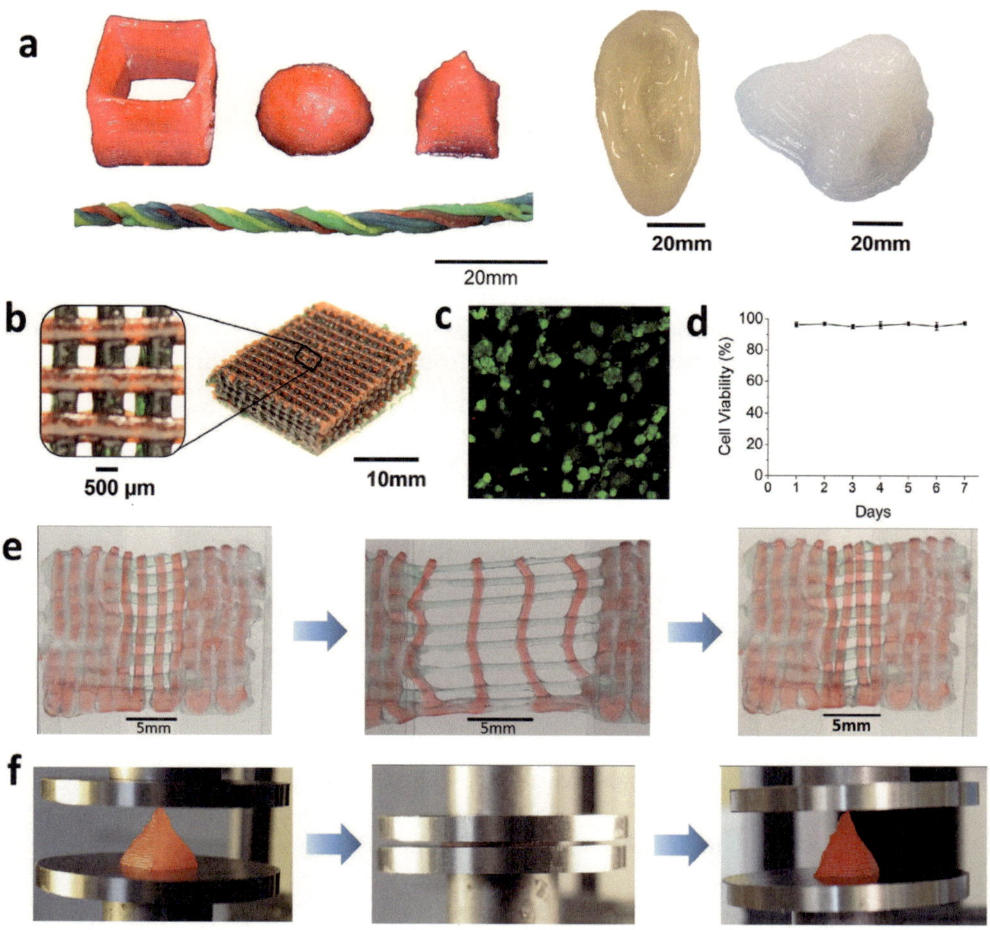

Figure 7. 3D printing of tough and biocompatible PEG–alginate–nanoclay hydrogels. (a) Various 3D constructs printed with the hydrogel (from left to right: hollow cube, hemisphere, pyramid, twisted bundle, the shape of an ear, and a nose. Non-toxic red food dye was added postprint on some samples for visibility). (b) A mesh printed with the tough and biocompatible hydrogel. The mesh was used to host HEK cells. (c) Live–dead assay of HEK cells in a collagen hydrogel infused into the 3D printed mesh of the PEG–alginate–nanoclay hydrogel. (d) The viability of the HEK cells through 7 d. (e) A printed bilayer mesh (top layer red, bottom layer green) is uniaxially stretched to three times its

initial length. Relaxation of the sample after stretching shows almost complete recovery of its original shape. (f) A printed pyramid undergoes a compressive strain of 95% while returning to its original form after relaxation. Reproduced with permission from ref. [71] © John Wiley and Sons.

Recent developments in 3D printing of hydrogels offer a potential to produce constructs with the higher structural organization, fine-tuned mechanical and chemical properties to control cell behavior and an environment that mimics *in vivo* tissue. 3D printing of hydrogels is promising and requires further development such that the hydrogels are easy and inexpensive to print, are favorable toward promoting cell viability, differentiation, migration, and cell–cell interactions, and are functionally versatile. However, hydrogels are soft, and their use for 3D printing largely depends on their viscosity, their structural integrity, and their ability to be cross-linked in a way such that the cells can be encapsulated.

4. Conclusion

Hydrogels have found extensive applicability in various fields of tissue engineering and regenerative medicine due to their underlying similarity to the native ECM. The role of hydrogels in regenerative medicine has progressed remarkably with their widespread use in peripheral nerve regeneration, tooth regeneration, and more recently in 3D printing. Long nerve gap repair, dentin-pulp complex reconstruction, and 3D printing of organs are few of the areas in regenerative medicine that are at the forefront. Understanding and development of functionally bioactive smart hydrogels could help tremendously in these regenerative therapies.

Acknowledgements

This work was supported by NSF-Inspire Program grant # DMR- 1344267.

Author details

Divya Bhatnagar[1*], Marcia Simon[2] and Miriam H. Rafailovich[3]

*Address all correspondence to: divya.bhatnagar22@gmail.com

1 New Jersey Center for Biomaterials, Rutgers University, Piscataway, NJ, USA

2 Department of Oral Biology and Pathology, Stony Brook School of Dental Medicine, Stony Brook, NY, USA

3 Department of Materials Science and Engineering, Stony Brook University, Stony Brook, NY, USA

References

[1] Tibbitt, M.W. and K.S. Anseth, *Hydrogels as extracellular matrix mimics for 3D cell culture.* Biotechnology and Bioengineering, 2009. 103(4): pp. 655–663.

[2] Hoffman, A.S., *Hydrogels for biomedical applications.* Advanced Drug Delivery Reviews, 2012. 64: pp. 18–23.

[3] Cabral, J. and S.C. Moratti, *Hydrogels for biomedical applications.* Future Medicinal Chemistry, 2011. 3(15): pp. 1877–1888.

[4] Nayak, S. and L.A. Lyon, *Soft nanotechnology with soft nanoparticles.* Angewandte Chemie International Edition, 2005. 44(47): pp. 7686–7708.

[5] Ghosh, K., et al., *Cell adaptation to a physiologically relevant ECM mimic with different viscoelastic properties.* Biomaterials, 2007. 28: pp. 671–679.

[6] Bhatnagar, D., et al., *Rheological Characterization of Novel HA-Pluronic Thermoreversible Hydrogels.* Journal of Chemical and Biological Interfaces, 2013. 1(2): pp. 93–99.

[7] Bhatnagar, D., et al., *Hyaluronic Acid and Gelatin Clay Composite Hydrogels: Substrates for Cell Adhesion and Controlled Drug Delivery.* Journal of Chemical and Biological Interfaces, 2014. 2(1): pp. 34–44.

[8] Hennink, W. and C.F. Van Nostrum, *Novel crosslinking methods to design hydrogels.* Advanced Drug Delivery Reviews, 2012. 64: pp. 223–236.

[9] Campoccia, D., et al., *Semisynthetic resorbable materials from hyaluronan esterification.* Biomaterials, 1998. 19(23): pp. 2101–2127.

[10] Prestwich, G.D., et al., *Controlled chemical modification of hyaluronic acid: synthesis, applications, and biodegradation of hydrazide derivatives.* Journal of Controlled Release, 1998. 53(1): pp. 93–103.

[11] Nayak, S. and L.A. Lyon, *Soft Nanotechnology with Soft Nanoparticles.* Angewantde Chemie International Edition , 2005. 44: pp. 7686–7708.

[12] Lee, K., *DJ Mooney Hydrogels for tissue engineering.* Chemistry Review, 1869. 101: pp. 2001.

[13] Ghosh, K., et al., *Rheological Characterization of in Situ Cross-Linkable Hyaluronan Hydrogels.* BioMacromolecules, 2005. 6(5): pp. 2857–2865.

[14] Rafailovich, M., D. Bhatnagar, and M.K. Cowman, *Nanocomposite hyaluronic acid-clay based hydrogels.* 2013, Google Patents.

[15] Watanabe, T., et al., *NMR studies on water and polymer diffusion in dextran gels. Influence of potassium ions on microstructure formation and gelation mechanism.* Magnetic Resonance in Medicine, 1996. 35(5): pp. 697–705.

[16] Huang, S. and X. Fu, *Naturally Derived Materials-Based Cell and Drug Delivery Systems in Skin Regeneration.* Journal of Controlled Release, 2010. 142(2): pp. 149–159.

[17] Jiang, J., et al., *Rheology of thermoreversible hydrogels from multiblock associating copolymers.* Macromolecules, 2008. 41(10): pp. 3646–3652.

[18] Eagland, D., N. Crowther, and C. Butler, *Complexation between polyoxyethylene and polymethacrylic acid—the importance of the molar mass of polyoxyethylene.* European Polymer Journal, 1994. 30(7): pp. 767–773.

[19] Mathur, A.M., et al., *Equilibrium swelling of poly (methacrylic acid-g-ethylene glycol) hydrogels: Effect of swelling medium and synthesis conditions.* Journal of Controlled Release, 1998. 54(2): pp. 177–184.

[20] Wichterle, O. and D. Lim, *Hydrophilic gels for biological use.* 1960.

[21] Gulrez, S.K., G.O. Phillips, and S. Al-Assaf, *Hydrogels: methods of preparation, characterisation and applications.* 2011: INTECH Open Access Publisher.

[22] Ratner, B.D., et al., *Biomaterials science: an introduction to materials in medicine.* 2004: Academic press.

[23] Yung, C., et al., *Transglutaminase crosslinked gelatin as a tissue engineering scaffold.* Journal of Biomedical Materials Research Part A, 2007. 83(4): pp. 1039–1046.

[24] Ito, A., et al., *Transglutaminase-mediated gelatin matrices incorporating cell adhesion factors as a biomaterial for tissue engineering.* Journal of Bioscience and Bioengineering, 2003. 95(2): pp. 196–199.

[25] Bhatnagar, D., et al., *Biomineralization on enzymatically cross-linked gelatin hydrogels in the absence of dexamethasone.* Journal of Materials Chemistry B, 2015.

[26] Bettini, R., P. Colombo, and N.A. Peppas, *Solubility effects on drug transport through pH-sensitive, swelling-controlled release systems: Transport of theophylline and metoclopramide monohydrochloride.* Journal of Controlled Release, 1995. 37(1): pp. 105–111.

[27] Cicek, H. and A. Tuncel, *Immobilization of α-chymotrypsin in thermally reversible isopropylacrylamide-hydroxyethylmethacrylate copolymer gel.* Journal of Polymer Science Part A: Polymer Chemistry, 1998. 36(4): pp. 543–552.

[28] Xu, D., et al., *Rheology of Poly (N-isopropylacrylamide)–Clay Nanocomposite Hydrogels.* Macromolecules, 2015. 48(3): pp. 840–846.

[29] Lombardi, J., et al., *Rheology of Glycated Poly (N-isopropylacrylamide)-Clay Nanogels.* Journal of Chemical and Biological Interfaces, 2014. 2(1): pp. 45–49.

[30] Haraguchi, K., et al., *Mechanism of forming organic/inorganic network structures during in-situ free-radical polymerization in PNIPA-clay nanocomposite hydrogels.* Macromolecules, 2005. 38(8): pp. 3482–3490.

[31] Haraguchi, K. and T. Takehisa, *Nanocomposite hydrogels: a unique organic-inorganic network structure with extraordinary mechanical, optical, and swelling/de-swelling properties.* Advanced Materials, 2002. 14(16): p. 1120.

[32] Miyazaki, S., et al., *Gelation mechanism of poly (N-isopropylacrylamide)-clay nanocomposite gels.* Macromolecules, 2007. 40(12): pp. 4287–4295.

[33] Haraguchi, K., et al., *Compositional effects on mechanical properties of nanocomposite hydrogels composed of poly (N, N-dimethylacrylamide) and clay.* Macromolecules, 2003. 36(15): pp. 5732–5741.

[34] Langer, R. and J. Vacanti, *Tissue engineering.* Science, 1993. 14(260(5110)): pp. 920–926.

[35] Slaughter, B.V., et al., *Hydrogels in Regenerative Medicine.* Advanced Materials, 2009. 21: pp. 3307–3329.

[36] Pabari, A., et al., *Recent advances in artificial nerve conduit design: strategies for the delivery of luminal fillers.* Journal of Controlled Release, 2011. 156(1): pp. 2–10.

[37] Kehoe, S., X. Zhang, and D. Boyd, *FDA approved guidance conduits and wraps for peripheral nerve injury: a review of materials and efficacy.* Injury, 2012. 43(5): pp. 553–572.

[38] Khaing, Z.Z. and C.E. Schmidt, *Advances in natural biomaterials for nerve tissue repair.* Neuroscience Letters, 2012. 519(2): pp. 103–114.

[39] Meek, M.F. and J.H. Coert, *US Food and Drug Administration/Conformit Europe-approved absorbable nerve conduits for clinical repair of peripheral and cranial nerves.* Annals of Plastic Surgery, 2008. 60(1): pp. 110–116.

[40] Archibald, S., et al., *Monkey median nerve repaired by nerve graft or collagen nerve guide tube.* The Journal of Neuroscience, 1995. 15(5): pp. 4109–4123.

[41] Archibald, S.J., et al., *A collagen-based nerve guide conduit for peripheral nerve repair: An electrophysiological study of nerve regeneration in rodents and nonhuman primates.* Journal of Comparative Neurology, 1991. 306(4): pp. 685–696.

[42] Farole, A. and B.T. Jamal, *A bioabsorbable collagen nerve cuff (NeuraGen) for repair of lingual and inferior alveolar nerve injuries: a case series.* Journal of Oral and Maxillofacial Surgery, 2008. 66(10): pp. 2058–2062.

[43] Li, S.-T., et al., *Peripheral nerve repair with collagen conduits.* Clinical materials, 1992. 9(3): pp. 195–200.

[44] Gu, X., et al., *Construction of tissue engineered nerve grafts and their application in peripheral nerve regeneration.* Progress in Neurobiology, 2011. 93(2): pp. 204–230.

[45] Phillips, J.B., et al., *Neural tissue engineering: a self-organizing collagen guidance conduit.* Tissue Engineering, 2005. 11(9-10): pp. 1611–1617.

[46] Sakai, Y., et al., *New artificial nerve conduits made with photocrosslinked hyaluronic acid for peripheral nerve regeneration.* Bio-Medical Materials and Engineering, 2007. 17(3): pp. 191–197.

[47] Jansen, K., et al., *A hyaluronan-based nerve guide: in vitro cytotoxicity, subcutaneous tissue reactions, and degradation in the rat.* Biomaterials, 2004. 25(3): pp. 483–489.

[48] Ornelas, L., et al., *Fibrin glue: an alternative technique for nerve coaptation-Part II. Nerve regeneration and histomorphometric assessment.* Journal of Reconstructive Microsurgery, 2006. 22(2): pp. 123–128.

[49] di Summa, P.G., et al., *Long-term in vivo regeneration of peripheral nerves through bioengineered nerve grafts.* Neuroscience, 2011. 181: pp. 278–291.

[50] Kalbermatten, D.F., et al., *New fibrin conduit for peripheral nerve repair.* Journal of Reconstructive Microsurgery, 2009. 25(1): p. 27–33.

[51] Pettersson, J., et al., *Biodegradable fibrin conduit promotes long-term regeneration after peripheral nerve injury in adult rats.* Journal of Plastic, Reconstructive & Aesthetic Surgery, 2010. 63(11): pp. 1893–1899.

[52] Pettersson, J., et al., *Muscle recovery after repair of short and long peripheral nerve gaps using fibrin conduits.* Neuroscience Letters, 2011. 500(1): pp. 41–46.

[53] Chen, M.B., F. Zhang, and W.C. Lineaweaver, *Luminal fillers in nerve conduits for peripheral nerve repair.* Annals of Plastic Surgery, 2006. 57(4): pp. 462–471.

[54] Carballo-Molina, O.A. and I. Velasco, *Hydrogels as scaffolds and delivery systems to enhance axonal regeneration after injuries.* Frontiers in Cellular Neuroscience, 2015. 9.

[55] Dodla, M.C. and R.V. Bellamkonda, *Differences between the effect of anisotropic and isotropic laminin and nerve growth factor presenting scaffolds on nerve regeneration across long peripheral nerve gaps.* Biomaterials, 2008. 29(1): pp. 33–46.

[56] Seckel, B., et al., *Hyaluronic acid through a new injectable nerve guide delivery system enhances peripheral nerve regeneration in the rat.* Journal of Neuroscience Research, 1995. 40(3): pp. 318–324.

[57] Wang, K.K., et al., *Hyaluronic acid enhances peripheral nerve regeneration in vivo.* Microsurgery, 1998. 18(4): pp. 270–275.

[58] Mohammad, J.A., et al., *Increased axonal regeneration through a biodegradable amnionic tube nerve conduit: effect of local delivery and incorporation of nerve growth factor/hyaluronic acid media.* Annals of Plastic Surgery, 2000. 44(1): pp. 59–64.

[59] Apel, P.J., et al., *Peripheral nerve regeneration using a keratin-based scaffold: long-term functional and histological outcomes in a mouse model.* The Journal of Hand Surgery, 2008. 33(9): pp. 1541–1547.

[60] Hill, P.S., et al., *Repair of peripheral nerve defects in rabbits using keratin hydrogel scaffolds.* Tissue Engineering Part A, 2011. 17(11–12): pp. 1499–1505.

[61] Yuan, Z., et al., *Biomaterial selection for tooth regeneration.* Tissue Engineering Part B: Reviews, 2011. 17(5): pp. 373–388.

[62] Laurenti, M. and M.-N. Abdallah, *Natural and synthetic hydrogels for periodontal tissue regeneration.* International Dental Journal of Students Research, 2015. 3(2): pp. 49–51.

[63] Kim, J.Y., et al., *Regeneration of dental-pulp-like tissue by chemotaxis-induced cell homing.* Tissue Engineering Part A, 2010. 16(10): pp. 3023–3031.

[64] Huang, G.T.-J., et al., *Stem/progenitor cell–mediated de novo regeneration of dental pulp with newly deposited continuous layer of dentin in an in vivo model.* Tissue Engineering Part A, 2009. 16(2): pp. 605–615.

[65] Dobie, K., et al., *Effects of alginate hydrogels and TGF-β1 on human dental pulp repair in vitro.* Connective Tissue Research, 2002. 43(2–3): pp. 387–390.

[66] Kwon, D.H., et al., *Evaluation of an injectable rhGDF-5/PLGA construct for minimally invasive periodontal regenerative procedures: a histological study in the dog.* Journal of Clinical Periodontology, 2010. 37(4): pp. 390–397.

[67] Highley, C.B., C.B. Rodell, and J.A. Burdick, *Direct 3D Printing of Shear-Thinning Hydrogels into Self-Healing Hydrogels.* Advanced Materials, 2015. 27(34): pp. 5075–5079.

[68] Wang, S., J.M. Lee, and W.Y. Yeong, *Smart hydrogels for 3D bioprinting.* International Journal of Bioprinting, 2015. 1(1).

[69] Kolesky, D.B., et al., *3D bioprinting of vascularized, heterogeneous cell-laden tissue constructs.* Advanced Materials, 2014. 26(19): pp. 3124–3130.

[70] Skardal, A., et al., *Photocrosslinkable hyaluronan-gelatin hydrogels for two-step bioprinting.* Tissue Engineering Part A, 2010. 16(8): pp. 2675–2685.

[71] Hong, S., et al., *3D P*rinting of Highly Stretchable and Tough Hydrogels into Complex, Cellularized Structures. Advanced Materials, 2015. 27(27): pp. 4035–4040.

Chapter 6

Nano-biomaterials in Antimicrobial Therapy

Pratima Parashar Pandey

Additional information is available at the end of the chapter

http://dx.doi.org/10.5772/61959

Abstract

Silver nanoparticles (AgNps) have attracted much interest in biomedical engineering, since they have excellent antimicrobial properties. Silver nanopolymer composites have applications in biochemical sensors, antimicrobial activity and drug delivery system. Silver nanoparticles are more effective than ionic homologues (Ag+) for their antimicrobial activity. Antimicrobial properties of silver nanoparticles are used by their incorporation into medical devices, tissues and other health-related products for skin pathologies to reduce the risk of contamination and to promote higher preventive infection control. Novel hybrid material thin films based on various polymeric systems with embedded silver nanoparticles were synthesized using various methods. The electrical, optical and plasmonic responses of AgNps onto thin layers of polymer composites show encapsulation of nanoparticles. The antibacterial activity of AgNps/polymer composites against various common bacteria is discussed in this chapter. The antibacterial activity of the synthesized hybrid materials was tested against various bacteria, commonly found in hospital environment. Silver nanostructures have especially been of interest because of contrast agents for biomedical image. Shunts used for hydrocephalous silicon elastomer grafted with hydrogel, polyvinylpyrrolidone (PVP) soaked in various antibiotics proved to be active for longer time.

Keywords: Polyvinylpyrrolidone, Silver nanoparticle, Antibacterial, Optical properties, Electrical properties, SEM

1. Introduction

Synthesis of nanoparticles is an interesting field in solid-state chemistry. Recently, much interest has been shown for ultrafine metal particles because they have unique properties that are different from bulk metals in optical property, catalytic activity and magnetic property. In particular, many studies have been devoted to silver, gold and copper colloids. Usually nanoparticles are spherical in shape. Nanoscale metal flakes are of interest in electronics, mechanics, electromagnetic, communication and stealth technology [1].

The silver antimicrobial effectiveness has been known for a long time. Over the past few years, the silver or silver salts have become important components to control the microbial proliferation. They are being currently incorporated in a wide variety of materials used in our daily lives, ranging from the textile and hospital areas to materials used in personal hygiene, such as deodorants and toothbrushes. A recent application is based on matrixes formed by collagen and bayberry tannin for the immobilization of silver nanoparticles. Various chemical forms of silver have been known for a long time as an effective antimicrobial agent with high activity against bacteria, viruses and fungi. Silver nanoparticles are most extensively studied in the recent years because of their possible usage as a strong bactericidal material with applications in the field of biomedical science. Well-developed large surface area to volume ratio of these particles provides better contact with the microorganisms and strong cytotoxicity to bacterial cells due to enough interaction with the functional groups on the bacterial cell surface [2].

Various approaches to control the size and shape of silver nanoparticles have provided many applications as antibacterial agents like medical devices, coatings for washing machines, food containers and portable water filters. In many of these applications, AgNps have been either embedded or attached on various inorganic and organic supports, such as filter materials made of zeolite, silica or fibre glass, natural macroporous materials, paper, carbon materials and various types of polymers for better contact and control for release of silver ions [3].

Various polymers such as polyglycidyl methacrylate-co-ethylene glycol dimethacrylate (polyGMA-co-EG-DMA) resins, polyvinyl alcohol (PVA) and polyvinylpyrrolidone (PVP) are used for encapsulation of silver nanoparticles.

2. Body

Santos et al. have shown the excellent antimicrobial effectiveness of AgNps stabilized by Pluronic™ F68 associated with other polymers such as PVA and polyvinylpyrrolidone. AgNps stabilized with PVP or PVA and co-stabilized with Pluronic™ F68 are effective against *Escherichia coli* and *Pseudomonas aeruginosa* microorganisms. The wide range of minimum inhibitory concentration (MIC) from 0.78% to 25% is achieved against bacteria [4].

Wang et al. have used polyvinylpyrrolidone, which is a good stabilizer for the silver nanoparticle due to its amine functional group. PVP was used as a dispersant to accelerate the reaction between silver ions and glucose and stabilized with H^+. The silver particles with the diameter <50 nm coordinate with the nitrogen in PVP, by forming a protection layer. The large silver particles, with the diameter of 500–1000 nm, coordinate with both nitrogen and oxygen in PVP molecules [5].

The control of shape and size of gold particles by introducing a small amount of salt into a *N,N*-dimethylformamide (DMF) solution containing poly(vinylpyrrolidone) or by changing the temperature is reported in literature. The shape of gold nanoparticles has been tuned in various forms like cubes, octahedrons and hexagonal nanosheets [6].

http://dx.doi.org/10.5772/61959

Jiang et al. have explored a new method to prepare colloidal five-twinned Au nanoparticles by solvothermal wet chemical method. In this process, hydrogen tetrachloroauric acid ($HAuCl_2$ $3H_2O$) was reduced by ethylene glycol (EG) to produce the desired shape of Au nanocrystals in the presence of poly(vinylpyrrolidone) molecules at 200ºC under autogenous pressure. They also found large, five-fold twinned Au particles with perfect decahedral shape [7].

Wang et al. have shown that Ag/PVP nanocomposite fibres can be electrospun from a PVP and Ag nanoparticles ethanol solution. The pyridyl group in PVP structure have strong affinity with metal particles by undergoing hydrogen bond [8].Therefore, many metals and semiconductor nanoparticles are prepared in the PVP solution.

Washio et al. have used the idea of kinetically controlled synthesis of metal nanocrystals by using hydroxyl end group of PVP. The kinetic control is a simple and versatile route for synthesis of metal nanocrystals with well-defined shapes. They have synthesized Ag nanoplates by heating an aqueous solution of silver salt ($AgNO_3$) and PVP in a capped vial. The hydroxyl end groups of PVP reduced $AgNO_3$ at a sufficiently slow rate so that the growth of Ag nanocrystals can be kinetically controlled, leading to the formation of triangular plates. This kinetically controlled synthesis gives Ag nanoplates of controllable edge lengths by varying the reaction time [9].

Polyethylene glycol (PEG) has a similar structure like polyethylene oxide (PEO). It is inexpensive, water-soluble and biocompatible. Therefore, these are used in the biomedical field for a long time. Tae et al. used these two substances (prodrug/silver) to provide a synergistic effect in both the biomedical and nanotechnology fields. Monocarboxylic ions and dicarboxylic acids are known to be very effective to stabilize silver surfaces for a long time. Vancomycin and sulfadiazine have long been known as highly active antibacterial drugs. Hence, PEO–vancomycin or PEO–sulpha drug conjugates as precursors of prodrugs, which must be effective to stabilize silver surface. Their active functional groups at the polymeric chain ends play an important role to stabilize metal surfaces by ionic–dipole or ionic–ionic interaction [10].

The polymer films of polystyrene (PS) and polymethyl methacrylate (PMMA) are technologically important for metal nanocomposites with controlled properties. Although these blend films usually result in phase separation, they are different from the bulk. This can be manipulated to produce different roughness surfaces providing different nucleation conditions for thin discontinuous metallic films. Ag nanoparticles are produced in polymer blends (PMMA/PS) on silica substrate by e-beam evaporation of metal in vacuum [11].

Bryaskova et al. have reported that hybrid materials showed a strong bactericidal effect against *E. coli*, *Staphylococcus aureus* and *P. aeruginosa* and therefore have potential applications in biotechnology and biomedical science [12].

Polymeric materials such as poly(vinyl pyrrolidone) carrying amine functional group are reported to be a good stabilizer for the silver nanoparticle. The PVP is used as a stabilizer for synthesis of silver nanoparticles using two different strategies based on thermal or chemical reduction of silver ions to silver nanoparticles. The synthesized AgNps/PVP were tested against etalon strains of three different groups of bacteria: *S. aureus* (gram-positive bacteria),

E. coli (gram-negative bacteria), *P. aeruginosa* (nonfermenting gram-negative bacteria), as well as against spores of *Bacillus subtilis* (*B. subtilis*) for antibacterial activity. Also, AgNps were tested in the presence of fungicidal activity against different yeasts and mould such as *Candida albicans, Candida krusei, Candida tropicalis, Candida glabrata* and *Aspergillus brasiliensis*. Hence, these hybrid materials show potential applications in biotechnology and biomedical science.

The chemical reduction of silver ions into stable AgNps in different polymer stabilizers is the most common method. Among them, polyvinylpyrrolidone has excellent chemical and physical properties. Therefore, it is a good material for coating or as an additive to different materials. Besides being a stabilizing agent, it also has an impact on the control of the reduction rate of the silver ions and the aggregation process of silver particles (Fig. 1). PVP is also used in other techniques for synthesis of AgNps as γ-irradiation synthesis and laser ablation [2].

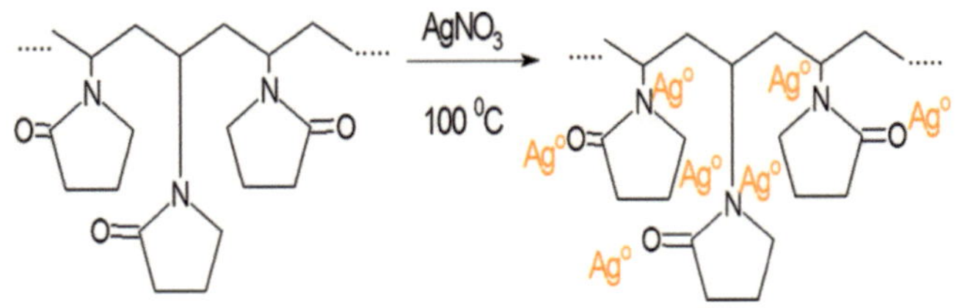

Figure 1. Chemical reduction of AgNps in PVP.

Irwin et al. have evaluated the antimicrobial properties of Ag-based nanoparticles using two solid-phase bioassays and found complete inhibition of the growth of bacteria like *Staphylococcus aureus, Salmonella typhimurium* or *Escherichia coli* on solid surfaces. Even after one week at 37°C on solid media, no growth was observed. At lower Np concentrations, visible colonies were observed, but they eventually ceased growing [13].

Dispersing silver nanoparticles homogeneously into a polymer matrix by ex situ methods is difficult because of the easy agglomeration of nanoparticles. Therefore, convenient and effective ways of preparing Ag nanoparticles in polymer materials are carried out by the author. Vacuum-deposited, thin discontinuous silver films on the composite of PVP and poly4-vinylpyridine (P4VP) is an in situ and eco-friendly method [14].

Vacuum-deposited, thin discontinuous silver films on Eudragit RL 100 (Eu) were also produced. Eudragit RL100 is insoluble in water at physiological pH values and capable of swelling, so they are good for the dispersion of active compounds. This polymer is being used for the enteric coating of tablets and for preparing controlled-release formulations. Hence, deposition of silver nanocluster on Eu would be useful for drug delivery system [15].

Various polymer systems and methods of dispersion of silver nanoparticles in them are discussed in this chapter. A comparison is drawn among them from the point of view of applications.

3. Experimental

Ivana et al. have prepared composites consisting of metallic silver nanoparticles and aminated poly(GMA-co-EGDMA) copolymer (G-NH2) by the reduction of silver ions with amino groups attached to copolymer at an elevated temperature. The content of silver in the Ag/G-NH2 composites was determined using inductively coupled plasma atomic emission measurements (ICP-AES Spectroflame 17 instrument). Bearing in mind that ICP-AES technique is not selective towards chemical state of silver, the measured value besides metallic silver includes traces of ionic silver in the composite. Further, ultraviolet-–visible (UV–Vis) reflection spectra, Fourier transform infrared spectroscopy (FTIR), transmission electron microscopy (TEM) and X-ray diffraction (XRD) were used for characterization of AgNps. The antimicrobial activity of Ag/G-NH2 composites was evaluated against gram-negative bacteria *E. coli* (ATCC 25922), gram-positive bacteria *S. aureus* (ATCC 25923) and yeast *C. albicans* (ATCC 24433) using the standard test method (ASTM E 2149-01). The percentages of microbial reduction (*R*, %) after 1 h of contact of Ag/G-NH2 composite, in concentration range from 50 to 1 mg/ml, with microbial cells were calculated using the following equation:

$$R = \left(C_0 - C\right) / C \; x \; 100$$

where C_0 (CFU, colony forming units) is the number of microbial colonies in the control sample (G-NH2 copolymer without AgNps) and C (CFU) is the number of microbial colonies in the Ag/G-NH2 composites [3].

Santos et al. have used the Turkevich method for metals nanoparticle synthesis with good antimicrobial properties. This method only involves benign substances, so it is suitable for applications in the human body. This method was applied with further modifications to allow incorporation of the polymeric stabilizers and the Pluronic™ co-stabilizer.

Wang et al. have used silver nitrate solution by dissolving in PVP, glucose and sodium hydroxide. Characterizations of the particles were achieved by TEM, infrared spectrum (IR), FTIR and extinction spectra by UV-2102 UV-Vis Spectrophotometer. All spectra were obtained from the particles immersed in water [5].

Washio et al. have applied reduction method by silver nitrate with PVP dissolved in water [9].

Tae et al. have used the PEO–sulphadiazine and the PEO–vancomycin to prepare nanosized silver particles in methanol using PEO–vacomycin prodrug as a stabilizing agent; silver nanoparticles were also obtained in aqueous media [10].

Ying et al. synthesized gold nanoparticles in DMF as both the solvent and reductant, with $HAuCl_4$ in the presence of PVP acting as the capping agent, and these had been effectively applied for the preparation of silver nanoparticles [6].

Peng et al. have prepared colloidal, five-twinned Au nanoparticles by a solvothermal wet chemical method. Ethylene glycol (EG) is used to reduce hydrogen tetrachloroauric acid

($HAuCl_4 \cdot 3H_2O$) to produce the five-twinned Au nanocrystals. The shape of a decahedron of Au crystal is produced in the presence of polyvinyl pyrrolidone at 200°C under the condition of autogenous pressure. TEM and field emission scanning electron microscopy (FE-SEM) have been used to observe the surface morphology of the produced crystals [7].

Wang et al. have prepared the silver nanoparticles by a known quantity of silver nitrate added to ethanol containing a predetermined quantity of PVP. The molar ratio of silver nitrate to a repeating unit of PVP was 1:10. Ethanol is widely used as a solvent for the electro-spinning of polymers. Thus, Ag nanoparticles in ethanol solution were prepared by reducing Ag+ ions in a PVP solution. It has been reported that ethanol alone could reduce Ag+ ions into Ag^0 under refluxing conditions without any external reducing agents [8].

Bryaskova et al. have prepared thin films based on PVA/tetraethyl orthosilicate (TEOS) embedded with silver nanoparticles using sol–gel method. Two different strategies were adopted for the synthesis of silver nanoparticles in PVA/TEOS matrix: one on reduction of the silver ions by thermal annealing and the other by synthesis of silver nanoparticles using PVA as a reducing agent [12].

Nanosilver polymer composite was prepared by vacuum evaporation and characterised by electrical and optical properties for PVP/AgNp and Eu/AgNp [14, 15].

4. Result and Analysis

Various polymeric systems and methods are found in the literature for encapsulation of nanoparticle in polymer matrix. Also, the size of the nanoparticle is controlled by modifying method as per their applications.

Polymers are widely used in the chemical synthesis of colloidal nanocrystals, and their roles are generally documented as steric stabilizers or capping agent [16 - 23]. Polyvinylpyrrolidone is an excellent dispersant, biocompatible polymer. In particular, PVP has received special attention because of its high chemical stability, no toxicity and excellent solubility in many polar solvents. The role of PVP in silver capping has received considerable attention due to the chemical structure of PVP that contains N atom that can protect the silver particles from growing and agglomerating. With the introduction of PVP into the process, silver ions coordinate with N or O in PVP, and a covered layer would generate on the surface of the particles. The layer inhibited the growth and agglomeration of the particles [5].

For the synthesis of silver nanoparticles, organic substances, such as ethanol, DMF and formaldehyde, can be used to reduce silver. DMF is a moderate reducing agent. Silver cation was reduced and the formed carbonic acid was decomposed easily:

$$HCONMe_2 + 2Ag^+ + H_2O \rightarrow 2Ag^0 + Me_2NCOOH + 2H^+$$

http://dx.doi.org/10.5772/61959

$$Me_2NCOOH \rightarrow CO_2 + Me_2NH$$

The reaction proceeds at a meaningful rate even at room temperature and in the dark. Moreover, the reaction can be conducted without taking any special care with regard to the presence of oxygen. Reduction of silver nitrate in DMF was readily observed by the appearance of a yellow colour, which deepened to greenish-black with time. In the UV–Vis spectrum (Fig. 2), a characteristic peak for Ag^0 at 415 nm appeared with the reduction of Ag^+ by DMF.

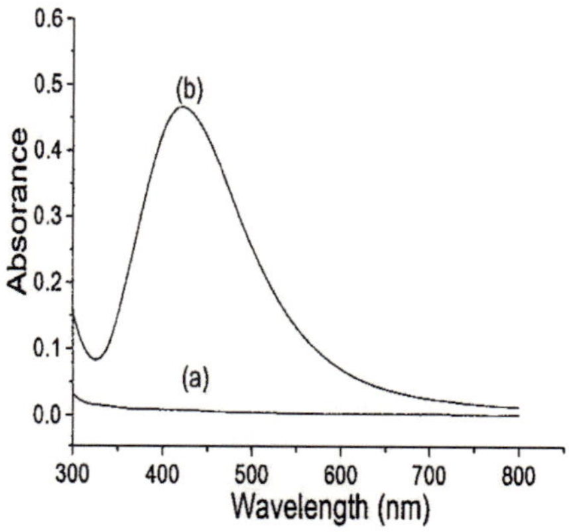

Figure 2. UV–Vis spectra for (a) $AgNO_3^-$.

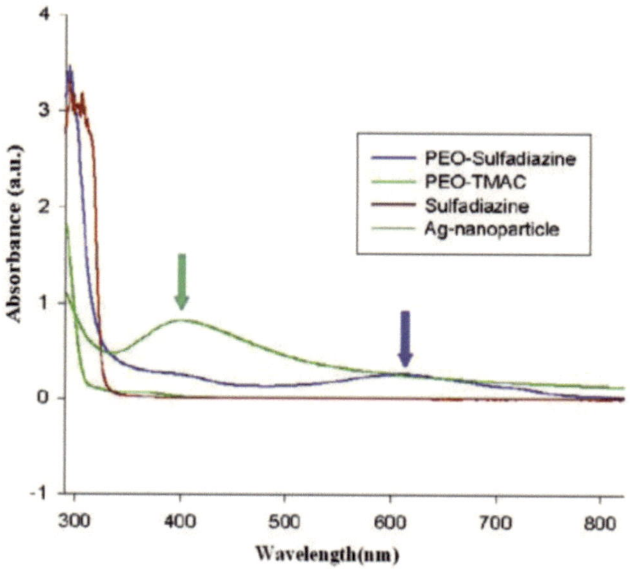

Figure 3. The electric spectral change of DMF system and (b) after 12 h at 60°C. PEO- stabilized Ag nanoparticles and the corresponding PEO-based polymers in methanol.

PVP, the molar ratio of vinylpyrrolidone unit to $AgNO_3$ (10/1), was used as a stabilizer. The Ag^+ ions and Ag particles were loosely bonded to the chain of PVP to make the system stable. It was found that the average diameter of silver particles changed with the ratio of PVP unit/Ag^+ [1].

PEO–sulfadiazine is used as a stabilizing agent, while reducing $AgNO_3$ with $NaBH_4$ in methanol shows interesting optical properties. These optical properties of nanosized metals are resulting from their surface plasmon resonances due to collective oscillation of conduction electrons in response to optical excitation. These properties are strongly dependent on size and shape. It has been observed that Cs symmetrical nanosized spheres have only one peak, whereas OH symmetrical cubic particles show three peaks. Also, three optical spectral peaks in the case of triangular silver nanoplates were observed at $\lambda_{max} = 335$, 470 and 690 nm as shown in Figure 3. The PEO–sulphadiazine conjugate exhibits the weak absorption peaks at $\lambda_{max} = 400$ and 620 nm. But, the broad absorption peak at $\lambda_{max} = 400$ nm in the case of PEO-sulphadiazine-stabilized Ag nanoparticles in water indicate that the nanoparticles are spherical particles exhibiting symmetrical cubic lattice structure [10].

Although PVP has been widely cited as a steric stabilizer or capping agent in the synthesis of various colloidal particles, Washio et al. have used the end groups of PVP as a reducing agent for the synthesis of metal nanoparticles. The repeating unit of PVP has been extensively investigated for its coordination capability in terms of functionality and reactivity. For commercially available PVP, their ends are terminated with the hydroxyl (–OH) group because of the involvement of water as a polymerization medium and the presence of hydrogen peroxide. The introduction of PVP accelerated the reaction between silver ions and glucose. PVP increases rate of converting ions into silver nanoparticles. If the rate of generating metal atoms is sufficiently high, then the final product takes the thermodynamically favoured shapes. As the reduction gradually decreases, the nucleation and growth turn into kinetic control and a range of shapes are produced. Metal nanostructures with sharp corners and edges are capable of generating maximum electromagnetic-field enhancement and hence make these nanoparticles attractive as substrates for surface enhanced Raman scattering (SERS) detection or other spectroscopic techniques [9].

Chen et al. have discussed that absorption of chemical species by the surface has a great impact on the surface energies. For the small particles, the ratio of surface free energy to the total energy yield about their structure. When the gold nanoparticles are prepared in the PVP solution, absorption of PVP and solvent molecules on the reduced gold particles takes place. The presence of NaOH or NaCl in the solution influenced the surface energy of various gold crystal planes due to the competitive absorption between the ions of the added salt and the original adsorbents. This influences the growth rate of various facets during the crystal growth. The evidence of the change of the surface energy of various crystal faces comes from the very few coexisting multi-twinned particles. This can be attributed to a kinetic occurrence, leading to different growth rates for the different crystal faces. However, the change of the surface energies by the adsorption of foreign species may be the origin of this kinetic problem. The adsorption abilities of chemical species vary with the temperature, and therefore, the shape can also be controlled by the reaction temperature [6].

PVP molecules play a very important role in the formation and evolution process of the five-twinned Au nanoparticles. It has been found that only large spherical Au particles were produced in the reaction solution without adding PVP. When PVP is added in the ethylene glycol solution, large triangular or hexagonal Au nanoplates as well as nanorods or nanowires were produced. Figure 4 shows small decahedral Au nanoparticles along with triangular and hexagonal shapes. Figure 5 is a typical TEM image of a decahedral Au nanoparticle. A nearly perfect pentagonal shape with each angle of about 108° can be clearly seen. Their sizes could be systematically controlled by changing the ratio of $HAuCl_4$/PVP. The authors suggest that PVP can act as a kinetic controller of the growth rates along various crystalline facet directions by adsorption and desorption processes [7].

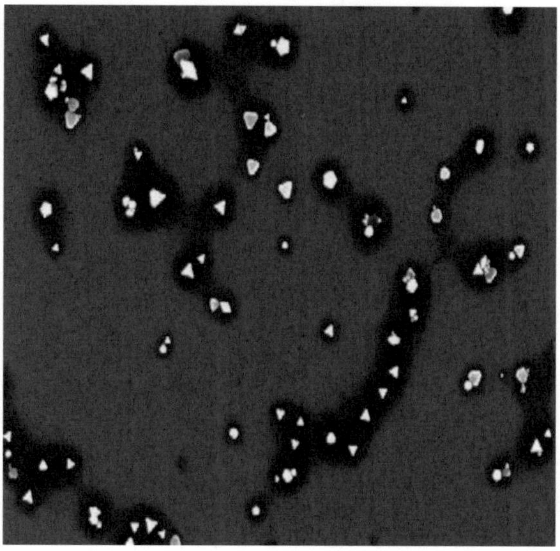

Figure 4. Decahedral Au nanoparticles.

The two major steps are involved in preparing PVP nanofibre film with homogeneously dispersed silver nanoparticles. In the first step, silver nanoparticles are prepared by reduction of $AgNO_3$ by ethanol under refluxing conditions in PVP ethanol solution. In the second step, the solution is electro-spun directly to prepare a PVP nanofibre film. It has been reported that ethanol could reduce Ag^+ ions into Ag^0 under refluxing conditions in the absence of any external reducing agents. The absorption spectra of the solution before and after refluxing for 5 h show remarkable change by reaching 410 nm. Silver nanoparticles exhibit a high optical absorbance due to the existence of discrete energy levels which was consistent with the result of the UV absorption spectrum [8].

Two different strategies were applied for producing antibacterial hybrid material/silver nanoparticles stabilized with PVP. In comparison to other techniques for synthesis of silver nanoparticles, thermal reduction of Ag^+ ions to Ag^0 can be performed at elevated temperature without using reducing agents. This was achieved by adding $AgNO_3$, the precursor for silver ions, to the PVP solution that led to the coordination of silver ions with –N or –O atoms from PVP, followed by boiling the PVP solution at 100°C for 60 min. In this case, PVP acts as a stabilizer that protects the silver nanoparticles from agglomeration and controls the rate of reduction. The optical absorption spectra clearly indicated the formation of silver nanoparticles by the appearance of strong absorption bands at 420 nm (Fig. 6a). Moreover, the absorption of silver nanoparticles becomes stronger and sharper with the increase in the silver concentration, which is an indication for he preparation of silver nanoparticles with narrower particles size distribution. This state is confirmed by TEM analysis that spherical silver nanoparticles with an average size of 6.4 ± 1.1 nm are homogeneously distributed in polymer matrix.

Figure 5. A typical TEM image of a decahedral Au nanoparticle.

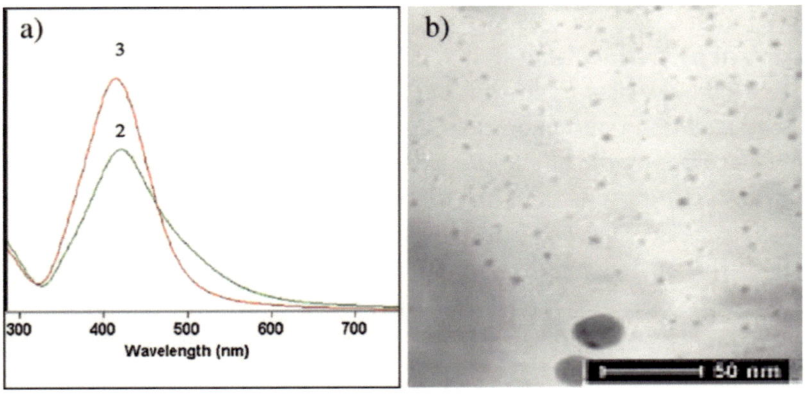

Figure 6. a) UV spectra of AgNps/ PVP obtained via thermal reduction at concentration of silver precursor: 2 for 1.96 mg/mL and 3 for 3.92 mg/mL. b) TEM of thermal-reduced silver nanoparticles stabilized by PVP at concentration of silver precursor 3.92 mg/mL.

In the second method, silver nanoparticles were prepared by adding a strong reduction agent as $NaBH_4$ to the PVP solution in the presence of $AgNO_3$. The optical absorption spectra of silver nanoparticles clearly show a band at 410 nm with peaks whose intensities increased with silver concentration (Fig. 7a). TEM analysis confirmed the formation of spherical nanoparticles with an average size of 13.8 ± 3.8 nm (Fig. 7b).

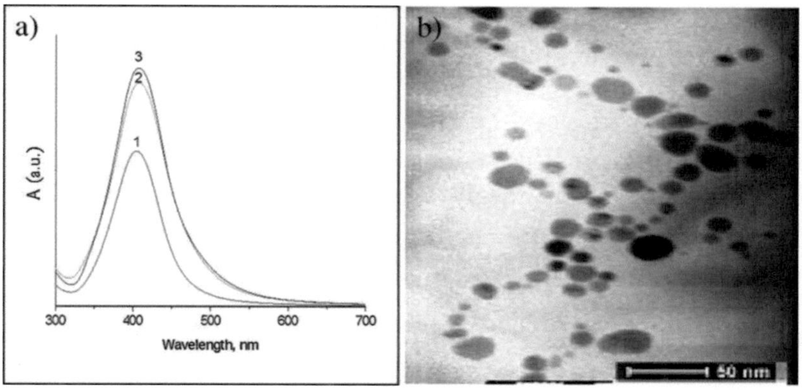

Figure 7. a) UV spectra of PVP/AgNps obtained via chemical reduction at concentration of silver precursor: 1. 0.98 mg/mL, 2. 1.96 mg/mL and 3. 3.92 mg/ mL. b) TEM of silver nanoparticles obtained via chemical reduction of $AgNO_3$ in the presence of NaBH4 at concentration of silver precursor 3.92 mg/mL.

FTIR spectra of pure PVP and AgNps/PVP show characteristic peaks for $-CH_2-H-CH$ groups of PVP observed in both structures. The shifting of peak for carbonyl group and –C–N bond of PVP was associated with the formation of coordination bonds between silver atoms and oxygen or nitrogen atoms in the PVP.

Antibacterial activity of AgNps/ PVP with three different concentrations of silver precursor in the range of 0.98, 1.96 and 3.92 mg/mL was tested against *S. aureus, E. coli, P. aeruginosa* and spores of *B. subtilis* by disk diffusion method (DDM). It was found that all AgNps stabilized with PVP exhibit bactericidal and sporocidic activity by the appearance of an inhibition zone in the range of 7.5–8.5 mm. Higher antibacterial activity was observed against both gram-negative *E. coli* and gram-positive *S. aureus,* as well as against spores of *B. subtilis,* but lowest antibacterial activity was found for *P. aeruginosa.* To confirm the bactericidal properties, AgNps/PVP (initial concentration of silver precursor 0.98 mg/mL) was tested using macro-dilution method. By this method, the MBC corresponding to the lowest concentration without detection of bacterial growth was determined for *E. coli* (ATCC 25922), *S. aureus* (ATCC 25923) and *P. aeruginosa* (ATCC 27853). These results clearly indicated the strong bactericidal activity demonstrated by the lack of any bacterial growth even at lowest silver concentration of 0.015 mg/mL [12].

Antibacterial PES–Ag membranes can be fabricated with PVP as dispersant in the casting solution. The effects of Ag loading and PVP (MW 360,000) on the thermal properties and contact angle revealed that membranes' contact angle decreased when Ag loading increased mainly due to hydrophilicity improvement. It was also observed that higher the concentration of PVP, the more hydrophilic the membrane. From the X-ray photoelectron spectroscopy (XPS) analysis, the PES–Ag membrane with 2.0 wt% of Ag and PVP displayed high Ag intensity, thus resulting in high activity in the antibacterial test against *E. coli*. Therefore, it is concluded that the presence of PVP in the PES–Ag membrane had significantly improved membrane performances in the antibacterial activity [19].

Irwin et al. have evaluated the antimicrobial properties of a biocompatible capping agent-based (keratin) AgNp using both solid- and solution-state media. It was found that on solid surfaces, keratin-based Nps completely inhibited the growth of *Staphylococcus aureus* and after several weeks at 37°C, no further growth was observed. At lower Np concentrations, formation of colony is found but ceased growing beyond a certain small size. Growth of small colonies on fresh media without Nps proceeded normally. These results imply that further cell division is limited due to the continued presence of AgNps on the solid surface. In liquid phase, it was found that growth always occurred but the t_m varied between 7 and >20 hrs (assuming a constant C_I) using either the citrate- ($[Np] \sim 3 \times 10^{-7}$ M) or keratin-based ($[Np] \sim 10^{-6}$ M) Nps. This delay was not related to the effect that Nps had on *S. aureus* k-values. To test the possibility that the Nps were effectively changing C_I bacteria via cell death, the probabilistic calculations assumed that the perturbations in t_m were due to C_I alone (i.e., with a fixed T). The observed large perturbations in t_m could only come about at concentrations where the probability for any growth occurring at all was small. This result indicates that much of the Np-induced change in t_m was due to a greatly increased value for the true microbiological lag time (T increased from ~ 1 to >15–20 hrs). In either solution or the solid state, a maximum perturbation was noticed only when the ratio of $[Np]:C_I$ (on a particle: cell basis) was about 10^{11}–10^{12}. The differences observed between the solid and liquid growth systems relates to obvious differences in the residence time of the Nps with respect to the bacterial cell membrane [13]. A high antimicrobial effectiveness against *E.coli* and *P. aeruginosa* by AgNps was found for all stabilizing polymers, and the inhibition value was found to be increased when Pluronic™ F68 was used as a co-stabilizer.. The use of AgNps co-stabilized by Pluronic™ F68 could help in the decrease of the resistance effect where the use of AgNps for burns treatment presented the onset of resistance against *Entero bacteriaceas, P. aeruginosa, Salmonella* and others [4].

Implantation of shunt for relief of hydrocephalus in neurosurgery is associated with infections. To prevent infection surface modified silicon elastomer polyvinylpyrrolidone grafted silicon elastomer (SEPVP) is used. PVP being a hydrogel, SEPVP soaked more in antibacterial solution results in more longevity in antibacterial activity [26].

A negative temperature coefficient of resistance (TCR) at high temperature followed by almost zero TCR at lower temperatures exhibited by the silver particulate film on polymer composite is the indication of film consisting of small particles separated by small distances. The overall resistance of film deposited on composite decreases with increase in thickness of silver films deposited on the composite. The dc resistances of the particulate film are exponentially dependent on intercluster separation or tunnelling length. Therefore, observed high resistances were attributed to very large interspacing and larger cluster sizes for PVP, which is reduced with the blending of P4VP with PVP and increasing thickness of silver film. Pyridine ring with nitrogen atom in the structure of P4VP causes the change in the morphology. Table 1 presents the resistance data for the silver films of various thickness deposited on polymer blend substrates held at 457 K with a rate of 0.4 nm/s [14].

Polymer PVP:P4VP	Thickness	R_{ts}	$R_{1\,hra}$	R_{rt}	$R_{0.5T}$	R_{fexp}
0:100	52 nm	1.90	26.87	27.17	32.8	72.35
50:50	86.2 nm	7.2	77.85	112.89	127.76	>1000
100:0	50 nm	15	123.3	>1000	–	–

All resistances are measured in MΩ/□.

R_{ts}: Resistance at the time of deposit, R_{1hra}: Resistance after 1 hr of annealing.

R_{rt}: Resistance at room temperature, $R_{0.5T}$: Resistance at time of exposure at 0.5 Torr.

R_{fexp}: Resistance on full exposure to atmosphere. a. Temperature: 457 K. b. Rate:- 0.4 nm/sec.

Table 1. Various resistances of silver particulate films deposited on polymer composite

In addition, optical properties of these films show the encapsulation of silver particles. The observed shifts of the resonance position to higher wavelength (red shift) for silver particles embedded in PVP/P4VP blends were correlated with changes of particle sizes and interseparation in silver clusters. It can be clearly seen (Fig. 8) that the plasmon resonance peak shifts towards the longer wavelength side for the PVP/P4VP, 50:50 (653), as compared to pure PVP (611 nm) for silver film thickness of 86 nm.

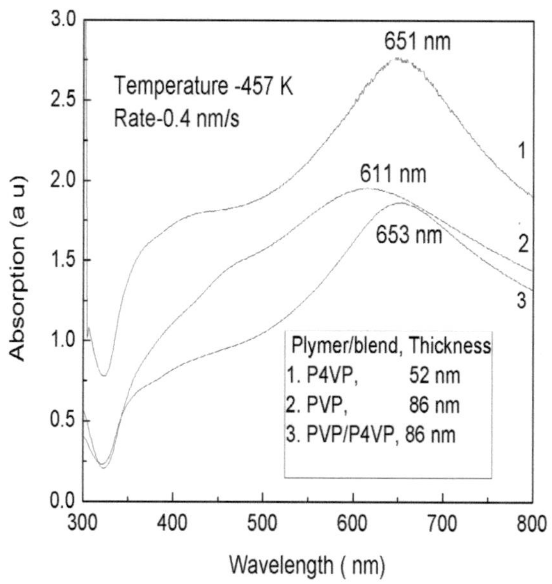

Figure 8. Optical absorption spectra for thicknesses 52 and 86 nm silver particulate films deposited on PVP/P4VP blends.

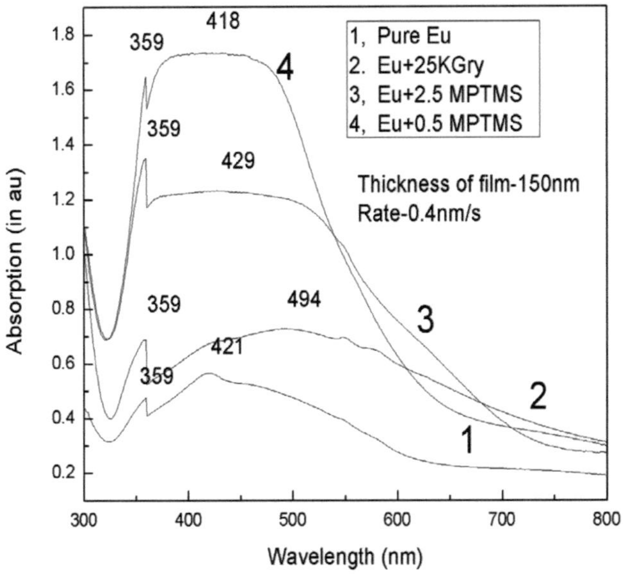

Figure 9. Optical absorption spectra for 150 nm silver particulate films deposited on pure Eu and its blends.

The optical absorption spectra recorded for 150 nm silver films deposited on polymeric blends of Eu held at 457 K at the deposition rate of 0.4 nm/s. For silver particles embedded in Eu blends, a higher wavelength red shift of the resonance position was found for irradiated Eu (494 nm) and Eu (+ 2.5 MPTMS) (429 nm) and negligible blue shift was found for Eu (+0.5MPTMS). These changes were correlated with changes of particle sizes and interseparation in silver clusters. It can be clearly seen (Fig. 9) that the plasmon resonance peak shifts towards the longer wavelength side as compared to pure Eu (421 nm). Also, there is an increase in intensity of absorbing peaks that signify the decrease of particle size. Irradiation on Eu released free radicals, which interact with silver clusters resulting in decrease in size. In addition, sedating with 2.5 methacryloxy-propil-trimethoxysilane (MPTMS) resulted in decrease in size of silver clusters.

5. Further Research

The effect of antimicrobial activity can be studied with vacuum-evaporated silver particulate films and a comparison can be drawn.

6. Conclusions

Polyvinylpyrrolidone is a very useful polymer for antimicrobial activity. It is the base for most of the application. The encapsulation of silver nanoparticles is generally characterized by optical absorption spectra. Therefore, this is an easy and economical method for detecting silver in polymer system.

Acknowledgements

The author is thankful to DST, India, for their financial support in carrying out this research work.

Author details

Pratima Parashar Pandey*

Address all correspondence to: pratimaparashara@rediffmail.com

Applied Sciences, CET, IILM, Greater Noida, India

References

[1] Gao YU, Yuzhaoun YU: Deposition of silver nanoparticles on montmorillonite platelets by chemical plating. J. Mater. Sci.2002; 37:5083–5087.

[2] Bryaskova R, Daniela P, Stanislav N, Kantardjiev T: Synthesis and comparative study on the antimicrobial activity of hybrid materials based on silver nanoparticles (AgNps) stabilized by polyvinylpyrrolidone (PVP). J. Chem. Biol.2011; 4:185–191.

[3] Ivana D, Vukoje ES, Dzˇunuzovic, Vesna V, Vodnik S, Dimitrijevic S, Phillip A, Jovan M, Nedeljkovic: Synthesis, characterization, and antimicrobial activity of poly (GMA-co-EGDMA) polymer decorated with silver nanoparticles. J. Mater. Sci.2014; 49:6838–6844.

[4] Santos CA D, Angela FJ, Adalberto P, Marcelo, Martins S: Antimicrobial effectiveness of silver nanoparticles co-stabilized by the bioactive copolymer pluronic F6. J. Nanobiotechnol.2012; 10:43–49.

[5] Hongshui W, Xueliang Q, Jianguo C, Xiaojian W, Shiyuan D: Mechanisms of PVP in the preparation of silver nanoparticles. Mater. Chem. Phys.2005; 94:449–453.

[6] Ying C, Xin G, Cha-Geng N, Zhi-Yuan J, Zhao-Xiong X, Chang- Jian L: Shape controlled growth of gold nanoparticles by a solution synthesis. Chem. Commun.2005; 4181–4183.

[7] Peng J, Jian-Jun Z, Rui L, Zhong-Lin W, Si-Shen Xie: Pol(vinyl pyrrolidone)- capped five-fold twinned gold particles with sizes from nanometres to micrometres. Nanotechnology.2006; 17:3533–3538.

[8] Yongzhi W, Yaoxian L, Songtao Y, Guangliang Z, Dongmin A, Wang C, Qingbiao Y, Xuesi C, Xiabin J, Yen W: A convenient route to polyvinyl pyrrolidone/silver nanocomposite by electrospinning. Nanotechnology.2006; 17:3304–3307.

[9] Isao W, Yujie X, Yadong Y, Younan X: Reduction by the end groups of poly (vinyl pyrrolidone): A new and versatile route to the kinetically controlled synthesis of Ag triangular nanoplates. Adv. Mater.2006; 18:1745–1749.

[10] Tae HK, Keunsuk K, Geon HP, Jin HC, Sangmi L, Ho-Jung K, Jae YL, Jungahn K: Preparation and characterization of water-dispersible silver nanoparticles stabilized by PEO-conjugated pro-drugs. Macromol Res.2009; 17(10):770–775.

[11] Pui˘soa J, Prosy˘cevasb I, Guobien˙eb A, Tamulevi˘ciusa S: Mater Sci Eng. B.2008; 149:230–236.

[12] Bryaskova R, Pencheva D, Kale GM, Lad U, Kantardjiev T: J. Colloid Interface Sci. 2010; 349(1):77–85.

[13] Irwin P, Martin J, Ly-Huong Nguyen LH, Yiping H, Gehring A, Yi Chen C: Antimicrobial activity of spherical silver nanoparticles prepared using biocompatible macromolecular capping agent: evidence for induction of a greatly prolonged bacterial lag phase. J. Nanobiotechnol.2010; 8:34–46.

[14] Parashar P: Synthesis of silver nanocomposite with poly(vinylpyrollidone) and poly(4-vinylpyridine) for antimicrobial activity. Advanced. Materials Res.2013; 772:9–14.

[15] Parashar P: Synthesis of silver nanocomposite with EUDRAGIT RL100 for the drug delivery system. Int. J. Edu. Appl. Sci. Res.2014; 3(1): 51–54.

[16] Duff DG, Baiker A, Edwards P: Langmuir.1993; 9:2301.

[17] Agt PY, Urbina RH, Elhsissen KT: J. Mater. Chem.1997; 7:293.

[18] Pastoriza-Santos I, Liz-Marzán LM: Nano Lett. 2002; 2:903.

[19] Chen J, Herrick T, Geissler M, Xia Y: J. Am. Chem. Soc.2004; 126(10):854.

[20] Xiong Y, Chen J, Wiley B, Xia Y, Aloni S, Yin Y: J. Am. Chem. Soc.2005; 127:7332.

[21] Sun Y, Xia Y: Science.2002; 298:2176.

[22] Wiley B, Herricks T, Sun Y, Xia Y: Nano Lett.2004; 4: 1733.

[23] Wiley B, Sun Y, Mayers B, Xia Y: Chem. Eur. J.2005; 11:455.

[24] Petroski JM, Wang ZL, Green TC, El-Sayed MA: Kinetically controlled growth and shape formation mechanism of platinum nanoparticles. J. Phy. Chem. B.1998; 102:3316– 3320.

[25] Basri H, Ismail AF, Aziz M: Assessing the effect of PVP of various molecular weights (MW) in PES–Ag membranes: Antimicrobial study using *E.Coli.* J. Sci Technol.2011; 3(2):59–67.

[26] Boelean JJ, Tan WF, Dankert J, Zaat SAJ: Antibacterial activity of antibiotic soaked polyvinylpyrrolidone-grafted silicon elastomer hydrocephalus shunt. J. Antimicrob. Chemother.2000; 45:221–224.

Section 3

Engineering of Biopolymers

Spectroscopic Characterization of Multilayered Functional Protective Polymers via Surface Modification with Organic Polymers against Highly Toxic Chemicals

Peter P. Ndibewu, Prince Ngobeni, Tina E. Lefakane and Taki E. Netshiozwi

Additional information is available at the end of the chapter

http://dx.doi.org/10.5772/62154

Abstract

Recent advances in biopolymers, including functional biomaterials for the manufacture of personal protective garments (PPGs) or equipment (PPE) have dramatically improved their efficiency and performance. Good and acceptable permeation characteristics, mechanical strength and durability are common attributes of these materials simultaneously without compromise for their cost-effectiveness and manufacturability. The comprehensive characterization of these materials and specimens' three-dimensionality with the endeavor to obtain the highest resistance to highly toxic agents such as nuclear, chemical and biological warfare agents is the must fulfilling aim in today's global interest in continuous development in this area.

Because energy absorption component seems to be important in considering quality the requirements related to the application of most protective materials (e.g. clothing), spectroscopy would seem to be the cornerstone to be considered for most analytical purposes to supplement the qualitative and quantitative assessment of polymeric materials. The major techniques worth mentioning include: scanning electron microscopy coupled to X-ray dispersive spectroscopy (SEM/EDS), atomic force microscopy (AFM), scattering-type near-field optical microscopy, scanning tunneling microscopy (STM), attenuated total reflectance Fourier transform spectroscopy (ATR-FT-IR), infrared spectroscopic ellipsometry, nano-FTIR absorption spectroscopy, near-field optical microscopy, and infrared vibrational nano-spectrosocopy. This chapter will discuss the importance of a particularly important natural polymer (cellulose) containing acetyl groups to form modifiable biopolymers (e.g. cellulose acetate polymers), doped to yield multi-layered functional protective materials (MFPMs) or composites (MFPCs). The ultimate aim seeks to provide critical insights into understanding the enhancement of their permeation characteristics against exposure to toxic industrial chemicals, including chlorine which is currently being used as a chemical warfare agent of choice in the Syrian conflict. MFPMs or MFPCs are a group of materials made from a combination of fiber or polymers together with varying amounts of additives possessing tailored physical and mechanical properties. Many of these materials should not only be durable but also must provide cost-competitive products in the manufacturability of personal protective garments (PPGs). The key advantages and disadvantages of available protective materials manufactured with synthetic polymers compared to biopolymeric ones based on the objective of achieving the highest quality, maximum protection, or both are presented. The chapter will also explain the fundamental differences of each material- and how biopolymers can potentially affect their design and the outcome of use. The challenges related to the cost and characterization for the purpose of facilitating correlation of different physical properties and morphological heterogeneities will be presented. The need for advanced characterization and analytical tools (such as spectroscopy and microscopy) shall be dealt with. This should pave the way in the critical understanding of how better permeation studies can be achieved from suitable biopolymer--based personal protective garments (PPGs). The reutilization of waste materials in the production of multilayered functional protective materials (MFPMs) or composites (MFPCs) have advantages to the economy, environment, and technology. The synthesis of targeted multi-layered metal-organic doped polymers via surface modification can be achieved in well planned experiments to yield products of acceptable permeation studies for industry use in the manufacture of PPGs and hence boost the third generation economy. The simple biopolymer preparation yet robust with intuitive practical applications in the industry will explore a strong scientific elucidation of mechanisms to explain or achieve the desired properties while overcoming the boundary of expensive needs for the development of the industry.

Keywords: Polymers, cellulose, functionalization, multilayer, personal protective garments, toxic chemicals, characterization, spectroscopy , microscopy

---

1. Introduction

According to Wang et al. [1], it has been documented that a fast depletion rate of non-renewable energy and environmental pollution caused by synthetic or petroleum-based polymers has motivated the utilization of naturally occurring polymers leading to the creation of a wide range of new materials with interesting physicochemical, mechanical and morphological characteristics. Cellulose [2] forms the structural component of the cell walls of plants and is by far the most abundant natural polymer (polysaccharide) on earth. Cellulose provides many manufactured products, including paper and paper products, rayon, linen, and cellulose nitrate (a constituent of nail polish). This natural abundance offers cellulose an attractive attention due to its renewability, wide availability, low cost, biocompatibility and biodegradability [1]. Cellulose acetate and rayon are known examples with many day-to-day applications ranging from thin-films to regenerated cellulose for particular usage such in packaging materials, e.g. cellophane. Rayon is the most important regenerated natural fiber [3], produced from cellulose derived from wood pulp. Cellulose acetate [4] has been widely produced from processed wood pulp (Figure 1) dictating the current market source with intense research focused on various other renewable materials as feedstock [5-7].

Cellulose can also be degraded with NaOH followed by carbon disulfide reaction to form cellulose xanthate and regenerated through precipitation in an acid medium- and then cold drawn to form spinnerette fibers for specific usage. Regenerated cellulose materials endowed with different functions and properties have been designed and fabricated in different forms,

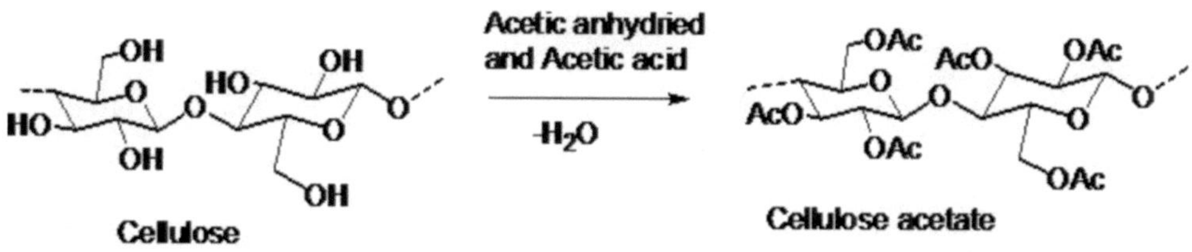

Figure 1. Reaction scheme for the synthesis of cellulose acetate - a modified natural polymer [8].

such as filaments, films/membranes, microspheres/beads, hydrogels/aerogels and bioplastics, for various applications and use in day-to-day demands. In this regard, the process of regeneration follows a physical process resulting in a number of novel regenerated cellulose materials employed for wide applications in textiles, packaging, biomedicine, water treatment, optical/electrical devices and agriculture and food materials. The 'ease-of-car' properties such as crease resistance can be imparted to cellulose by various cross—linking agents [3]. The most common are di-(N-hydroxymethyl) ureas. In the presence of Lewis acid, these reagents bridge hydroxy groups on adjacent polycellulose chains as shown in reaction scheme 1.

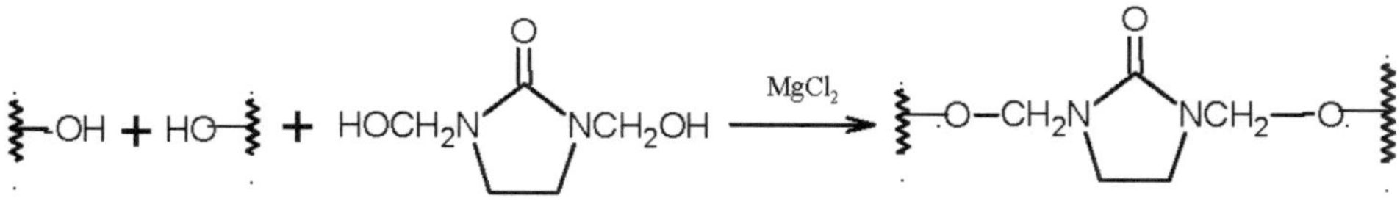

Scheme 1. Bridging of hydroxy groups on adjacent polycellulose chains [3].

However, many processes have been developed to improve the fiber characteristics of cellulose. For example, treatment with strong aqueous NaOH (mercerization) alters the strength, surface character, and dyebility of cellulose during manufacturing.

Another interesting research area that expands its scope of development mainly in the fields of tissue engineering for many years is in the processing of organic-inorganic hybrids and photoactive polymer nanocomposites. In this area, processing strategies with very strong scientific interest and several applications in the nanocomposites field are based on the development of interpenetrating networks and sol–gel processed materials [9-16]. Other processing strategies of polymer nanocomposites that should be mentioned here include the microwave-assisted processes [17, 18], frontal polymerization [19-24] and processing of foams [25, 26] and aerogels [27-29]. In all these processes, the presence of nanofillers and the interaction with the matrix represent again relevant factors for the processing behavior and the final properties of the nanocomposites obtained, which can be exploited in the manufacture of protective polymeric materials such as personal protective garments (PPGs).

Amidst all these recent advances in biopolymers, including functional biomaterials, their usage for the manufacture of personal protective garments (PPGs) or equipment (PPE) have dramatically improved their efficiency and performance. Thus from the research point of view, generating wide impact and benefits of the regenerated cellulose materials to the society. Good and acceptable permeation characteristics, mechanical strength, and durability are common attributes of these materials simultaneously without compromise for their cost-effectiveness and manufacturability. The comprehensive characterization of these materials and specimens three-dimensionality with the endeavor to obtain highest resistance to highly toxic agents such as nuclear, chemical and biological warfare agents is the must fulfilling aim in today's global interest in the continuous development in this area.

The reutilization of waste materials in the production of multilayered functional protective materials (MFPMs) or composites (MFPCs) has advantages to the economy, environment, and technology. The synthesis of targeted multi-layered metal-organic doped polymers via surface modification can be achieved in well planned experiments to yield products of acceptable permeation studies for industry use in the manufacture of PPGs and hence boost the third generation economy. The simple biopolymer preparation yet robust with intuitive practical applications in the industry will explore a strong scientific elucidation of mechanisms to explain or achieve the desired properties while overcoming the boundary of expensive needs for the development of the industry. Apart from the 'green' methodologies of material processing and the resultant properties and functions, with emphasis on the regenerated cellulose materials and the composite materials, this chapter also emphasizes other composite blending procedures. Wang et al. [1] reported that the latter processes followed new intensive developments resulting in environment-friendly biopolymerization steps. This avoided consuming chemicals because most of the agents (solvents, coagulants, etc.) may be recycled and reused, with no accompanying chemical reaction. These authors pointed out that regenerated cellulose varies in different shapes, such as powder, fibers, films, hydrogels, and spheres, especially the applications of 'green' cellulose solvents in dissolution and regeneration, leading to sustainable development, environmental preservation and energy conservations.

Although new polymers have been developed from complex polymeric blends, generally through multilayered linkages, for the purpose of improving their surface function and performance, their molecular self-assembly has proportionately led to very complex organic and inorganic molecular interactions at even nanoscales [30]. Hence, the understanding and control of such materials during the manufacturability of protective equipment, including personal protective garments (PPGs) have been impeded by difficulties in their characterization using conventional analytical techniques [31]. Various state-of-the-art spectroscopic and hyphenated techniques are currently available for both qualitative and qualitative characterizations of polymeric protective materials (PPMs) to directly resolve nanoscale morphology and associated intermolecular interactions for the systematic control of functionality in multicomponent systems and manufacturability. These include, but not limited to techniques such as:

1. Scanning electron microscopy coupled to X-ray dispersive spectroscopy (SEM/EDS) for surface imaging [32-35] and structural elucidation.
2. Atomic force microscopy (AFM) [36] and scattering-type near-field optical microscopy [37, 38] for nanoscale morphology and nanoscale-resolved subsurface imaging.

3. Scanning tunneling microscopy (STM) [39] for associated intermolecular interactions.
4. Attenuated total reflectance Fourier transform spectroscopy (ATR-FT-IR) [40, 41].
5. Infrared spectroscopic ellipsometry [42] and nano-FTIR absorption spectroscopy [43, 44] for apertureless near-field optical microscopy [45, 46], and infrared vibrational nanospectrosocopy [47].

This chapter fundamentally addresses these challenges by first overviewing the chemistry of an important natural polymer, cellulose, critically discussing the potential of its use to form modifiable biopolymers (e.g. cellulose acetate polymers), doped to yield multi-layered functional protective materials (MFPMs) or composites (MFPCs). The mention of regenerated cellulose processing is being made. The ultimate aim of the chapter seeks to provide critical insights into understanding the acetylation mechanisms for the enhancement of their permeation characteristics against exposure to toxic industrial chemicals, including chlorine which is currently being used as a chemical warfare agent of choice in the Syrian conflict. MFPMs or MFPCs are a group of materials made from a combination of fibers or polymers together with varying amounts of additives possessing tailored physical and mechanical properties. Many of these materials should not only be durable but must provide cost-competitive products in the manufacturability of personal protective garments (PPGs). The key advantages and disadvantages of available protective materials manufactured with synthetic polymers compared to biopolymeric ones based on the objective of achieving the highest quality, maximum protection, or both are presented. The chapter will also explain the fundamental differences of each material and how biopolymers can potentially affect their design and the outcome of use. The challenges related to the cost and characterization for the purpose of facilitating correlation of different physical properties and morphological heterogeneities are presented. This should pave the way in the critical understanding of how better permeation studies can be achieved from suitable cellulosic biopolymer--based personal protective garments (PPGs).

2. Overview of natural polymer chemistry

2.1. Natural polymers

Polymers are large molecules constructed by the repetitive bonding of many smaller molecules called monomers (or monomer residues) [48]. Although synthetic polymers appear in many forms that are familiar to the consumer, biopolymers constitute all living organisms [7]. Biopolymers are responsible for the structural and functional chemistry of all plants and animals. The polymeric sugars, known as polysaccharides, are an important component of the cell membranes of plants. The chemical formula of this type of sugar is illustrated in Figure 2b.

Wool is a fibrous insoluble animal protein known as keratin. In its natural form, it has an α-keratin structure and is a classic example of the protein α-helix. A variety of groups that make up a polypeptide provide many possibilities for inter- and intramolecular interactions. Hydrogen bonding, steric repulsions, van der Waals attraction, and solvation contribute to the

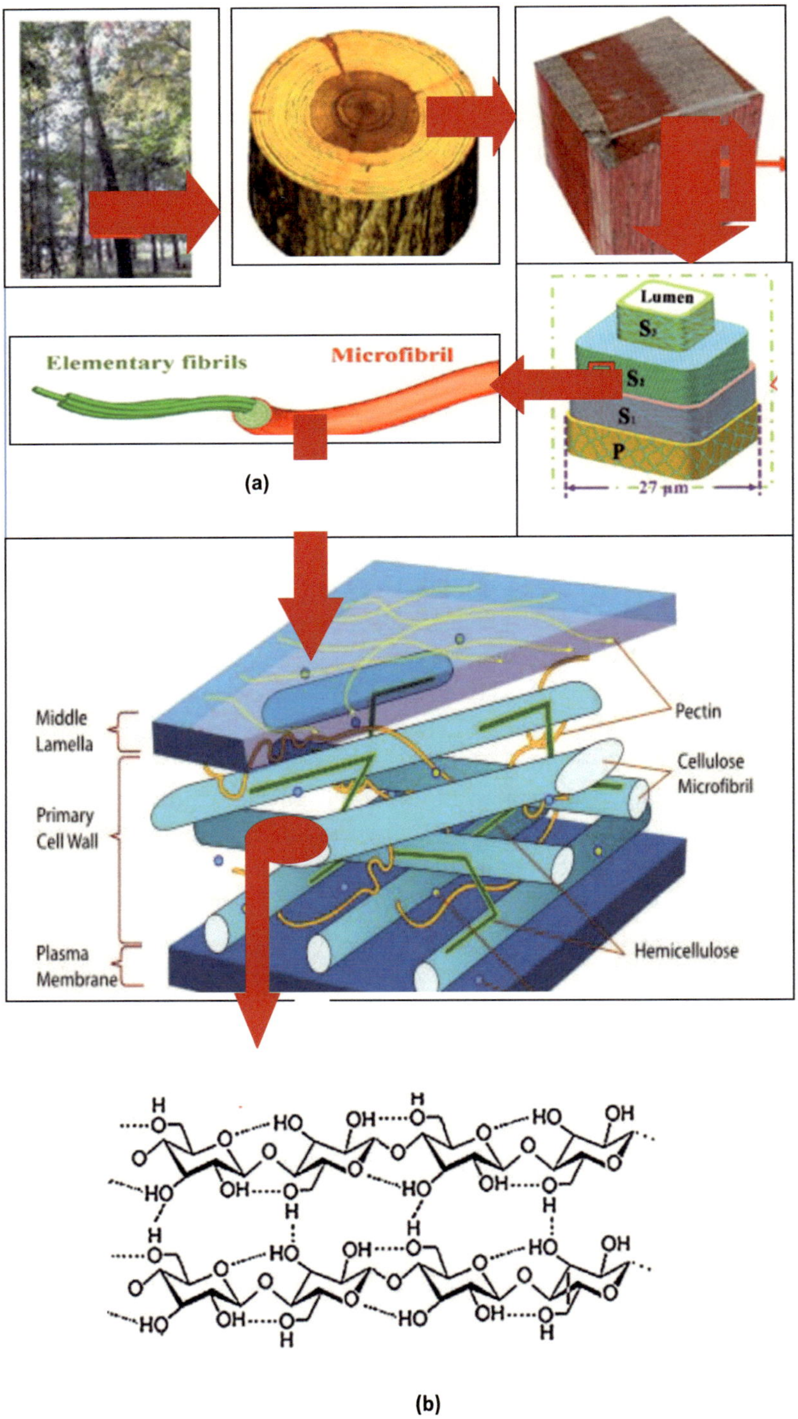

Figure 2. (a) Hierarchical structure from plant to cellulose chains, Adapted from Peponi et al. (2014). (b) A strand of cellulose (Nelson and Cox, 2013).

three-dimensional conformations of these types of proteins. The peptide bond (C-H) is the major factor in determining peptide conformation (Figure 3). Rotations are most important at the single bonds connected to the α-carbon atoms.

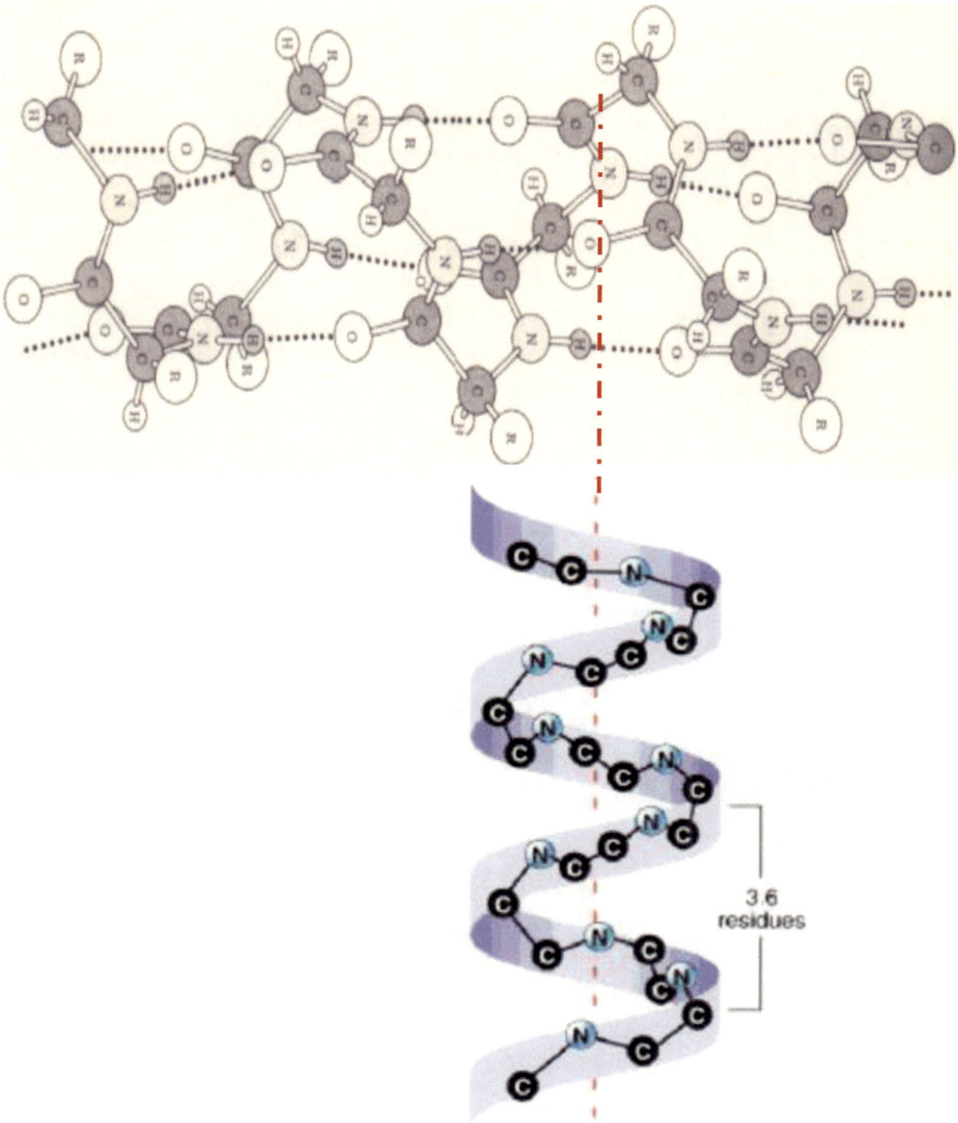

Figure 3. A schematic representation of a protein α-helix showing possibility of rotations at the single bonds connected to the α-carbon atoms, Adapted from [3, 51].

The reshaping of wool fibers usually involves the reductive cleavage of the disulfide bonds and the formation of new cross-links involving disulfides or other groups. These cross-linking modifications are used to impart permanent press to wool fabric (Scheme 2).

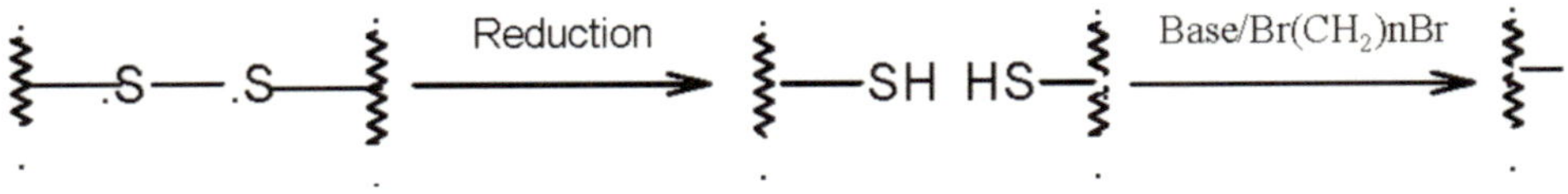

Scheme 2. Cross-linking modifications to impart pressability to wool fabric.

The flexible protein structure of wool leads to fabric with excellent handling qualities. When exposed to moisture and heat, the tendency to shrink creates a problem. However, blends of wool and synthetic fibers are currently the most effective approach to shrink-resistant "wool" fabrics.

Pure pulp from cotton (Figure 4), with the scientific name as *Gossypium herbaceum spp.* [52] is widely available in the arid regions of the world, particularly Africa. This natural polymer is very rich in α-cellulose consisting of 95 – 98% [53]. Cellulose is the most widely available biopolymer in this class (cotton) and accounts for approximately 50 % of total linear β-1,4-polyglucoside exhibiting a strong fibre structure that is quite versatile [54]. Cotton fibers are also cross-linked for crease resistance by using epichlorohydrin (3-chloropropylene oxide) or the diepoxide of butadiene. Cotton fabrics retain their strength whether wet or dry. They have excellent wearability and are pleasing in appearance and to the touch. However, the resistance of these fibers against toxic chemicals can be possibly improved through surface modifications.

Figure 4. A photograph of fiber from the *G. herbaceum* species represents the purest natural form of cellulose, containing more than 90% of this polysaccharide [55].

Cellulosic fibers from natural protein (silk) are most widely spread out in southern Africa, and elsewhere [56]. Silk is a fibrous protein produced by insects, including those from sources such as the common *Bombyx mori* or the wild *Antheraea pernyi* and *Antheraea assama.* The "pleated sheet" structure of silk fibroin (Figure 5) has a repeating sequence of glycine at every other position with alanine or serine in between. The sheet sections interact through dispersion forces.

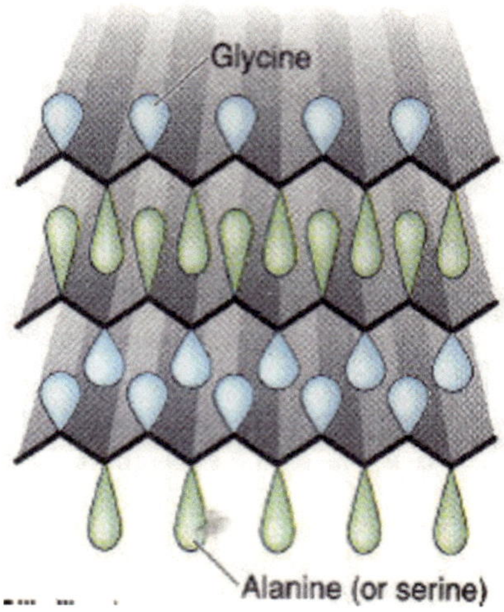

Figure 5. A schematic representation of silk fibroin, Adapted from [57].

The polymer chains tightly pack together in pleated sheet conformations that result in a strong fiber. Interestingly, silk fabric does not have good wear resistance, and because of its high price it is luxury fabric. All these fibrous materials can be very good basic raw biomaterials for manufacturing cellulose acetate biopolymers. The cocoons of various silkworms (moth larvae), after appropriate chemical treatment, provide the fibroin used for most silk fabric. More than 70 % of the average composition of fibroin is due to the small amino acids glycine, alanine, and serine.

Cotton is not the only source of cellulose polymers but just one of the available sources. Non-wood bioresources has always constituted an enormous supply potential for cellulosic biopolymers for centuries [58] and would continue to be potentially so [59] should exploitation follow sustainable principles. A polymeric form of glucosamine known as chitin is a major component of the exoskeleton (shell) of many insects [3].

2.2. Cellulosic biomaterials

Cellulose, $(C_6H_{10}O_5)n$ is an organic compound (polysaccharide) with the chemical structure presented in Figures 6A and B. This molecule is the major chemical component of the cell wall fiber, contributing to 40 % - 45 % of the wood's dry weight [60]. This polysaccharide is composed of linear chains of $\beta(1 \rightarrow 4)$ D-glucose units held together by ß-1,4-glycosidic bonds

(Figure 7). In bleached kraft pulps prepared from native wood, there is possibly a degree of polymerization from 10,000 to 1,000 units [61]. The D-anhydroglucopyranose unit is endowed with hydroxyl (-OH) groups at positions C_2, C_3, and C_6. These groups are capable of undergoing the typical reactions known for primary and secondary alcohols [62].

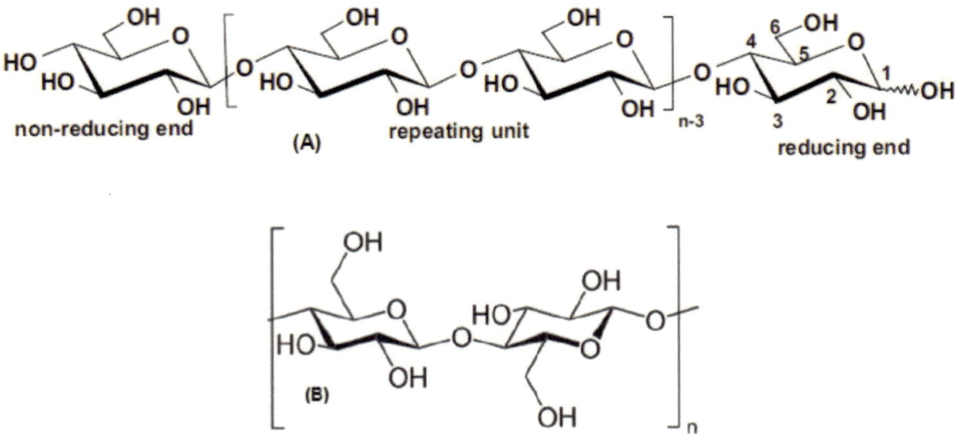

Figure 6. Molecular structure of cellulose (A): Chain of several hundred to more than 10,000 β(1 → 4) linked D-glucose units [60, 63], (B): polymeric polysaccharide [54].

On one hand, cellulose exhibits a strong tendency to form intra- and inter-molecular hydrogen bonds via the hydroxyl groups as shown in Figure 7. On the other hand, the linear cellulose chains tend to stiffen the entire straight chain, thus, promoting aggregation into a crystalline structure that gives the cellulose molecule a multitude of partially crystalline fiber structures and morphologies [60].

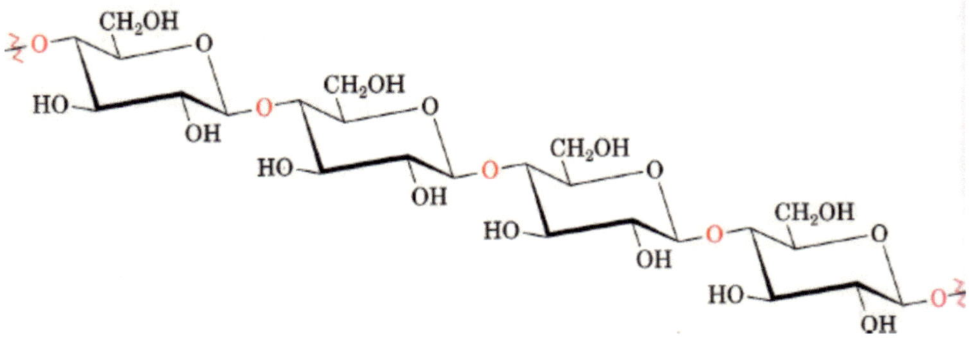

Figure 7. Cellulose, a 1,4′-O-(β-D-glucopyranoside) polymer, Adapted from [2, 48].

A schematic representation of a strand of cellulose (conformation I_α) showing the hydrogen bonds (thin lines) within and between cellulose molecules [50] is illustrated in Figure 8.

http://dx.doi.org/10.5772/62154

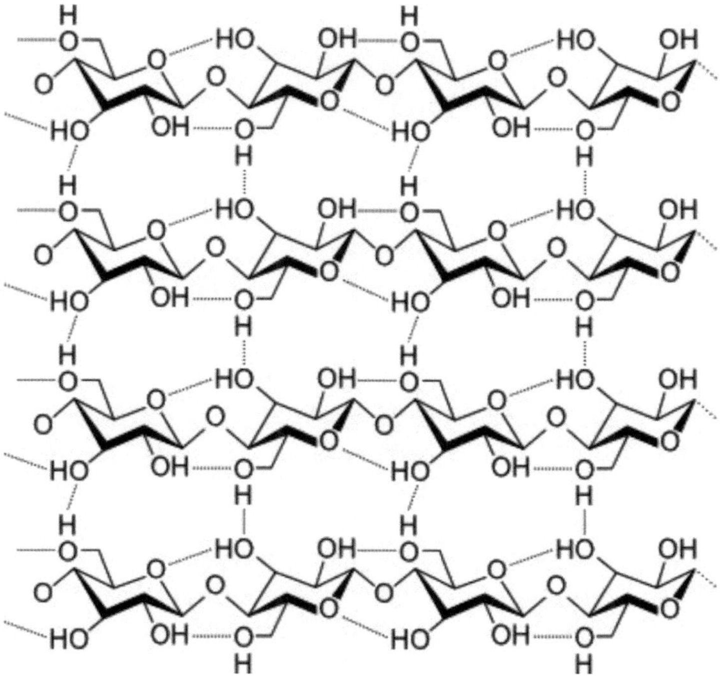

Figure 8. A schematic representation of a strand of cellulose (conformation I_{α}) showing the hydrogen bonds (thin lines) within and between cellulose molecules [50].

This readily bioavailable resource thus offers a great potential for continuously applied polymer research because, as already mentioned, it is the most abundant organic chemical on earth. It is reported in the literature [64] that more than 50% of the carbon in plant occurs in the cellulose support structure of stems and leaves; wood is largely cellulose, and cotton is more than 90% cellulose. It is worth mentioning that the great strength of wood is due largely to the H bonds between cellulose chains [57]. Cellulose is composed of long unbranched chains from 100 to 10,000 glucose molecules. Groups of these chains, held together by the hydrogen bonding between – OH groups on adjacent chains, are twisted into rope-like structures that make cellulose tough and fibrous [64]. Note that the absorbent properties of cotton and paper towels are due to capillary action and formation of hydrogen bonds between water molecules and –OH groups on cellulose chains.

2.3. Cellulose polymers

Cellulose acetate (Figure 9) is one of the oldest man-made macromolecules used extensively in the textile and polymer industries [63]. It has an inherent advantage in that the starting material, cellulose (Figure 7), is a renewable natural resource [66] widely produced from processed wood pulp (Figure 2a) dictating the current market source with intense research focused on various other renewable materials as feedstock [5-7].

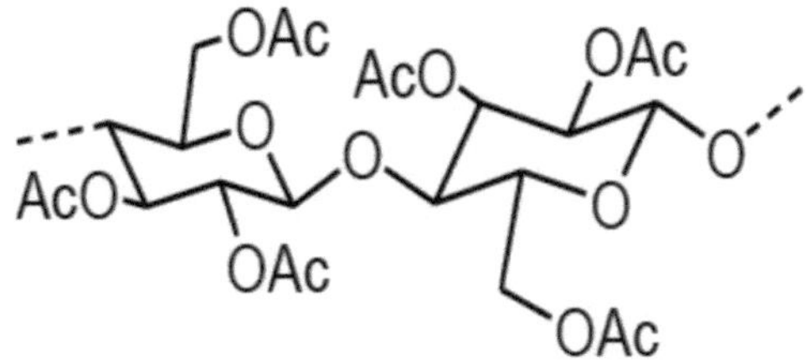

Figure 9. Cellulose acetate molecule [4].

Cellulose acetate is currently being used in fiber or film processing; however, it has been reported [67] to have high glass transition temperature which limits its thermal processing. Current applications of cellulose acetate include textiles, cigarette tow, lacquers, cellulose films, and in food packaging [68, 69]. Because it is nontoxic, it possesses a great potential to be used in new blends for the manufacture of a variety of biomaterials with very high potential eco-friendly characteristics. An example of such application(s) is in the manufacture of protective materials, e.g. personal protective garments. Cellulose acetate is broadly classified into two types:- cellulose diacetate and cellulose triacetate with acetyl values of approximately 55% (degree of substitution: 2.4) and 61% (degree of substitution: 2.9), respectively [67]. This type of biopolymer, sourced from a variety of renewable materials could undoubtedly revolutionize the polymer industry in just the near future owing to its versatility in chemical modification for better physicochemical qualities.

Cellulose acetate gives fabrics a silk-like appearance [8] (Figure 8) and it can be blended easily with fibers from other materials, providing room for the manipulation of the physico-chemical properties for improvement via surface modification with organic polymers and/or metal-organic supramolecules. Its texture is soft and cool against the skin, is naturally absorptive and breathable, has good drape and is excellent at combating static cling [70]. Cellulose acetate is also frequently used for linings in suits or coats, for formal wear including wedding gowns [71].

Two conventional acetylation techniques have successfully been employed in the fibrous and solution processes [65] to fully acetylate purified plant-derived cellulose-based mechanical pulp. The reaction media for acetylating pulp by the fibrous process consist of xylenes, acetic anhydride, and H_2SO_4. Xylene acts as a non-solvent, diluent. Cellulose acetate is isolated from the remainder of the acetylated components by differential solubility in dichloromethane/methanol (9:1,v/v) [65].

Through this approach, it is possible to modify cellulosic fiber using raw materials that contain acetyl groups capable of forming cellulose acetate polymers. Such raw materials used in the manufacture of these types of biopolymers are usually acetic acid and acetic anhydride. Acetic acid is one of the simplest organic acids; a main component of vinegar can be prepared naturally or synthetically from chemical processes. Acetic anhydride is produced by combining two acetic acid molecules with the removal of a water molecule. Thereafter, it follows a comprehensive evaluation by measuring key physicochemical parameters characterizing the polymer and some chemical/instrumental analyses are also conducted to complement the physicochemical measurements in the evaluation protocol.

Thus, the biopolymer preparation protocol is simple (Fig. 10) and entails that cellulosic fibre, acetic acid, and acetic anhydride should be mixed together and reacted to form cellulose acetate polymers. This process is aided by the addition of a small amount of H_2SO_4 which is subsequently neutralized during processing. The unique properties [8, 72] of the biopolymer (cellulose acetate) enable a great variety of end-use applications to be investigated in a number of permeation studies against selected Industrial toxic material, nuclear, biological and chemical agents.

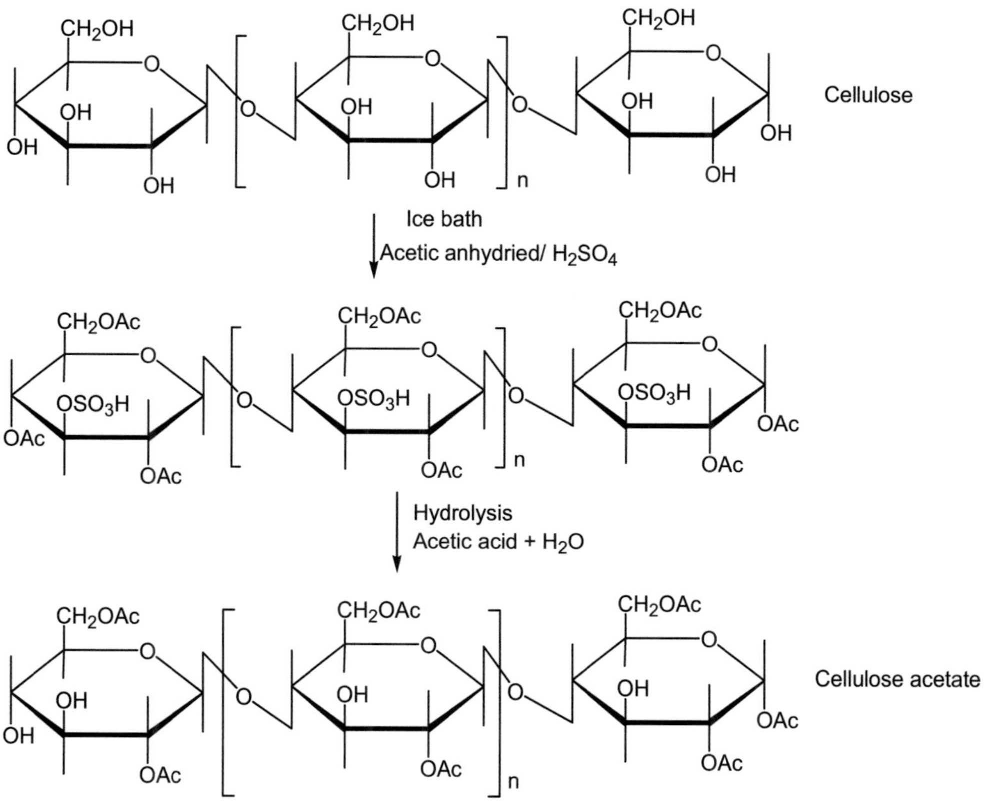

Figure 10. A simple preparation protocol of cellulose acetate polymer [73].

2.4. Biopolymers and crystallinity

A particular polymer is not made up of a single molecular species but is a mixture of macromolecules with similar structures and molecular weights that exhibit some average characteristic properties. The monomer-polymer term has probably been erred since time immemorial before humans employed it to such an advantage today. Biological macromolecules are nothing more than condensation polymers created by nature's reaction chemistry and improved through evolution [57]. Remarkably, these molecules are today's greatest proof of the versatility of carbon and its kingdom of atomic partners. Polymers, therefore, do not crystallize in exactly the same way that "pure" compounds do. The packing of identical molecules required to form precise crystal structures is only partially attained. Yet some polymers have many of the physical characteristics of crystals said to be crystalline. For example, long segments of linear polymer chains are oriented in a regular manner with respect to another. Such crystalline regions of a polymer are referred to as crystallites [74]. Structurally, amorphous and noncrystalline regions lie between the crystallites and constitute defects in the crystalline structure [72].

Biopolymers that have polar functional groups show considerable tendency to be crystalline. This would seem a positive character that favors surface modification to meet specific demands, for example, to create "creuse" blends with considerable resistance to nuclear, chemical and biological warfare agents. During processing and manufacturability, orientation can be aided by alignment of dipoles on different chains. Another good example is proteins, constituting a group of the many natural polymers that can crystallize because of their regular helical structure. In this regard, the order imposed by hydrogen bonds is further inducement to malleable crystallinity. Because polar groups are not, however, a necessary prerequisite to crystallization, crystalline polyethylene and polypropylene can be prepared interlayered with biopolymer to yield a desired strength and wearability. The van der Waals interactions between the long hydrocarbon chains provide sufficient total attractive energy to account for the high degree of regularity within the polymers. Sight should not be lost, though, that irregularities such as branch points, comonomer units, and cross-links lead to amorphous polymers. In other words, crystalline polymers usually are relatively strong and nonelastic and have characteristic melting points, such as synthetic fibers (e.g. nylon and Dacron). Because amorphous polymers do not have true melting points but instead have glass transition temperatures at which the rigid and glasslike material becomes a viscous liquid as the temperature is raised, multilayered surface modifications via organic or biopolymers can be envisaged to enhance the functional protective properties.

Each type of protein has its own amino acid composition, specific number and proportion of the different amino acids. The forces responsible for protein shapes (varying from long rods to undulating sheets, baskets with deep crevices to Y-shaped blobs) are the same bonding and intermolecular forces that operate for all molecules in nature. The –SH ends of two cysteine R groups often form an –S-S- bond, a covalent disulfide bridge that brings distant parts of the chain together as clearly illustrated in Figure 11 [57].

Polar and ionic R groups usually protrude into the aqueous fluid, interacting with water through ionic-dipole forces and H bonds: sometimes securing the chain's bends through an acid ($-COO^-$) R group lying near a basic ($-NH_3^+$) one to form an electrostatic salt bridge. This allows for an enormous potential for multilayer intercalation with other functional organic biomaterials to obtain desirable characteristics and qualities. The helical and sheet-like segments, thus, arise from H H bonds between the C=O of one peptide bond and the N-H of another. Other H bonds act to keep distant portions of the chain near each other. Nonpolar R groups usually congregate through dispersion forces within the protein interior. As such, fibrous proteins have relatively simple amino acid compositions and structures shaped like

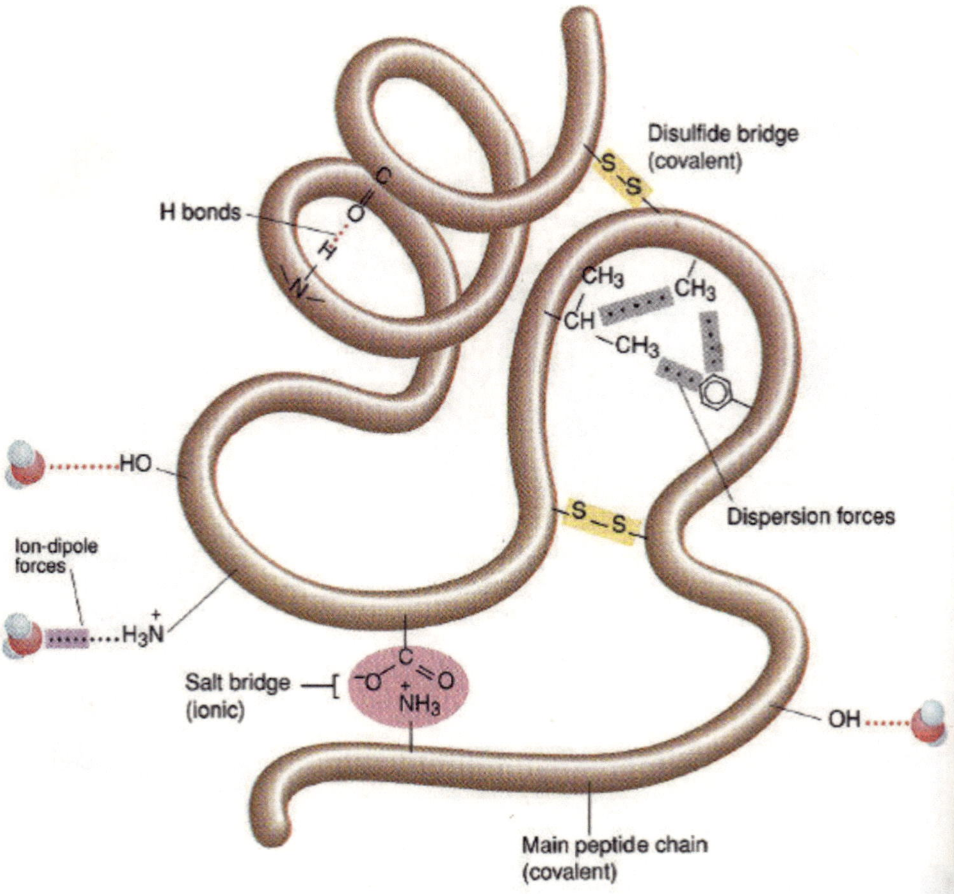

Figure 11. Schematic depiction of the forces operating in many fibrous proteins, Adapted from [57].

extended helices or sheets. These are key components of hair, wool, skin, and connective tissue materials that require strength and flexibility.

Finally but not the least, spatial configurations (Scheme 3(a)) of polymer molecules have a critical effect upon the physical properties of the polymer. The structure of natural rubber is an example to illustrate this. In this case, three spatial arrangements can result from polymerization process:

1. Atactic, in which the configurations are random along the polymer chain.

2. Isotactic, in which the configurations at each chiral center are identical.

3. Syndiotactic, in which the configurations at each chiral center alternate.

If the polypropylene stereoisomers have been prepared, the regular structures of the isotactic and syndiotactic polymers lead to hard crystalline materials. The random repeating configurations along the atactic polymer chain result in soft, amorphous, elastic material.

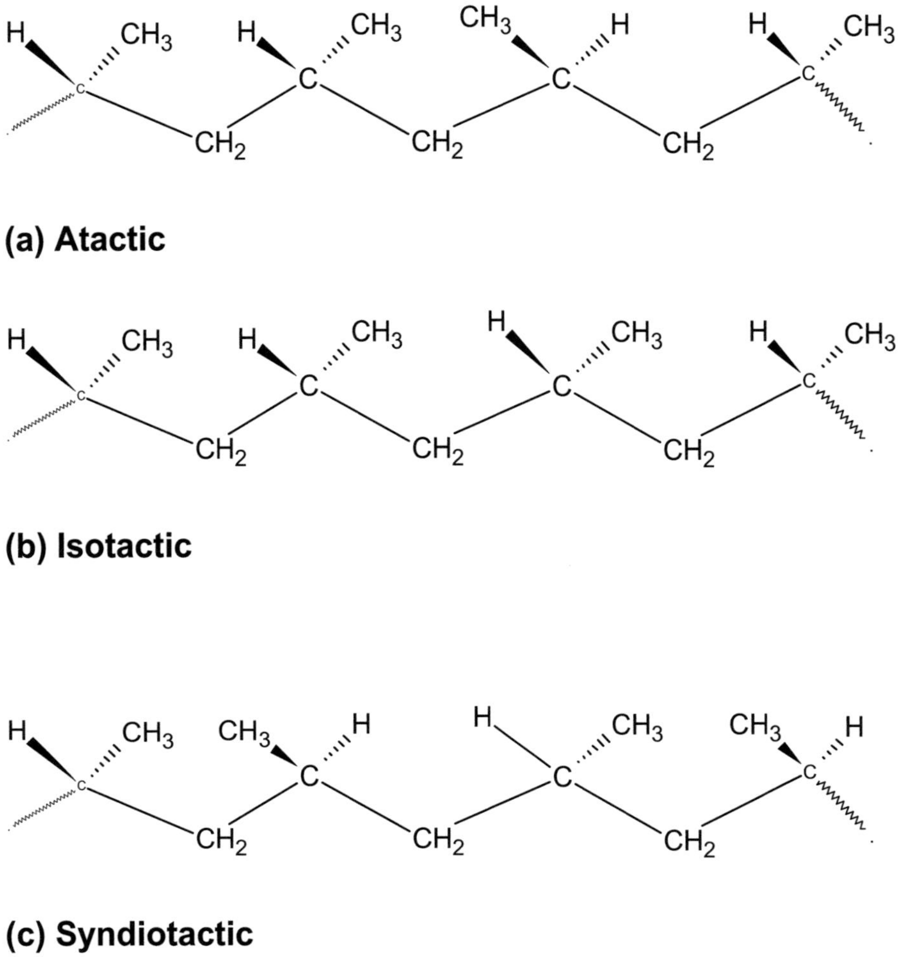

Scheme 3. Configurations of polypropylene: (a): atactic, (b): isotactic, and (c): syndiotactic [3].

Further processing using biopolymer copolymerization under well- monitored conditions could further improve not only the resistance of such polymers against both toxic industrial and chemical agents but also may not create great wearability and manufacturability.

3. Polyelectrolytic polymers

3.1. Polymeric Ionic Liquids (PILs)

Polyelectrolytes constitute a class of polymers with repeating units bearing an electrolyte group. The main moieties of polyelectrolytes are polycations and polyanions. In an aqueous medium, these groups are susceptible to dissociate yielding charged polymers. As such, polyelectrolyte properties are similar to both electrolytes in the form of salts and usual high molecular weight polymeric compounds. Thus, polyelectrolytes are sometimes called poly-salts (Figure 12). However, theoretical approaches [75-77] to describing the statistical properties of these charged macromolecules differ profoundly from those of both their electrically neutral and synthetic counterparts [78], while technological and industrial fields exploit their unique properties for a wide range of applications.

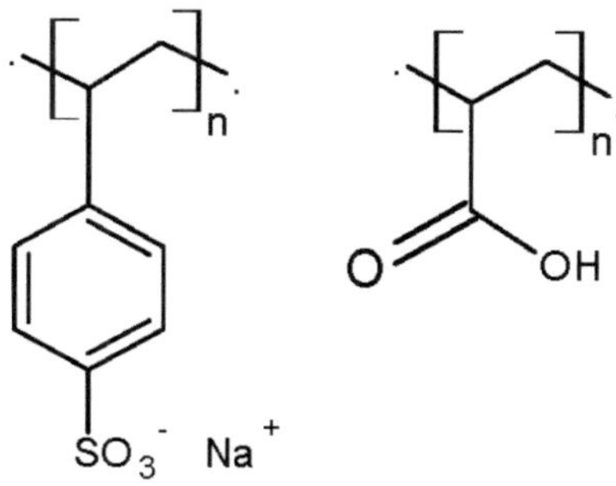

Figure 12. Example of a polyelectrolyte molecule, a molecule in which a substantial portion of the constitutional units has ionizable or ionic groups, or both [77]. The chemical structures of two synthetic polyelectrolytes: To the left is poly(sodium styrene sulfonate) (PSS), and to the right is polyacrylic acid (PAA).

As illustrated in Figure 12, PSS is a 'strong' polyelectrolyte (fully charged in solution), whereas PAA is a 'weak' polyelectrolyte (partially charged). Like polymers, their solutions are often viscous. Charged molecular chains, commonly present in soft matter systems, play a fundamental role in determining the structure, stability and the interactions of various molecular assemblies. Both natural and synthetic polyelectrolytes are used in a variety of industries. The good examples to be discussed in this chapter are those of natural polyelectrolytes (e.g. liquid cellulose). For instance, polypeptides (already discussed in Section 2.4), glycosaminoglycans, and DNA are good examples of natural polyelectrolytes.

The main factor strongly affecting the physical properties of polyelectrolyte solutions is the degree of charging. The dissociation of polyelectrolytes would release counter-ions directly affecting the solution's ionic strength and the Debye length. Consequentially, other properties such as electrical conductivity are affected.

If a polyelectrolyte completely dissociates in solution with reasonable pH values yields, it is referred to as very 'strong'. In contrast, polyelectrolytes that partially dissociate at immediate pH with a dissociation constant (*pKa or pKb*) in the range of ~2 to ~10 are termed 'weak'. Electrochemically, it can be seen that, weak polyelectrolytes are not always fully charged in solution. Moreover, their fractional charge can be modified by changing the solution *pH*, counterion concentration, or the ionic strength. These properties have been exploited in several industrial applications. The introduction of new ionic moieties, cations and anions, extend the properties and classical applications of the polyelectrolytes. Ndibewu et al. [79] used two locally available ionic polyelectrolytes (PP1 and PP2) to study the polymeric cementation mechanistic types on two South African subgrade soils (black cotton soil: BCS and reddish chert soil: RCS) during soil composite pavement compaction processes. The compositional physicochemical properties (*pH* and electrical conductivity, mS) of the two polyelectrolytes measured by preparing a 0.03% (v/v) solution at ambient temperature are presented for illustration in Table 1.

Ionic-lignosulfonate based polyelectrolytes (Stabilizers)	pH	Conductivity (mS)	Turbidity (NTU)	Total Dissolved Solutes (TDS in mg/L)
PP1	4.31±0.02	1.016±0.017	0.58±0.03	0.51±0.03
PP2	3.85±0.01	0.400±0.052	1.57±0.03	0.21±0.01

Table 1. Compositional chemical-physical properties of the enzyme-based formulations measured by preparing a 0.03% (v/v) solution at ambient temperature

In this study, the authors were able to access (via infrared synchrotron light spectroscopy and micro-imaging spectroscopy) the strength enhancement properties (plasticity index) affected by the use of the polyelectrolytes. Although, it wasn't possible for them to describe the exact mechanisms through which polyelectrolytes achieved this kind of polyelectrolytic polymeric-enhanced surface networking, new molecular and complex bonding rearrangements were prominent in more performing microcrystalline formations (Figure 13), demonstrating the polymer character of this important group of natural polyelectrolytes.

PILs have many applications, mostly related to modifying flow and stability properties of aqueous solutions and gels. For instance, they can be used to destabilize a colloidal suspension and to initiate flocculation (precipitation). These properties, if so well desirably monitored, can yield very interesting sought properties obtained from blends with their synthetic counter-parts.

Mecerreyes [78] in his review article has discussed the two main approaches for synthesizing ionic liquids (Figure 14) or polyelectrolytes, named polymeric ionic liquids (PILs) in analogy to their monomeric constituents. This author listed a few examples herein included: cations such as imidazolium, pyridinium, and pyrrolidonium and anions such as hexafluorophosphate, triflates, and amidotriflates. These new liquid macromolecules are now giving rise to a new family of functional polymers with particular properties and new applications. The first part of this chapter has focused on the renewable sources of these PILs while synthetic aspects of PILs and the main aspects related to their physico-chemical properties are being discussed. There are a few technological applications of these polymers, namely: polymer electrolytes used in electrochemical devices, use as building blocks in materials science, nanocomposites, gas membranes, innovative anion-sensitive materials and smart surfaces. There are also a countless set of applications in different fields such as energy, environment, optoelectronics, analytical chemistry, biotechnology and catalysis. It would be interesting to explore other possibilities targeting their synthesis and blends in the manufacture of personal protective materials.

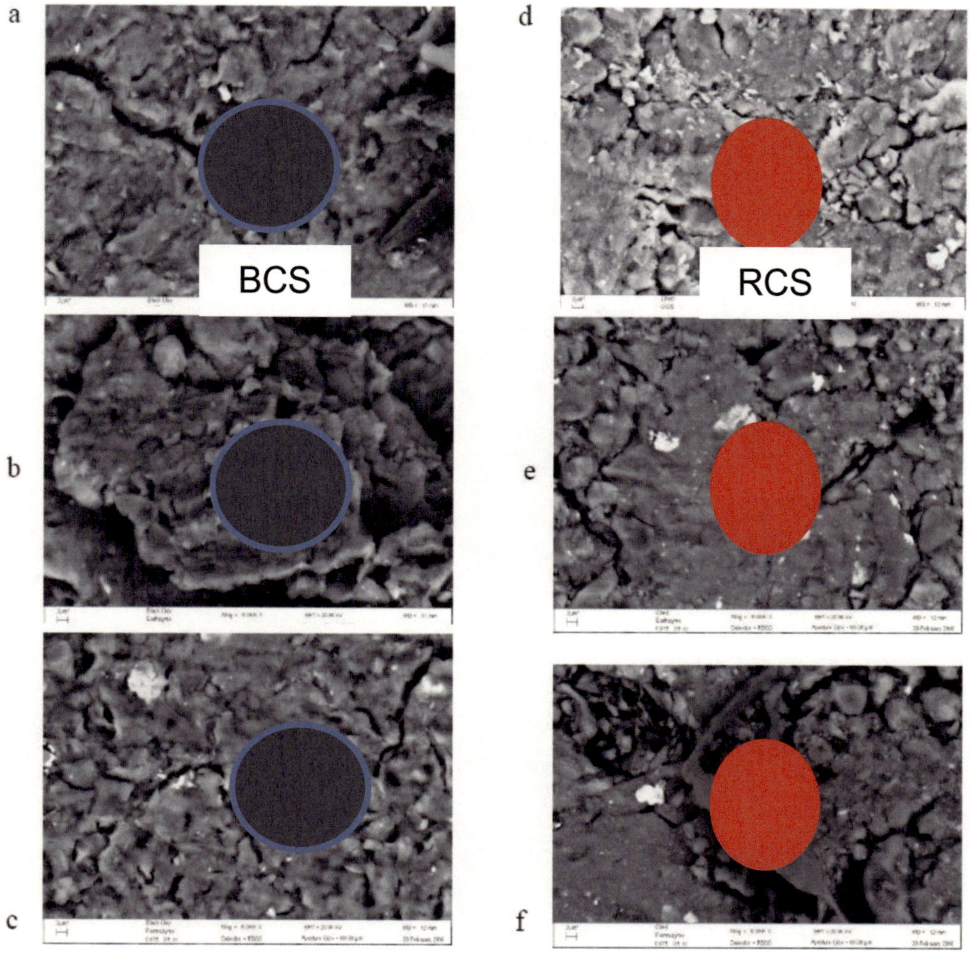

Figure 13. SEM micrographs of (a) black soil composite specimen (BCS), (b) stabilized with PP1 (BCS-PP1) or (c) stabilized with PP2 (BCS-PP2) and (d) red chert soil composite specimen (RCS), (e) stabilized with PP2 (RCS-PP1) or (f) stabilized with PP2 (RCS-PP2) [79].

The potential use of PILs in blends or complex composites of new clothing with high resistance against nuclear, chemical and biological agents is because PILs can potentially impart a surface charge to neutral particles, enabling them to be dispersed in aqueous solution or mixtures (superplasticizer). Elsewhere, they are often used as thickeners, emulsifiers, conditioners, clarifying agents, and even drag reducers. They are used in water treatment and for oil recovery. Many soaps, shampoos, and cosmetics incorporate polyelectrolytes. Some of the polyelectrolytes that appear on food labels are pectin, carrageenan, alginates, and carboxymethyl cellulose. All but the last are of natural origin. Finally, they are used in a variety of protective or surface coating components, e.g. cement.

PILS are also being investigated for biochemical and medical applications exploiting the high solubility characteristics. For example, much research currently focuses on using biocompat-

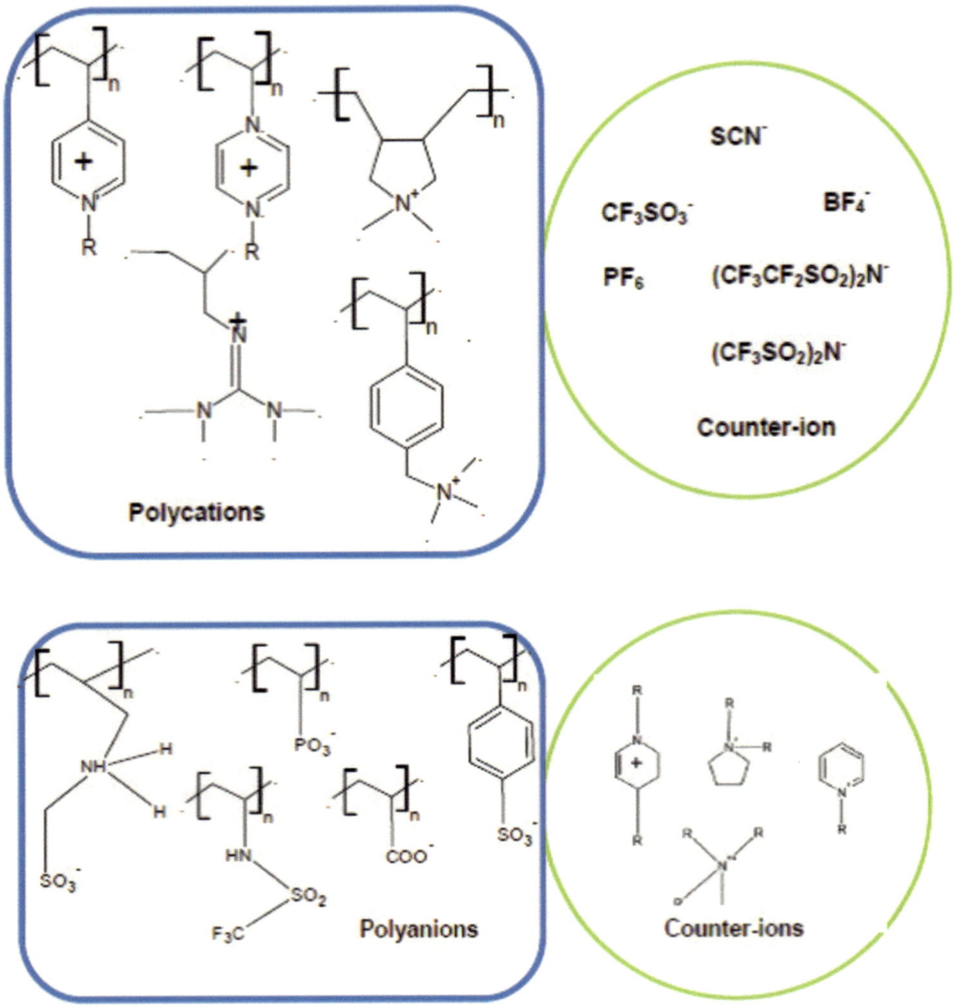

Figure 14. Typical chemical structures of polymeric ionic liquids [80].

ible polyelectrolytes for applications in: implant coatings, controlled-drug release, and many other applications. Recently, the biocompatibility and biodegradability of macroporous materials composed of polyelectrolyte complexes have been described in which the material is presented to exhibit excellent proliferation of mammalian cells [81]. Perhaps, the approach to incorporate PILs into synthetic polymer blends for desired properties and characteristics could exploit the recent processing strategies of polyelectrolyte multilayers (PEMs) utilized in the formation of new types of materials for fuel cells. These thin films are prepared using a technique known as the layer-by-layer (LbL) deposition. In the technique, a suitable charged growth substrate is dipped back and forth between dilute polyelectrolyte solution baths that are positively and negatively charged. The mechanism of polyelectrolyte deposition during which the surface are reversibly charged is adsorption. This allows for a gradual but controlled build-up of electrostatically cross-linked films of polycation-polyanion layers. This procedure sounds feasible in creating intercalated layers in cellulose-based polymeric protective garments of any controlled thickness to stop the penetration of toxic or chemical agents. The advantage is that thickness control of such films down to single-nanometer scale maybe possible. LbL films can also be constructed by substituting charged species such as nanoparticles or clay platelets [82] in place of or in addition to one of the polyelectrolytes. LbL deposition has also been accomplished using hydrogen bonding instead of electrostatics. Another interesting avenue to explore in this regard could be polyelectrolyte adsorption.

The main benefits of using water-based processes in PILs-cellulose-based garment manufacturing would be reasonably less costs, the utilization of particular chemical properties of the film for further modification, such as the synthesis of inner surface metal-organic, nanoparticles, or porosity phase transitions to create anti-reflective coatings, chemical shutters, and superhydrophobic surface coatings with environmental benefits to biodegrade.

As illustrated in Figure 14, Mecerreyes [80] has also reported that PILs present some of the unique properties of ILs (ionic conductivity, thermal stability, tuneable solution properties and chemical stability) together with the intrinsic polymer properties that are unique. This uniqueness is premised on the ground that most PILs are not soluble in water but in polar organic solvents. This is in contrary to classic polyelectrolytes, usually water soluble while dissociating in aqueous solutions and making the polymers charged to form polysalts. This is due to the hydrophobic character of the counter-ion and the reduced columbic interactions.

Thus, the preparation of inorganic–organic nanocomposites/polymeric blends offer a potential route for producing functional materials that combine the best features of inorganics (e.g. improved mechanical strength, thermal stability) with the properties of PILs.

4. Functional protective materials

4.1. Functional polymers

Because the relatively high persistence length of the cellulose molecular chain conformation and their close packing through numerous hydrogen bonds have made the dissolution of cellulose - a difficult process, the development of new polymeric compounds (e.g. regenerated cellulose) using environment-friendly low-cost solvents for cellulose regeneration is essential for the successful utilization of the cellulose as a component of polymeric materials with a wide applications including the use in the manufacture of protective materials such as PPGs.

Regenerated celluloses have drawn attention owing to their ease to be fabricated. Also, they are biocompatible and biodegradable while thermal and chemically stable.

It should be noticed, though that there is still some persistence difficulty in dissolving this natural polymer thus widening the field of current research into using them in preparing "green" solvents for further polymerization processes.

The different solvents used in the preparation of regenerated cellulose confer to the regenerated newer functional properties (physicochemical and mechanical). These solvents have been extensively discussed elsewhere by Wang et al. [1]. It is worth mentioning a few of these solvents here including: N-methylmorpholine-N-oxide (NMMO), LiCl/N, N-dimethylacetamide (LiCl/DMAc), NaOH aqueous solution, alkali/urea, NaOH/thiourea aqueous solution, and tetra butyl ammonium fluoride/dimethyl sulfoxide system.

5. Toxic industrial chemicals and their potential use by terrorists

5.1. Toxic industrial chemicals

The industrial revolution brought prosperity and improved quality of humankind, along with it the use of chemicals and complex technologies. As a result, industrial chemicals are produced in large quantities, transported and stored for making petroleum, textiles, plastics, fertilizers, pesticides, herbicides and other products for peaceful use [84]. Some of these chemicals such phosgene are utilized for making isocyanates, perfumes and fumigants; cyanogen chloride for making herbicides, dyes and vitamins; and hydrogen cyanide for making plastics, pesticides and sanitizers. Phosgene and cyanogen chloride were used during World War II as chemical weapons. These served to illustrate that some of these industrial chemicals are toxic. According to some researchers [84-86], a substance is toxic if it has a median lethal concentration (LC_{50}) in air of more than 200 parts per million (ppm) but not more than 20 mg/L of mist, fume or dust when administered by continuous inhalation for 1 hour (or less, if death occurs within 1 hour) to rats weighing between 200 and 300 g. It should be noted that the toxic effects of chemical, biological, radiological, nuclear, explosive and other toxic industrial chemicals (TICs) are dose dependent and depends on the mode of exposure such as inhalation through the lungs, ingestion, injection through puncture and absorption through skin.

Interestingly, a substantial use of chemicals is essential to meet the social and economic goals of the world community,, and today's best practice demonstrates that chemicals are used widely in a cost-effective manner and with a certain degree of safety. However, past events such as the 1984 accidental release of methyl isocyanate in Bhopal, India, resulted in 3 000 deaths [87] and according to Sohrabji [88], it contaminated underground water within a 3.5 km radius around the affected industrial buildings. The recent Tianjin hazardous chemical storage facility explosions [89] on the 23 August 2015 resulted in 123 people confirmed to have died and another 50 reported to be missing and most of them being fire fighters. This incident has demonstrated that more work need to be done, to ensure the environmentally sound management of toxic chemicals within the principles of sustainable development. According to the United Nations Environment Programme (UNEP) Agenda 21 [90], there are two challenges, particularly in developing countries, which are: (a) a lack of sufficient scientific information for the assessment of risks associated with the use and storage of a large number of chemicals, and (b) a lack of resources for the assessment of chemicals for which data are available. However, countries such as Germany, Switzerland and China [91], in partnership with the International Co-operation and Assistance Division of the Organisation for the

http://dx.doi.org/10.5772/62154

Prohibition of Chemical Weapon Convention (OPCW) are addressing some of these challenges. They are developing and implementing program such as 'Promoting Chemical Safety Management Training' [92] and the internship program in organic chemistry, respectively, which are aimed at developing skills of personnel from developing countries.

In the current political situation, many governments of the world are afraid that if some of these toxic industrial chemicals fall on the hands of terrorist, or deliberately released, they may have some serious effects on the exposed individuals. A detailed list of toxic industrial chemicals was compiled by International Task Force 25: 1998 Industrial Chemicals Final Report 1998 [93]. As an example, chlorine is commonly used in water treatment facilities, and paper and cloth manufacturing industries and it is often transported by road. According to the United Nations Security Council, Report No. S/2007/314, insurgents in Iraqi have increasingly used chlorine in improvised devices for the purpose of harming unprotected people. Historically, chlorine was used by Germans in 1915 [94, 95] against the British forces, and it was released against the military in Iraq in 2007. Furthermore, chlorine has attracted some attention in both the Syria and Iraqi conflicts, where the parties involved are accusing each other of using chlorine gas as a chemical weapon [94].

5.2. Chemical and biological warfare agents

Chemical warfare agents (CWAs) are chemical substances, whether gaseous, liquid or solid, which are used because of their direct toxic effects on humans, animals or plants. In military operations, CWAs are intended to kill, seriously injure, or incapacitate the victims and are particularly effective because of their extremely high toxicity [85, 97]. They have been classified into nerve, blister or vesicants, choking or lachrymators, blood, tear, vomiting and incapacitating agents [98] based on mode of their effect on humans [99].

Dr Gerhard Schrader of I.G Farben in Germany discovered the nerve agents in the 1936 [100]. The first nerve agent to be discovered was Tabun, followed by Sarin and Soman, their chemical structures are shown in figure 15.

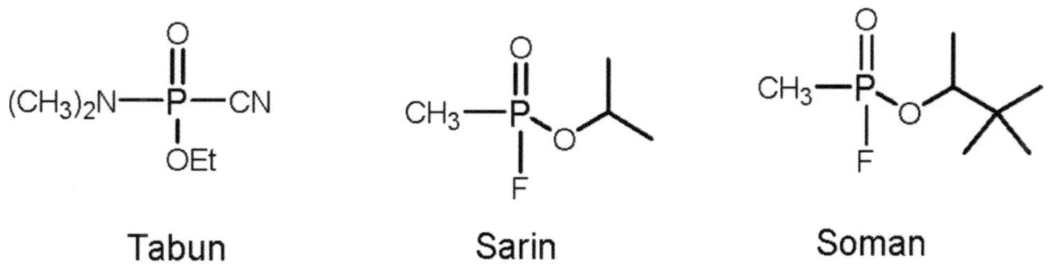

Figure 15. Selected nerve chemical warfare agents.

More ominously, governments such as those of Germany, Britain, the United States, the former U.S.S.R and Japan invested in the development of delivery system technology comprising artillery shells, unguided gas rockets and the Mark 116 'Weteye' air- dropped gas bombs [101]. The Chemical Weapon Convention (CWC) prohibits the development, production, stockpiling, and use of chemical weapons, except for the purpose not prohibited by the CWC such as industrial, agricultural, pharmaceutical and other peaceful purpose [102].

The most widely published chemical terror incident involving nerve agent Sarin occurred in Japan [103]. A Japanese religious doomsday sect used Sarin against unprotected people in Matsumoto resulting in the death of seven people [104]. Sarin was again used in terror attack by the same religious sect on 20 March 1995 Tokyo subway resulting in approximately 5 500 people seeking medical attention [105, 106]. Unfortunately, the first responders also become victims. Seemingly, the terror attack with CWAs seems to be an on-going activity in Syria, where multiple incidents involving Sarin gas were reported and some investigated by the OPCW [107-109]. The most disturbing threat according to Almemar [96] is the increasing amount of evidence of chlorine gas attacks, by the Islamic State of Iraq and Syria (ISIS) in Iraq and Syria against Kurdish Peshmerga.

Biological warfare agents refer to the military or terrorist employment of microbial or other biological agents such as bacteria, viruses, fungi and toxins to intend or inflict death, temporary incapacitation or permanent harm in humans, animals and plants [110]. This is important in case of filoviruses, Marburg and Ebola; both these agents pose a serious threat as lethal pathogens and their use by terrorist will results in fear and panic [111]. Contemporary biological warfare has its roots in World War I and the post-war era. According to Guillemin [110], the German military spy agents mounted an international sabotage by infecting horses and mules that were being shipped to Britain and France with glanders and anthrax from neutral nations such as the United States, Norway, Spain and Romania. This resulted in entire shiploads of animals being sick and killed. After the tragic terror events of 11 September 2001 in the United States, an unknown perpetrator mailed four letters containing Anthrax spores to unsuspecting victims in Florida, New York and two senators in Washington DC [112]. As a result, 5 of the victims eventually died of inhalation of anthrax wheras 17 others who contracted a cutaneous form of the diseases were successfully treated [112].

Similar to chemical agents, governments around the world are worried that if some of the biological agents, such as anthrax, small pox, plague, tularemia, Ebola, Marburg and others, may find their way into the hands of terrorists, such attack may cause a huge economic loss [113]. According to Aken and Hammond [114], the members of the Japanese cult, Aum Shinrikyo reportedly tried unsuccessfully to get their hands on Ebola virus during an outbreak of the diseases in the Democratic Republic of Congo (formally known as Zaire) in the early 1990s. They attempted using a complex published method for cultivating polio; however, only highly trained experts would be able to master the technique [114]. Thus, with the current developments in biotechnology, there is a possibility that in the near future, such groups or interested states could theoretically, cultivate viruses in the laboratory [115].

In conclusion, perhaps the human suffering and economic impacts resulting from biological agents were recently illustrated in West African countries (Guinea, Liberia, Sierra Leone, Senegal and Nigeria) with the outbreak of Ebola in March 2014. As of 2 October 2014, according to the World Health Organisation (WHO) [116], 7157 of Ebola infections and 3330 deaths were suspected, with only 3953 infections that were positively confirmed by laboratory results.

5.3. Protection against biochemical warfare agents

The two incidents namely: terror attack in Tokyo subway and the Ebola outbreak in West African countries have demonstrated the importance of protecting the first responders and health workers. In the Tokyo subway incident, it was reported [117] that the first group of responders and health-care workers involved in the initial response were not wearing proper personal protective equipment (PPE), and surely not knowing that a highly toxic nerve agent was released to unsuspecting passengers. As a result, 135 of the 1364 fire department personnel who responded to the incident experienced secondary exposure while transporting victims to emergency facilities [118]. Lastly, some the first firefighting responders in the Tianjin series of chemical explosions were also reported to be the victims of the secondary explosions [89].

According to Bray [111], in Africa, Ebola outbreaks resulted in a large number of infections among health practitioners, such as doctors and nurses, mainly because of the inability to employ proper infection control measures, such as proper use of gloves, gowns and masks. Furthermore, people handling the deceased, conducting funerals, traditional healers, mid-wives, mothers, humanitarian and aid workers may also become the victims [119, 120], which may illustrate the importance of training the above groups in the techniques of using personal protective garments correctly.

According to DuPont [121], safe and reliable protective clothing is essential to prevent spreading infection and protect against chemicals. They have recommended different over-the-counter garments that are tailor-made to offer protection against biological contaminated liquids and dust that first responders should consider.

5.4. Personal protective garments against industrial toxic chemicals

For any operation, proper protective clothing is required. However, it comes with numerous variables to be considered such as weight, comfort, level of protection, and duration of protection. Furthermore, it is dependent on the nature of the challenges to be encountered, for example, CWAs or TICs. The commonly used chemical protective clothing are carbon-based hooded suits (one- or two-piece), worn together with gloves and boots and some even wear the carbon-based undergarments [122, 123]. Tugara et al. [124] reported that the elimination of one or more of the ensemble components will reduce and improve weight, logistic concerns, and cost reduction while increasing comfort. More researchers are considering alternative materials such as membranes in the form of films, electro-spun nanofibers, super-hydrophobic materials, fabrics and others [125, 124]. When evaluating the membranes relative to protective clothing, the main properties to be considered for some applications, for instance, the military would include the following: high strength, antiballistic performance, fire-retardant characteristics, vapor absorption and liquid barrier characteristics [124]. So far, the spun-bond and melt blown non-woven materials used for protective clothing offer low cost, improved barrier properties, impermeability to particulate matter, medium strength and comfort [124, 126, 127].

There are different types of materials used as personal protective garments (PPGs). Mostly they are found as multilayers to provide multiple functions. The classic types of the multilayered PPGs consist of the following: (a) air-permeable materials that usually consist of a woven shell fabric, a layer of sorptive material (for example, activated carbon-impregnated foam or a carbon-loaded non-woven felt), and a liner fabric. At low hydrostatic pressures, liquids can easily penetrate permeable materials; hence the shell fabrics are usually coated with Quarpel and other fluoro-polymers to improve liquid repellency [124, 126]; (b) semi-permeable materials that consist of two types of membranes, the porous as well as the solution-diffusion membranes; and (c) impermeable materials that provide better protection against TICs and CWAs. However, because they prohibit transmission of sweat from the body to the environment, they should be cautiously used with the appropriate microclimate cooling/heating system (Rao, 2006; SANS 5101, 2004). Some examples of the impermeable PPGs are: 4H by the Safety 4 Company, Tychem by DuPont [121], and CPF3 by Kappler [126,127].

Types of fibres	Composition	Commercial products	Protection from mechanical hazards (stab, ballistic, puncture, chainsaw, etc.)	Protection from flame, fire and thermal hazards	Protection from chemical and biological hazards
Aramid fibres	p-aramid (poly-para-phenylene terephthalamide)	Kevlar (DuPont), Twaron (Acordis), and Technora (Teijin)	++ Tenacity: 194.3~229.6 cN/tex Modulus: 4061.8~9713 cN/tex	++ LOI: 25–28 T*: 190°C	+
	m-aramid (poly(meta-phenyleneisophthalamide))	Nomex, Kermel and Corex	++ Tenacity: 33.6~63.6 cN/tex Modulus: 106.0~123.6 cN/tex	+ LOI: 30 T*: 205°C	+
Polyamide imide (PAI) fibre and Polyimide	Polyimide (PI)	Torlon (Solvay) P-84® (Evonik)[8]	+ Tenacity: 37.1 cN/tex	++ LOI: 38 T*: 260°C	+ Poor in alkali
PBI fibres	Polybenzimidazole	PBI (Celanese)	++ Tenacity: 23.8 cN/tex Modulus: 282.6 cN/tex	++ LOI 41 T*: 250°C	+
PEN, PBO and PIPD fibres	PEN (polyethylene-2,6-naphthalate) PBO (p-phenylene-2,6—benzobisoxazole)	Zylon (Toyobo), M5 (Akzo Nobel) Novel yarn	++ (excellent abrasion) Tenacity: 370.9 cN/tex Modulus: 11479 cN/tex	++ LOI: 68 T*: 288~315°C	++ Fair in acid
UHMW-PE fibres[9,10]	Ultra-high Molecule Weight Polyethylene, Gel spun	Dyneema (DSM), Spectra (Honeywell), Tekmilon (Mitsui)	++ abrasion Tenacity: 264.9 cN/tex Modulus: 12362 cN/tex	– LOI 41 T*: 121°C	++
LCP fibre[11]	Polyarylate, liquid crystal polymer fibre	Vectran (Kuraray)	++ (excellent abrasion) Tenacity: 247.2~256.1 cN/tex modulus: 61.8~88.3 cN/tex	++ LOI 37 T*: >392°C	++ Good in solvent
PPS fibres	Polyphenylene sulphide, Crystalline thermoplastic fibre	Ryton® (Amoco), Procon® (Evonik), Toray PPS® (Toray)	+ Tenacity: 30.9~39.7 cN/tex modulus: 61.8~88.3 cN/tex	+ LOI: 34 T*: 288~315°C	++
PEEK fibre	Polyetheretherketone, Semi-crystalline thermoplastic fibre	ZYEX, ZEUS	++ Tenacity: 17.7 cN/tex modulus: 70.6 cN/tex	++ LOI: 95 T*: 260°C	++

Table 2. Properties of high-performance fibres used for protective clothing [128]

There has been an extensive research work conducted to improve the properties of the materials while trying not to compromise the protection capacity. The material industry has expanded and adventured into many alternatives that are available to meet the individual's requirements. There are selectively permeable materials (SPMs) that have the combined properties of the impermeable and semipermeable materials; they are extremely thin, and lightweight, and consist of flexible protective barrier materials offering resistance to CWAs and selected TICs. Their protection mechanism is based on selective solution/ diffusion process instead of adsorption process that is used by the activated carbon materials [126, 129]. Other material types consist of self-detoxification abilities in which the agent-reactive catalysts are incorporated to reduce the hazard by either hydrolyzing the chemical or trapping it [124, 126].

In general, the list of different types of PPG materials does not end here; there are many more technologies that are also focusing on improving the protection of the PPGs. Some of these technologies include super-repellency functions and electronic fabrics with sensors, for example to monitor physiological parameters [125, 127]. It all depends on what functions one chooses to focus on, ultimately for each requirement, there will be a suitable technology.

Mao [128] recently reviewed the properties of high-performance textile materials available in the form of fiber, film, membrane and liquid (herein presented in Table 2), used for protective clothing. These materials are reported to have exceptional mechanical, fire resistance and chemical resistance.

6. Spectroscopy and permeation tests

6.1. Permeation tests for evaluation of chemical protective materials

Several tests for the evaluation of chemical protective materials have been evaluated [130-135]. This includes that of the mustard gas and a matrix thereof has been compiled, especially in the textile industry [136]. However, most of the materials tested either are expensive or need further improvement. The matrix of materials evaluated will automatically increase the chance of developing readily available ones, also for use by civilians. The challenge remains though in the sampling and detection systems developed so far. A variety of sampling and detection systems have been used to meet the analytical needs of a particular testing scenario described in detail by other authors [137, 138]. Test cell designs prescribed by standard methods have remained basically fixed for over two decades [130, 131]. This has led to the continuous questioning of the analytical reliability for these standard cell designs [139- 141] noticeably with the desire for further improvement. In this regard, some authors [142] have extensively discussed the efforts to improve current nerve guide conduits (NGCs). Their article focuses on research to improve material selection, geometric structures, and incorporation of cells and chemical cues. In this report, they suggested that the advanced NGC model suggested by Hudson and Evans [143] is considered the most widely accepted based on which model the above six major components need to be considered and integrated in an advanced NGC.

Despite all these developments, there is still a general consensus that a comprehensive description of performance criteria for permeation test cells is lacking [144] in the literature. This suggests that there is an urgent need to define scientifically acceptable procedures and criteria addressing a test cell design with minimum attributes that ensure its reliability for the evaluation of chemical protective materials. In this line of thinking, Verwolf et al. [144] emphasized the vapor challenges as one of the more problematic than liquid permeates. These authors state that flowing vapor streams exiting the feed side require precision in its generation and proper monitoring of standard atmospheres of the test chemical(s). This operation is often not an easy task. This can result in large volumes of contaminated air to be disposed of. Furthermore, there would be the requirement of decontamination of the vapor flow path and vent stream upon the completion of a test run. As a practical example, highly toxic compounds such as chemical weapon agents (CWAs) could make this approach practically and economically not feasible or even prohibitive. Finally, the delivery of a standard vapor to the feed side of the test material may not be taken for granted, because the transfer conduit and connectors could interact with the analyte via sorption phenomena or via a chemical reaction, and/or undesirable partitioning in stagnant regions.

6.2. Resistance of protective materials and spectroscopic characterization

The quality of chemical protective materials is generally judged by the breakthrough time [145- 149]. This parameter describes the time from the beginning of the exposure until the chemical appears on the inside of the glove [148]. As an example, the time range is from 15 minutes to more than 24 hours for different pesticides through different types of glove material [150, 151] Furthermore, the resistance against penetration of a given toxic chemicals through a protective material will depend on chemical-specific physico-chemical properties [128]. This parameter is very important during manufacturability and testing or decontamination exercise. It could be thoroughly evaluated using permeation studies [152] for various toxic chemicals against the protective materials or personal protective garments used for this purpose.

From most of the literature reports and published results [128], the overall outcome is that composite materials and hybrid structures used in protective clothing and personal protection equipment (helmet, knee protection panel, gloves, protection curtains, protection tent and shield, etc.) would be those with unique properties, and should be highly resistant to toxic chemicals. Continuous research should focus on the development of composite materials and hybrid structures usually employed to provide multiple functions in addition to resistance requirements. The results on stress–strain relationship and deformation characteristics during development and manufacturability can no longer be over-emphasized here. This then also places lots of emphasis on simultaneous analytical capabilities and techniques to accompany efforts in the development of traditional protective equipment to highly desired ones such as in the decontamination after any chemical spillage. Because the energy absorption component seems to be important in considering the quality requirements related to the application of most protective materials (e.g. clothing), spectroscopy would seem to be the cornerstone to be considered for most analytical purpose. This is because qualities (aesthetic properties, clothing comfort and human mobility) of protective materials can successfully be assessed or quantitatively evaluated employing many already discussed spectroscopic techniques in the introduction of this chapter such as scanning electron microscopy coupled to X-ray dispersive spectroscopy (SEM/EDS) for surface imaging and structural elucidation; atomic force microscopy (AFM) and scanning tunnelling microscopy for both morphology, and surface-resolved property or associated intermolecular interactions; scattering-type near-field optical microscopy for nanoscale morphology and nanoscale-resolved subsurface imaging, attenuated total reflectance Fourier transform spectroscopy (ATR-FT-IR) for functionalized surface identifications, infrared spectroscopic ellipsometry and nano-FTIR absorption spectroscopy for apertureless near-field optical microscopy, and infrared vibrational nanospectrosocopy for nanoscale composite network formations.

http://dx.doi.org/10.5772/62154

7. General conclusion

7.1. General outlook

The general outlook is a peaceful world, free from not only hunger and poverty but wars and natural disasters. Perhaps amongst the challenges that man must tackle to create a safer environment is not only the need to develop new products to suit specific needs using more sustainable technologies is his ability to solve both yesterday's pollution that include toxic chemicals dumps into the environment and quickly act to stop further chemical accidental spillages. Since man is at the center of protection against nuclear and chemical attack whether conscious or unconscious, his safety is critically important. These projects continue research into the future focusing on developing far more performance protective materials, e.g. polymeric materials of various composite blends with required both mechanical strength and chemical resistance for common use and in the manufacture of personal protective equipment. To realize this sustainable world, we are facing the emerging challenges in agriculture, forestry, academia, government and industry. Thus, stakeholders (governments and researchers) should find the need here to work together for the common goal. From the 'green' perspective, cellulose is the most abundant natural and renewable resource, and is a prime candidate for replacing the petroleum-based products to expand to new applications without polluting the earth. We, as well as have indicated the inherent advantages of safety, biocompatibility and biodegradability of cellulose owing to the fact that cellulose is intrinsically a part of plants and animals (tunicates), and thus, a renewable biomaterial. Co-workers in their works have demonstrated the versatility of the cellulose natural polymer. The abundance of the OH groups endows cellulose with the affinity to inorganic/organic substances, leading to the preparation of hybrid materials, and expanding the potential application of cellulose as a functional material. This chapter has further expanded on the diversity and richness of regenerated cellulose materials fabricated via physical dissolution. Numerous citations have been exploited in putting this chapter together pointing out how in the last decades, regeneration have been astonishing, demonstrating promising potentials in textiles, packaging, biomedicine, water treatment, optical/electrical devices, agriculture and food fields. From one of the literature source, the general agreement is that physical dissolution and regeneration process are environmentally friendly by avoiding the consumption of chemicals because most of the agents can be recycled and reused, and the nature of cellulose is retained. As a result, no chemical reaction occurs, promising to bring another Green Revolution to the comprehensive utilization of cellulose-like natural resources. Therefore, we herein agree to it that cellulose-polymeric focused research impact and benefits are truly fascinating to society. This considers the physical processes in the preparation of new materials via environment-friendly technologies as substitution to the petroleum-based materials.

The overall outlook of functional polymers against toxic chemicals broadly discussed in this chapter can be coined in the following problem statement 'Explorations in cellulose (natural polymer) as a renewable resource may improve the world to be greener and more sustainable in the future, particularly in terms of functional polymers processing (e.g. usage in PPGs and PPE against toxic chemicals) if current efforts encourage further investigations of cellulose-based materials; that will increase a deeper understanding of the new mechanisms involved in the cellulose dissolution, regeneration and blends tunable with various applications owing to improved mechanical strength and physicochemical properties'. To achieve these goals, a spectroscopic and microscopic approach that combines newly developed simple to advanced hyphenated techniques will surely provide an efficient tool to supplement the existing ones for the assessment of structural integrity and durability of polymeric materials; and to determine differences between designs and manufacture deficiencies towards improvement. Sample preparation may affect the quality of results and their interpretation thus, higher-quality results and reliability will depend on the quality and appropriate sample preparation techniques. An understanding of the impact of sample preparation on the results is vital.

7.2. Conclusion and future perspectives

This chapter started with an overview of natural polymer chemistry with focus on cellulose and cellulose acetate. Polyelectrolytic materials, a typical class of polymeric ionic liquids (PILs), have been discussed to stimulate interest in the cellulose polymers and regenerated cellulose. It is herein shown how modified/functionalized cellulose employing 'green' procedures and strategies owe a great potential in producing new polymeric materials with improved mechanical strength and physicochemical properties against toxic chemical penetration. A subsection has dealt with toxic industrial chemicals and their potential treat in terrorism or bioterrorism. In this section, emphasis has been placed on the resistance of current protective materials to deal with this dreadful threat to world peace and development. Before providing an overview of the authors' outlook on the cellulose polymers and future research perspectives; key analytical features to supplement the understanding of both structural and morphological integrity and the mechanisms imparting enhanced manufacturability/wearability are briefly discussed.

With no intention to claim for any exhaustive review, the chapter has attempted to provide a broad vision on the importance of natural polymers (e.g. cellulose), their regeneration via 'green' routes, processing, blends via multilayering strategies that have been accomplished to date, to stimulate the increasing interests in cellulose research and development. Advantages and disadvantages co-exist, no matter what technologies one invents; however, it is worth noting that the novel "green" cellulose solvents as well as the physical dissolution/regeneration techniques open up a completely new avenue to create novel enticing materials with desired properties and functions. The current challenge to effectively deal with any incident of toxic chemical spillage while protecting the recue agents, as well as preventing environmental disasters constituted the backbone of motivations to undertake this project of writing this chapter.

In view of the literature sources reviewed in this chapter, further explorations in cellulose-based polymeric research could give a full perspective view of the preparation of the environment-friendly cellulose materials and their potential use in personal protective garments (PPGs) against nuclear and chemical agents. New cellulose functionalized materials or regenerated solvents modified for specific properties could also add hope to decontamination exercises when dealing with cases of usage of chemical war agents'. On a final note, information provided in this chapter is far from being conclusive, but simply a calling for more research interest and creativity into making the world a more peaceful environment if the clean-up of toxic chemicals or decontamination of nuclear and chemical agents were dealt with effectively using appropriate protective materials and clothing.

7.3. Recommendations

This chapter has limitations, and we apologize for any during readership. We humbly realize that not all that the audiences would expect could be provided in such a short chapter. The first conclusive recommendation would be further readership through the exhaustive list of references and citations. Second, research should continue to provide in the nearest future a validated permeation test cell(s) for many known toxic industrial chemicals and/or specific nuclear and chemical agents of immediate treats to mankind. Moreover, research is encouraged in the creation of new cellulose-based polymeric materials modified with organic/synthetic counterparts to acquire the most desired characteristics for a wider use in protective materials in general, and for use in the manufacture of personal protective garments (PPGs) against toxic chemicals in particular. Testing of current materials should be done in ambient temperatures to assess the impact of environmental factors such as humidity and air velocity on their permeation breakthrough rates, as well as penetration resistance for specific chemicals.

Experimental challenges during permeation studies lie in the difficulty to determine exactly the start of the chemical breakthrough. This is because of factors such as variations in the temperature settings, the evaporation of the agent exposed and the chemical bonding and interlocking within the material that interferes with the detection of breakthrough [136]. In some types of materials, the permeation time is directly proportional to the material thickness, whereas in some, the permeation time depends on the composition of the material. For example: aluminum foil and activated carbon cloth are materials that can be used for the manufacture of personal protective clothing and whose permeation time would depend much on their chemical or physical composition. In view of the above, it is highly recommended to use non-toxic chemicals for initial testing and experiments before an attempt to simulate 'real' situation experimental designs. Of interest would be a need to compile a world compendium of existing polymers prepared and tested or investigated under set conditions to serve future research purpose. A deeper understanding of the new or old mechanisms involved in the cellulose dissolution and regeneration/processing will not only make the regenerated cellulose materials more functional, but also more reproducible in view of their particular applications in this particular example.

The development of new analytical capabilities, as well as new method development is encouraged to explore the nano- and macrostructure of polymer integrity and durability. This could be specific, starting with cellulose-based polymers or regenerated solvents. There should be an increasing interest for investigations of the renewable resources with 'limitless' abundance or which can be non-competitively produced. Nanostructured polymers and nanocomposites have been gaining popularity in the last two decades due to their exciting bulk and surface properties. Indeed, nanocomposites based on polymeric matrices could yield new properties of particular interest. A number of new nanoparticles with extraordinary properties have encouraged the enlargement of the polymer markets. To name a few, carbon nanotubes, graphenes, as well as nanoclays, nanocellulose, metals and ceramics have created new and exciting possibilities. Unfortunately, the successful development of these new materials would strongly depend on the scale-up of reliable processing technologies. Organic–inorganic hybrids are also an interesting research area since many years mainly in the fields of tissue engineering and photoactive polymer nanocomposites. This area can be explored for the development of new interpenetrating networks through sol–gel processes.

Current industry trends in the processing of polymers focus on traditional synthetic polymers with only targeted composite blends or nanocomposites using some particular techniques and approaches, proprietary to industries. The modernization of processing infrastructures at transformation companies must be encouraged to involve research, including fundamental research, as well as collaborative research between industry-based research and development (R & D). Technical efforts are dedicated to the modification and updating of the current equipment and real innovations in new polymer creation and processing strategies are relevant in conquering most challenges raised in this chapter and elsewhere.

Acknowledgements

We are indebted to the Tshwane University of Technology for their financial support to this endeavor, courtesy of the Executive Dean of the Faculty of Science. T. E. Lefakane acknowledges Protechnik Laboratories, a Division of Armscor SOC Ltd, a division of the South African National Defence Research and Development Board, whose financial support granted to the mentioned author has permitted extensive laboratory investigations that generated interesting results upon which valuable insights into drafting this chapter have been partly sourced.

Author details

Peter P. Ndibewu[1*], Prince Ngobeni[1], Tina E. Lefakane[2] and Taki E. Netshiozwi[2]

*Address all correspondence to: NdibewuP@tut.ac.za

1 Tshwane University of Technology, Department of Chemistry, Arcadia Campus, Pretoria, Republic of South Africa

2 Protechnik Laboratories, Pretoria, Republic of South Africa

References

[1] Wang, S., Lu, A. & Zhang, L. 2015. Recent advances in regenerated cellulose materials, Progr Polym Sci., 5 – 38.

[2] De Van, T. V. & Kadla, J. 2013. Cellulose - Fundamental Aspects, Cellulose – Biomass Conversion, In: Vande Ven, T. & Kadla, J. InTech 978-953-51-1172-6http://www.intechopen.com/books/cellulose,fundamental-aspects [Accessed 10/09/2015].

[3] Pine, S. H., Hendrickson, J. B. Cram & Hammond, D. J. 1982. Organic Chemistry, International Student Edition, 4th ed., McGraw-Hill, New York, USA, p. 960.

[4] Zepnik, S., Hildebrand, T., Kabasci, S., Ra-dusch, H. J. & Wodke, T. 2013. Cellulose Acetate for Thermoplastic Foam Extrusion, Cellulose – Biomass Conversion, In:Van De Ven, T. & Kadla, J. (Eds.), InTech. ISBN: 978-953-51-1172-6 http://www.intechopen.com/books/cellulose-biomass-conversion, [Accessed 10/09/2015].

[5] Bogati, D. R. 2011. Cellulose Based Biochemicals and Their Applications, Saimaa University of Applied Sciences, Imatra Unit of Technology, Degree Programme in Chemical Engineering, p. 37.

[6] Biswas, A., Saha, B. C., Lawton, J. W., Shogren, R. L. & Willet, J. L. 2006. Process for obtaining cellulose acetate from agricultural by-products. Carbohydr Polym., 64:134-7.

[7] Hurter, R. W. 1998. Will Nonwoods Become an Important Fibre Resources for North America? World Wood Summit, Chicago, Illinois, USA, Summit Proceedings, August 31 - September 2.

[8] GAMA (Global Acetate Manufacturers Association at http://www.acetateweb.com). 2014. Cellulose acetate polymer, environmentally degradable material made from a modified natural polymer (cellulose), 1 – 6.

[9] Stamatopoulou, C., Klonos, P., Koutsoumpis, S., Gun'ko, V., Pissis, P. & Karabanova, L., J. 2014. Polym. Sci. Pol. Phys. 52,397–408, http://dx.doi.org/10.1002/ polb., 23427[Accessed 27/08/2015].

[10] Zhao, L., Zhou, Z. L., Guo, Z., Gibson, G., Brug, J. A., Lam, S., Pei, J. & Mao, S. S. 2012. J. Mater. Res., 27:639–52.

[11] Poologasundarampillai, G., Yu, B., Tsigkou, O., Valliant, E., Yue, S., Lee, P. D., Hamilton, R. W., Stevens, M. M., Kasuga, T. & Jones, J. R. 2012. Soft Matter., 8:4822–32, http://dx.doi.org/10.1039/C2SM00033D[Accessed 27/08/2015].

[12] Kuo, S. W. & Chang, F. C., Prog. 2011. Polym. Sci., 36:1649–96.

[13] Cai, J., Kimura, S., Wada, M., Kuga, S. & Zhang, L. 2008. Cellulose aero-gels from aqueous alkali hydroxide-urea solution, ChemSusChem, 1:149–54.

[14] Vendamme, R., Onoue, S. Y., Nakao, A. & Kunitake, T. 2006. Nat. Mater. 5:494–501, http://dx.doi.org/10.1038/nmat1655. [Accessed 27/08/2015].

[15] Valle, K., Belleville, P., Pereira, F. & Sanchez, C. 2006. Nat. Mater. 5:107–111, http://dx.doi.org/10.1038/nmat1570 [Accessed 27/08/2015].

[16] Novak, B. M. 1993. Adv. Mater. 5:422–433, http://dx.doi.org/10.1002/ adma.19930050603 [Accessed 27/08/2015].

[17] Hatori, S., Matsuzaki, R. & Todoroki, A. 2014. Compos. Sci. Technol., 92:9–15.

[18] Lin, W., Moon, K. S. & Wong, C. P. 2009. Adv. Mater. 21:2421–2424, http://dx.doi.org/10.1002/adma.200803548 [Accessed 25/08/2015].

[19] Alzari, V., Nuvoli, D., Scognamillo, S., Piccininii, M., Kenny, J. M., Malucelli & Mariani, A. 2011. J. Mater. Chem. 21:16544–16549, http:// dx.doi.org/10.1039/C1JM12104A [Accessed 25/08/2015].

[20] Alzari, V., Nuvoli, D., Scognamillo, S., Piccininii, M., Gioffredi, E., Malucelli, G., Marceddu, Sechi, M., Sanna, V. & Mariani, A. 2011. J. Mater. Chem. 21:8727–33, http://dx.doi.org/10.1039/C1JM11076D [Accessed 25/08/2015].

[21] Fang, Y., Chen, L., Wang, C. F. & Chen, S., 2010. J. Polym. Sci. Chem. 48:2170–77, http://dx.doi.org/10.1002/pola.23986 [Accessed 25/08/2015].

[22] Mariani, A., Alzari, V., Monticelli, O., Pojman, J. A. & Caria, G., 2007. J. Polym. Sci. Chem. 45:4514–21, http://dx.doi.org/10.1002/pola.22185[Accessed 25/08/2015].

[23] Mariani, A., Bidalii, S., Caria, G., Monticelli, O., Russo, S. & Kenny, J. M., 2007. J. Polym. Sci. Chem. 45:2204–11, http://dx.doi.org/10.1002/pola.21987 [Accessed 24/08/2015].

[24] Chen, S., Sui, J., Chen, L., John, A. & Pojman, J. A. 2005. J. Polym. Sci. Chem. 43:1670–80, http://dx.doi.org/10.1002/pola.20628 [Accessed 24/08/2015].

[25] Mittal, V. 2014. Polymer Nanocomposite Foams, CRC Press, Taylor & Francis Group, Boca Raton.

[26] Pielichowski, K., Njuguna, J., Michalowski, S. In: Pandey, K.R. Reddy, A. K. Mohanty & Misra, M. (Eds.), Handbook of Polymer Nanocomposites. Processing, Performance and Application, Volume A: Layered Silicates, Springer, Berlin, Heidelberg, 2014, pp. 453–79, ISBN: 4 978-3-642-38648-0.

[27] Longo, S., Mauro, M., Daniel, C., Galimberti, M., Guerra, G. 2013. Front. Chem., 13:1–28, http://dx.doi.org/10.3389/fchem.2013.00028 [Accessed 25/08/2015].

[28] Hernandez, R. A., Nogales, A., M. Sprung, M., Mijangos, C. & Ezquerra, T. A. 2014. J. Chem. Phys., 140,024909, http://dx.doi.org/10.1063/1.4861043 [Accessed 25/08/2015].

[29] Boday, D. J., Keng, P.Y., Muriithi, B., Pyun, J. & Loy, D. A. 2010. J. Mater. Chem., 20:6863–65, http://dx.doi.org/10.1039/C0JM01448F [Accessed 25/08/2015].

[30] Liu, X. & Mallapragada, S. K. 2011. Bioinspired Synthesis of Organic/Inorganic Nanocomposite Materials Mediated by Biomolecules, on Biomimetics, InTech. ISBN: 978-953-307-271-5, http://www.intechopen.com/books/on-biomimetics/bioinspired-synthesis-of-organicinorganic-nanocomposite-materials-mediated-by-biomolecules [Accessed 0/09/2015].

[31] Pollard, B., Muller, E. A., Hinrichs, K. & Raschke, M. 2014. Vibrational nano-spectroscopic imaging correlating structure with intermolecular coupling and dynamics, Nat. Commun., p. 3587.

[32] Mamani, J. B., Costa-Filho, A. J., Cornejo, D. R., Vieira, E. D. & Gamarra, L. F. 2013. Synthesis and characterization of magnetite nanoparticles coated with lauric acid, Mater Charact., 81:28–36.

[33] Aßmann, A. & Wendt, M. 2003. The identification of soft X-ray lines, Spectrochim., Acta, 58(4):711–16.

[34] Cross, W. M., Duke, E. F., Kellar, J. J. & Johnson. 2001. Scanning electron microscopy and other techniques to investigate low-strength concrete. Transport Research Record 1775. TRB, National Research Council, Washington, DC, p 10.

[35] Newbury, D. E. & Williams, D. B. 2000. The electronic microscope: the materials characterization tool of the millennium. Acta Mater., 48:323.

[36] Knoll, A., Magerle, R. & Krausch, G. 2001. Tapping mode atomic force microscopy on polymers: where is the true sample surface? Macromolecules, 34:4159–65.

[37] U.S. Department of Energy. 2014. AFM combined with infrared synchrotron light improves spatial resolution of infrared spectroscopy. http://www.azonano.com/news.aspx?newsID=30072 [Accessed 01/08/2015].

[38] Taubner, T., Keilmann, F. & Hillenbrand, R. 2005. Nanoscale-resolved subsurface imaging by scattering-type near-field optical microscopy. Opt. Express 13:8893–99.

[39] Levinson, N. M., Fried, S. D. & Boxer, S. G. 2012. Solvent-induced infrared frequency shifts in aromatic nitriles are quantitatively described by the vibrational Stark effect. J. Phys. Chem., B, 116:10470–6.

[40] Ando, H., Yoshizaki, T., Aoki, A. & Yamakawa, H. 1993. Mean-square electric dipole moment of oligo- and poly(methyl methacrylate)s in dilute solution, Macromolecules, 30:6199 –207.

[41] Hamanoue, K., Nakajima, K. & Amano, M. 1983. Matrix effects on absorption spectra of 9-cyano-10-anthryloxy anion and its use for probing a microscopic polarity of poly (methyl methacrylate). Polym.Photochem., 3:407–19.

[42] Rosu, D. M. et al. 2009. Molecular orientation in octanedithiol and hexadecanethiol monolayers on Ga-As and Au measured by infrared spectroscopic ellipsometry. Langmuir, 25:919–23.

[43] Huth, F. et al. 2012. Nano-FTIR absorption spectroscopy of molecular fingerprints at 20 nm spatial resolution. Nano Lett. 12:3973–3978.

[44] Decamp, M. F. et al. 2005. Amide I vibrational dynamics of N-methylacetamide in polar solvents: the role of electrostatic interactions. J. Phys. Chem. B., 109:11016–26.

[45] Raschke, M. B. et al. 2005. Apertureless near-field vibrational imaging of block-copolymer nanostructures with ultrahigh spatial resolution. Chemphyschem. 6:2197–203.

[46] Raschke, M. B. & Lienau, C. 2003. Apertureless near-field optical microscopy: tip-sample coupling in elastic light scattering. Appl. Phys. Lett., 83:5089–91.

[47] Xu, X. G., Rang, M., Craig, I. M. & Raschke, M. B. 2012. Pushing the sample-size limit of infrared Vibrational nanospectrsocopy: from monolayer toward single molecule sensitivity. J. Phys. Chem. Lett., 3:1836–41.

[48] Mcmurry, J. & Simanek, E. 2007. Fundamentals of Organic Chemistry, Thomson Brooks/Cole, Thomson Higher Education, 6th ed., Belmont, USA, p. 460.

[49] Peponi, L., Puglia, D., Torre, L., Valentini, L. & Kenny, J. M. 2014. Processing of nanostructured polymers and advanced polymeric based nanocomposites. Mater Sci and Eng R Rep.,, 85:1–46.

[50] Nelson, D. L. & Cox, M. M. 2013. Lehninger Principles of Biochemistry, 6th ed., (W H Freeman Publishers), New York, USA. p. 1158.

[51] Dumas, P. 2012. Biological and biomedical applications of synchrotron infrared microspectroscopy, Science at Synchrotron, Lecture Notes, Pretoria, S. Africa, 9-13 February 2009.

[52] Sepasal, 2014. Gossypium herbaceum. Survey of Economic Plants for Arid and Semi-Arid Lands (SEPASAL) database. Royal Botanic Gardens, Kew, Richmond, United Kingdom, http://www.kew.org/ ceb/sepasal/ [Accessed March 2014].

[53] Wakelyn, P. J., Bertoniere, N. R., French, A. D., Zeronian, S. H., Nevell, T. P., Thibodeaux, D. P., Blanchard, E. J, Calamari, T. A., Triplett, B. A., Bragg, C. K., Welch, C. M., Timpa, J. D., Franklin, W. E., Reinhardt, R. M. & T. L. Vigo. 1998. Cotton Fibres, Marcel Dekker, Inc., Jerusalem, Israel, Vol. 15.

[54] Nishiyama, Y., Sugiyayma, J., Chanzy, H. & Langan, P. 2003. Crystal structure and hydrogen bonding system in cellulose Iα from synchrotron X-ray and neutron fibrious diffraction. J. Am. Chem. Soc., 125:14300-6.

[55] Piotrowski, S. & Carus, M. 2011. Multi-criteria evaluation of lignocellulosic niche crops for use in biorefinery processes. Nova-Institut GmbH, Hürth, Germany, 5.

[56] Mhuka, V., Dube, S. & Nindi, M. M. 2013. Chemical, structural and thermal properties of Gonometa postica silk fibroin, a potential biomaterial. Int J of BiolMacromol., 52:305–11.

[57] Silverberg, M., Chemistry: The Molecular Nature of Matter and Change, Mosby Year Book Inc., St. Louis, Missouri, USA, pp. 626 - 30.

[58] Casey, J. P. 1960. Pulp and Paper: Chemistry and Chemical Technology, 2nd ed., Vol. I, Interscience Publishers, New York.

[59] Soo, C. M., Daud, W. R. W., Leh, C. P. 2012. "Cellulose acetate in paper". BioResources, 7:5333-45.

[60] Klemm, D., Heublein, B., Fink, H. P. & Bohn, A. 2005. Cellulose: fascinating biopolymer and sustainable raw material. Angew Chem Int Ed., 44,22:3358 –93.

[61] Updegraff, D. M. 1969. "Semimicro determination of cellulose in biological materials". Anall Biochem 32:420–4.

[62] Suumakki, A. & Westermak, U. 1996. Chemical characterization of the surface layers of unbleached pine [Pinus] and birch [Betula] kraft pulp fibers. J. Pulp Pap. Sci., 22 J43-7.

[63] Nishiyana, Y.; Langan, P. & Chanzy, H. 2002. "Crystal structure and hydrogen-bonding system in cellulose Iβ from synchrotron X-ray and neutron fiber diffraction". J Am Chem Soc., 124: 9074–82.

[64] Buell, P. & Girard, J. 2003. Chemistry Fundamentals: An Environmental Perspective, 2nd ed., Jones & Barlett Publishers, Sudbury, USA, pp. 417 –8.

[65] Barkalow, D. G., Rowell, R. M. & Yound, R. A. 1989. A new approach for the production of cellulose acetate: acetylation of mechanical pulp with subsequent isolation of cellulose acetate by differential solubility, Jl of Appl Polym Sci., 37:1009 -18.

[66] Fatehi, P. 2013. Production of Biofuels from Cellulose of Woody Biomass, Cellulose – Biomass Conversion In: Van De Ven, T. & Kadla, J. (Eds.), InTech. ISBN: 978-953-51-1172-6. http://www.intechopen.com/books/cellulose-biomass-conversion, [Accessed 10/09/2015].

[67] Golova, L., Makarov, I., Kuznetsova, L., Plotnikova, E. & Kulichikhin, V. 2013. In: Van De Ven, T. & Godbout, L. (Eds.), Structure - Properties Interrelationships in Multicomponent Solutions Based on Cellulose and Fibers Spun Therefrom, Cellulose - Fundamental Aspects, InTech. ISBN 978-953-51-1183-2. http://www.intechopen.com/books/cellulose-fundamental-aspects [Accessed 11/09/2015].

[68] AZOM. 2013. http://www.azom.com/article.aspx?ArticleID=1461 (Updated: 06/11/2013) [Accessed 04/03/2014].

[69] Kalia, S., Dufresne, A., Cherian, B. M., Kaith, B. S., Averous, L., Njuguna, J. & Nassiopoulos, E. 2011. Cellulose-based bio- and nanocomposites: a review. Int J of Polym Sci., 1 – 36.

[70] Gardner et al., 1994. Compostability of cellulose acetate films", J of Appl Polym Sci., 52:1477-88.

[71] Kroon-Batenburg, L. M. J., Kroon, J. & Nordholt, M. G. 1986. Polym. Commun., 27:290.

[72] El-Ashhab, F., Sheha, L., Abdalkhalek, M. & Khalaf, H. A. 2013. The influence of gamma irradiation on the intrinsic properties of cellulose acetate polymer. J Assoc Arab Univ Basic Appl Sci., 14:46–50.

[73] Berins, M. L. 2000. SPI Plastics Engineering Handbook of the Society of the Plastics Industry, Inc., Kluwer Academic Publishers, 5th ed., p. 272.

[74] Pereira, D. C., De Faria, D. L. A. & Constantino, V. R. L. 2006. Cu(II) hydroxy salts: characterization of layered compounds by vibrational spectroscopy. J Braz Chem Soc., 17:1651–7.

[75] Kudaibergenov, S. 2012. Novel macroporous amphoteric gels: preparation and characterization. Express Polym Lett., 6:346.

[76] Jenkins, A. D., Kratochvil, P., Stepto, R. F. T. & Suter, U. W. 1996. Glossary of basic terms in polymer science (IUPAC Recommendations 1996) Pure Appl Chem., 68

[77] De Gennes, P-G., 1979. Scaling Concepts in Polymer Physics. Cornell University Press. ISBN: 0-8014-1203-X [Accessed May 2015].

[78] Lefakane, T.E., Ndibewu, P.P. & Netshiozwi, T.E. 2015. Characterization, chemical phase identification and performance evaluation of omnipotent polymers used in the manufacturing of personal protective garments. Polym Res J., 9.

[79] Ndibewu, P. P, Mgangira, B. M. & Mccrincle, R. I. 2015. FT-IR spectroscopy combined infrared synchrotron light improves micro- and macroscopic study of pavement subgrade composite materials stabilized using polyelectrolytic polymers. Polym Res J.,

[80] Mecerresyes. 2011. Polymeric ionic liquids: broadening the properties and applications of polyelectrolytes. Prog in Polym Sci., 36:1629–48.

[81] Berillo, D. Elowsson, L., Kirsebom, H. 2012. Oxidized Dextran as crosslinker for chitosan cryogel scaffolds and formation of polyelectrolyte complexes between chitosan and gelatin. Macromol Biosci. 12:1090.

[82] Lee, G. S., Lee, Y-Jo; Y., K. Byung. 2001. Layer-by-layer assembly of zeolite crystals on glass with polyelectrolytes as ionic inkers". J Am Cheml Soc., 123: 9769–79.

[83] Small, L. 2002. Toxic industrial chemical: A future weapons of mass destruction threat. A thesis presented to the Faculty of the US Army Command and General Staff College in partial fulfilment of the requirements for the degree Master of Military Art and Science, Boston, Massachusetts, Boston University.

[84] Patnaik P, 2007. Comprehensive Guide to the Hazardous Properties of Chemical Substances John Wiley & Sons, New York, pp. 3-5. ISBN-13: 978-0-471-71458-3.

[85] Munro, N.B., Talmage, S.S., Griffin, G.D., Waters, L.V., Watson, A.P., King, J.F., & HAUSSCHILD, V. 1999. The sources, fate and toxicity of chemical warfare agent degradation products. Environ Health Perspect., 107: 933–74.

[86] Obryan, T.R., & Ross, R.H. 1988. Chemical scoring system for hazard and exposure identification. *J Toxicol Environ Health,* 1: 119 -34.

[87] Mannan, M.S., West, H.R., Krishna, K., Alddeb, A.A., Keren, N., Sarsf, R.F., Liu, Y., & Gentile, M. 2005. The Legacy of Bhopal: the impact over the last 20 years and future direction. J Loss Prev Process Ind., 18:218 - 24.

[88] Sohrabji, S. 2014. Bhopal thirty years later no relief still for gas plant disaster victims. In: *West India Newspaper,* 5 December 2014. Retrieve 24 August 2015.

[89] Ryan, F. 2015. Tianjin explosion: warehouse Handled toxic chemicals without licence- reports. In: The Guardian, 18 August 2015, Retrieved 22 August 2015.

[90] Kačer, P., Švrček, J., Syslová, K., Václavík, J., Pavlík, D., Červený, J., & Kuzma, M. 2012. Vapor phase hydrogen peroxide – method for decontamination of surfaces and working areas from organic pollutants, organic pollutants ten years after the Stockholm convention - environmental and analytical update. In: UZYN, T. (Ed.), InTech. ISBN: 978-953-307-917-2). http://www.intechopen.com/books/organic-pollutants-ten-years-after-the-stockholm-convention-environmentaland-analytical-update/vapor-phase-hydrogen-peroxide-method-for-decontamination-of-surfaces-and-workingareas-rom-organic [Accessed 11/09/2015].

[91] OPCW Secretariat 2015, "Invitation to apply for two chemical engineering internships for African countries to be held at Beijing University of Chemical Technology, China, Report No. S/1245/2015.

[92] OPCW Technical Secretariat Report s/1275/2015.

[93] Fatah, A.A., Barrett, J.A., Arcilesi, R.D., Ewing, K.E., Latting, C.H., & Helinski, M.S. 2000. Guide for the selection of chemical agent and toxic industrial material detection equipment for emergency first responders. US National Institute of Justice Guide 100-00, June, 6 – 14.

[94] Petr. 2012. Prepared from the International Task Force 25: 1998 Industrial Chemicals Final Report 1998.

[95] Jones, E. 2014. Terror weapons: The British experience of gas and its treatment in the First World War, In: War in History, 21: 355 –75.

[96] Almemar, Z. 2015. Uniting Against a Common Enemy: Formation of International Alliances Against CBRN Terrorism, CBRNe, Washington, DC, May.

[97] Sharm, N., & Kakkar R. 2013. Recent advancement on warfare agents/ metal oxides surface chemistry and their simulation. Adv Mater Lett., 4:508 –21.

[98] Yeshua, I. 1990. *Chemical warfare: A Family Defense Ma*nual. Centre for Educational Technology, Ramat Aviv, Israel.

[99] Hoenig, S.L. 2007. Compendium of Chemical Warfare Agents. Springer, New York. ISBN-10-0-387-34626-0.

[100] Schrader, G. 1942. (a) German Patent 720 577, 1942, (b) US Patent 2 336 302, 1942.

[101] Langford, R. E. 2004. Introduction to Chemical Weapons of Mass Destruction. John Wiley & Sons, Canada, pp. 34 - 36.

[102] Budiman, H. 2007. Analysis and identification of spiking chemical compounds related to chemical weapon convention in unknown water samples using gas chromatography and gas chromatography electron ionization mass spectrometry. Indo J Chem., 7: 297-302.

[103] Kaplan, D.E., & Marshall, A. 1996. The Cult at the End of the World: The Incredible Story of Aum, Arrow Books, London, pp. 141 - 252. ISBN: 0-09-972851-6.

[104] Larson, R. C., Metzeger, M. D., & Cahn, M. F. 2006. Responding to emergencies: lesson learned and the need for analysis. Interfaces, 36: 486–501.

[105] Brackett, D.W. 1996. Holy Terror Armageddon in Tokyo. Weatherhill, New York, pp. 121-2. ISBN: 0-8348-0353-4.

[106] Pangi, R. 2002. After the attack: the psychological consequences of terrorism.In: Perspectives on Preparedness, 7. August.

[107] Johnson, S. 2013. A high degree of confidence. In: CBRNe World, August, 8(4),: 26-29.

[108] Sellström, A., Cairns, S, & Barbeschi, M. 2013. United Nations mission to investigate allegations of the use of chemical weapons in the Syria Arab Republic. September.

[109] Robinson, J.P. 2013. Alleged use of chemical weapons in Syria. In: Harvard Sussex Program Occasional Paper. June 26, 1-42.

[110] Guillemin, J. 2005. Biological Weapons: From the Invention of State –Sponsored Programs to Contemporary Bioterrorism. Columbia University Press, New York, pp. 1 - 5.

[111] Bray, M. 2003. Defence against filoviruses used as biological weapons. Antiviral Res., 57:53 – 60.

[112] Krenzelok, E.P. 2003. Biological and Chemical Terrorism: A Pharmacy Preparedness Guide In: American Society of Health System Pharmacists. Pp. 1-7. ISBN 1-58528-048-8.

[113] Croddy, E.A. 2005. Encyclopedia of Weapons of Mass Destruction. California, ABC-CLIO, pp. 18-12. ISBN: 1-85109-495-4.

[114] Aken, J., & Hammond, E. 2003. Genetic engineering and biological weapons. EMBO Rep. S57–60. doi. 10.1038/sj.embor.embor860.

[115] Mahy, B.W. 2003. An overview on the use of a viral pathogen as a bioterrorism agent: why small pox. Antiviral Res., 57: 1 - 5.

[116] WHO. 2014. Ebola viruses. Technical Report, 7157 Ebola. October 2014.

[117] Okumura, S., Okumura, T., Ishimatsu, T., Miura, K., Maekawa, H., & Naito, T. 2005. Clinical review: Tokyo protecting the health care worker during a chemical mass casualty event: An important issue of continuing relevance. Acad Emerg Med., 9: 397 - 400.

[118] Okumura, T., Suzuki, K., Fukuda, A., Kohama, A., Takasu, N., Ishimatsu, S., & Hinohara, S. 1998. The Tokyo subway Sarin attack: disaster management. Part I. Community emergency response.Acad Emerg Med., 5:618 –7.

[119] WHO. 2014. Ebola and Marburg virus disease epidemics: preparedness, alert, control and evaluation. In: World Health Organization. August 2014.

[120] WHO. 2015. Health workers Ebola infections in Guinea, Liberia and Sierra Leone. A preliminary report In: World Health Organization 01 May 2015.

[121] DUPONT. 2005. Technical information on biological hazards. protection from contact with biological agents (brochure).

[122] Yuan, Z., Chen, H. & Zhang, J. 2008. Facile method to prepare lotus-leaf like superhydrophobic polyvinyl chloride film.Appl Surf Sci., 254:1593-8.

[123] Lodewyckx, P. 2006. Adsorption of Chemical Warfare Agents. Elsevier, New York.

[124] Tugara, U., Kendall, R.J., Singh, V., LALAGIRI M. & Ramkumar, S.S. 2012. Chapter 12: advances in materials for chemical, biological, radiological and nuclear (CBRN) protective clothing. In: SPARKS, E. (Eds.), Advances in Military Textiles and Personal Equipment. Woohead Publishing, Sawston.

[125] Truong, Q. & Wilusz, E. 2013. Chapter 13: advances in chemical and biological protective clothing. In: Smart Textiles for Protection. Woodhead Publishing, Cambridge.

[126] Gugliuzza, A. & Drioli, E. 2013. A review on membrane engineering for innovation in wearable fabrics and protective textiles. J Membr Sci., 444: 350-75.

[127] Boopathi, M., Singh, B. & Vijayaraghavan, R. 2008. A review on NBC body protective, Open Textile J., 1:1-8.

[128] Mao, N. 2014. High Performance Textiles for Protective Clothing, Elsevier Ltd.

[129] Sun, G., Worley, S.D. & Broughton, R.M. 2008. Chapter 12: self-decontaminating materials for chemical biological protective clothing. In: Wilusz, E. (Eds.), Military Textiles. Woohead Publishing, Cambridge.

[130] Wang, P. H., Tseng, I. L. & Hsu, S.H. 2011. Review: bioengineering approaches for guided peripheral nerve regeneration. J Med Biol Eng., 31:151–60.

[131] Valmikinathan, C. M. & Bellamkonda, R. V. 2011. Peripheral nerve regeneration. Compr Biomater., 5:421–34.

[132] Deumens, R., Bozkurt, A., Meek, M. F., Marcus, M. A. E., Joosten, E. A. & Weis, J. 2010. Repairing injured peripheral nerves: bridging the gap. Prog Neurobiol., 92:245–76.

[133] Gu, X., Ding, F., Yang, Y. & Liu, J. 2011. Construction of tissue engineered nerve grafts and their application in peripheral nerve regeneration. Prog Neurobiol., 93:204–30.

[134] Bellamkonda, R. V. 2006. Peripheral nerve regeneration: an opinion on channels, scaffolds and anisotropy. Biomaterials, 27:3515–8.

[135] Schlosshauer, B., Dressmann, L, Schaller, H. E. & Sinis, N. 2006. Synthetic nerve guide implants in humans: a comprehensive survey. Neurosurgery, 59:740–8.

[136] Joy, R. J. T. 1997. Historical aspects of medical defence against chemical warfare. In: SIDELL et al., Medical Aspects of Chemical Warfare and Biological Warfare. Office of the Surgeon General, Washington, DC., USA.

[137] Chao, K. P., Lai, J. S. & Lin, H. C. 2007. Comparison of permeation resistance of protective gloves to organic solvents with ISO, ASTM and EN standard methods. Polym Test., 26:1090–9.

[138] Henry, N.W. 2003. History of permeation test cells. In: NELSON, C. N. & HENRY, N. W. (Eds.), Performance of Protective Clothing: Issues and Priorities for the 21st Century: Seventh Volume.

[139] Huang, P. H. & Kacker, R. 2003. Repeatability and reproducibility standard deviations in the measurement of trace moisture generated using permeation tubes. J Res Natl Inst Stand Technol., 108:235–240.

[140] Anna, D. H., Zellers, E. T. & Sulewski, R. 1998. ASTM F739 method for testing the permeation resistance of protective clothing materials: critical analysis with proposed changes in procedure and test-cell design. Am Ind Hyg J., 59:544–56.

[141] Bromwich, D. 1998. The validation of a permeation cell for testing chemical protective clothing. Am Ind Hyg J., 59:842–51.

[142] Hung, W. C., Lin, L. H, Tsen, W. C, Shie, H. S., Chiu, H. S., Thomas, C. K., Yang, T. C. K., & Chen, C. C. 2015. Permeation of biological compounds through porous poly(L-lactic acid) (PLLA) microtube array membranes (MTAMs). Eur Polym J., 6:166–73.

[143] Hudson, T. W. & Evans, G. R. 1999. Engineering strategies for peripheral nerve repair. Clin Plast Surg., 26:617–28.

[144] Verwolf, A., Sherry, O., Farwell, S. O., Zhongtao, C. & Smith. 2009. Performance-based design of permeation test cells for reliable evaluation of chemical protective materials. Polym Test., 28:437–5.

[145] Creely, K. S. & Cherrie, J, W. 2001. A novel method of assessing the effectiveness of protective gloves—results from a pilot study. Ann Occup Hyg., 45:137–43.

[146] Guo, C., Stone, J. & Stahr, H. M. & Shelley, M. 2001. Effects of exposure time, material type, and granular pesticide on glove contamination. Arch Environ Contam Toxicol., 41:529–36.

[147] Krzeminska, S. & Szczecinska, K. 2001. Proposal for a method for testing resistance of clothing and gloves to penetration by pesticides, Ann Agric Environ Med., 8:145–50.

[148] Ehntholt, D. J., Cerundolo, D. L., Bodek, I., Schwope, A. D., Royer, M. D. & Nielsen, A. P. 1990. A test method for the evaluation of protective glove materials used in agricultural pesticide operations. Am Ind Hyg Assoc J., 51:462–8.

[149] Archibald, B. A., Solomon, K. R. & Stephenson, G. R. 1995. Estimation of pesticide exposure to greenhouse applicators using video imaging and other assessment techniques. Am Ind Hyg Assoc J., 56:226–35.

[150] Chao, K. P., Wang, P., Cheng, C. P. & Tang, P.Y. 2011. Assessment of skin exposure to N, Ndimethylformamide and methyl ethylketone through chemical protective gloves and decontamination of gloves and decontamination of gloves for reuse puposes. Sci Total Environ., 409:1024–32.

[151] Nielsen, J. B. & Andersen, H. R. 2001. Dermal in vitro penetration of methiocarb, paclobutrazol and pirimicarb - influence of nonylphenylethoxylate and protective gloves. Environ Health Perspect, 129–32.

[152] Gao, P., El-Ayouby, N. & Wassell, J. T. 2005. Change in permeation parameters and the decontamination efficacy of three chemical protective gloves after repeated exposures to solvents and thermal decontaminations. Am J Ind Med., 47:131–43.

Chapter 8

A Novel Application of Oceanic Biopolymers — Strategic Regulation of Polymer Characteristics for Membrane Technology in Separation Engineering

Keita Kashima, Ryuhei Nomoto and Masanao Imai

Additional information is available at the end of the chapter

http://dx.doi.org/10.5772/61998

Abstract

Membranes prepared from oceanic biopolymers have a high potential in membrane separation processes and water purification. It is anticipated to result in more biocompatible and lower-cost materials compared with artificial polymers. This chapter describes the excellent performance of oceanic biopolymer membranes in separation engineering and the regulation factors controlling membrane properties. In particular, chitosan and alginate were picked up as intelligent membrane materials to provide the promised molecular size recognition and other membrane properties. Future prospective strategies for a simple methodology for preparing stable membranes from oceanic biopolymers and the development of selective separation processing were reviewed.

Keywords: Membrane, oceanic biopolymer, chitosan, alginate, mechanical strength, mass transfer characteristics

1. Introduction

1.1. Overview of oceanic biopolymers in membrane separation technology

Membrane separation technology has been applied in various fields, such as the chemical industry, food production, pharmaceutical products, environmental sciences, and water purification, because it can be operated without heating and residual toxicity. Its energy cost was lower than conventional thermal separation technologies [1–4]. It can be employed for broad components by selecting optimum membranes.

Development of oceanic biopolymer membranes such as chitosan and alginate has exponentially increased due to increases in the demand for biocompatibility, environment adaptability, renewability, and selective separation ability [5–6]. In the field of membrane science, most membranes have been made from petroleum-based polymers used as raw materials for the membrane body. Petroleum-based synthetic polymers such as polyethylene, polyamide, polyimide, and polysulfonate have been used as the main materials of the membrane body for a long time. The practical use of biopolymer membranes was less than synthetic polymer membranes, because controlling the physical and chemical properties of a biopolymer produced from natural bioresources was difficult in practical membrane applications. Cellulose-based membranes have been extensively used in practical applications since the development of the anisotropic cellulose acetate membrane in 1963 by Loeb and Sourirajan [7]. Applications of biopolymer materials have increased recently for membrane sciences [8]. Recent studies of biopolymer membranes are summarized in Table 1 referred to previous literatures [9-22]. In the past five years, oceanic biopolymers have attracted attention for their chemical functional ability. Investigations of oceanic biopolymer membranes, especially chitosan and alginate, for various applications have drastically increased.

Authors	Year	Membrane materials	Application	Coments	Ref.
Torres et al.	2010	eggshell	biomedical application	regenerative medicine	9
Kashima et al.	2010	calcium alginate	membrane separation	mass transfer characteristics	10
Kashima and Imai	2011	calcium alginate	membrane separation	mass transfer characteristics	11
Michalak and Mucha	2012	polylactid acid, dibutyrylchitin, chitosan	controlled release	drug delivery system	12
Wu and Imai	2013	pullulan and κ-carrageenan	membrane separation	dyes removal	13
Lakra et al.	2013	chitosan or cellulose brended polyethersulfone	membrane separation	food industry	14
Moraes et al.	2013	chitosan and alginate	membrane separation	environment science	15
Nomoto and Imai	2014	chitosan	membrane separation	mass transfer characteristics	16
Ma et al.	2014	chitosan	biomedical application	guided bone regeneration	17
Uragami et al.	2015	chitosan	membrane separation	pervaporative dehydration	18
Alias et al.	2015	chitosan blended SiO_2	membrane separation	proton battery	19
Puspasari et al.	2015	cellulose	membrane separation	nanofiltration	20
Livazovic et al.	2015	cellulose	membrane separation	water treatment	21
Zhang et al.	2015	calcium alginate and polyacrylamide	membrane separation	nanofiltration	22

Kashima, Nomoto and Imai 2015.

Table 1. Recent investigations of biopolymer membrane and applications.

The common aim of membrane separation technology is to separate a target component from a mixture with the aid of permeation through the membrane, while rejecting other components. The key characteristics for practical use are as follows: mechanical strength, permeation flux, diffusivity of molecule, and micro-/nanostructure. This chapter describes the dominant factors in controlling important properties for practical membrane separation applications for oceanic polymer membrane, especially those of chitosan and alginate.

2. Chitosan membrane

Chitosan is a well-known sustainable and biocompatible oceanic biomaterial. It has been attracted in creating new polymer materials for broad application due to its nontoxicity, biocompatibility, and biodegradability [5]. Chitosan is refined by removing an acetyl group from chitin, which is mainly produced in oceanic bioresources such as the shells of crabs and shrimps. Chitin is the second most abundant natural polymer in nature, after cellulose [23]. Chitosan is generally discarded as industrial waste around the world [24]. It is strongly expected to be useful as a biocompatible and reactive material for making functional gel membranes. Recent studies of various chitosan membranes are summarized in Table 2 and Table 3 referred to previous literatures [5-6, 12, 15-19, 24-46].

2.1. Chemical composition and membrane formation

Chitosan is a heteropolymer obtained by alkaline deacetylation of chitin at the C-2 position as shown in Figure 1 (a) and (b). Chitosan is generally defined by a deacetylation degree (*DD* [%]) of 60–100%. The deacetylation degree is estimated as follows:

$$DD = \frac{n_{\text{GS}}}{n_{\text{GS}} + n_{\text{AGS}}} \times 100 \quad (1)$$

Here, n_{GS} (n_{AGS}) is the molar number of glucosamine residues (acetylglucosamine residues) in the molecular chain. Chitosan has great potential for chemical modification because its molecular chain has a rich amino group and a hydroxyl group in the glucosamine residues [29].

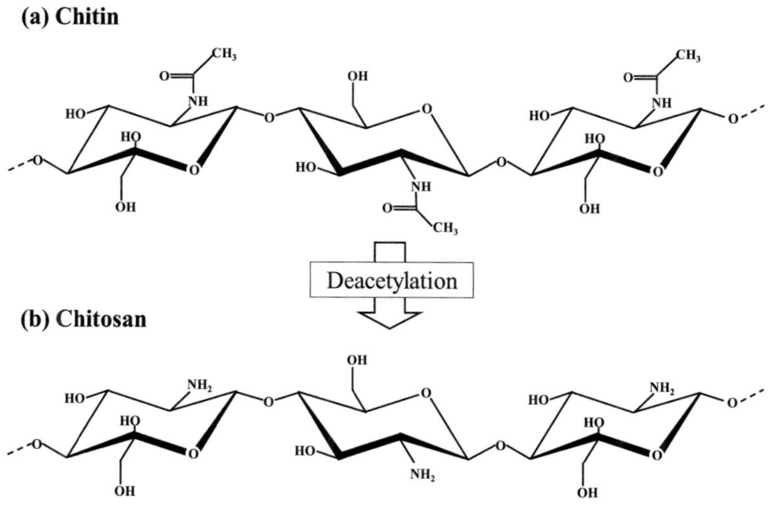

Figure 1. Chemical structure of chitosan showing β-1,4-D-glycoside linkage. (a) Chitin molecular chain consisting of poly-β-1,4-D-*N*-acetylglucosamine residues, (b) chitosan molecular chain consisting of poly-β-1,4-D-glucosamine residues.

The beneficial effects of chitosan are as a dietary fiber, such as its inhibition of fat digestibility [47] and its reduction of cholesterols [48]. The biodegradability and biocompatibility of chitosan are also suitable for biomedical applications [49].

Authors	Year	Membrane material	Application	Raguration factor	Controled properties of membrane	Coments	Ref.
Michalak and Mucha	2012	chitosan blended polylactid acid and dibutyrylchitin	controlled release	-	-	drug delivery system	12
Tasselli et al.	2013	chitosan	membrane separation	concentration of glutataldehyde as a cross-linker	mechanical strength, swelling ratio, adsorption of dye	hollow fiber membrane	25
Moraes et al.	2013	alginate blended chitosan	membrane separation	blended chitosan	swelling ratio, thickness, adsorption of herbicide	water treatment	15
Sun et al.	2014	chitosan	antibacterial medium	concentration of gallic acid	antibacterial activity, mechanical strength	food packaging film	26
Wanichapichart et al.	2014	chitosan blended polysulfone	membrane separation	mixing ratio of chitosan and polysulfone	water permeation coefficient, rejection of fluoride and cadmium	water treatment	27
Ma et al.	2014	chitosan containing chitin whisker	biomedical application	containing ratio of chitin whisker	mechanical strength, antibacterial activity	wound dressing	5
Bai et al.	2014	chitosan containing nanotube bearing sulfonate polyelectrolyte	fuel cell	containing ratio of nanotube	water uptake, proton conductivity, methanol permeability	direct methanol fuel cell	24
Nomoto and Imai	2014	chitosan	membrane separation	concentration of NaOH for neutralization of chitosan	mechanical strength, porosity, water permeation flux, effective diffusion coefficient	mass transfer characteristics	16
Shenvi et al.	2014	chitosan supported on PPEES membrane	membrane separation	pH of tripolyphosphate solution as a cross-linker	water flux, water uptake, salts rejection	water treatment	28
Han et al.	2014	alginate and chitosan	biomedical application	-	-	wound dressing	29
Karim et al.	2014	chitosan containing cellulose nanocrystals	membrane separation	with or without cross-linking by gluteraldehyde	dyes rejection, porosity, mechanical strength	dyes purification	30
Dudek et al.	2014	chitosan containing ion oxide nanoparticles	membrane separation	containing ratio of iron oxide, additive sulphuric acid or glutaraldehyde	water permeation coefficient of water and ethanol	pervaporative dehydration	31
Kumar et al.	2014	polysulfone blended chitosan and functionalized chitosan	membrane separation	functionalization of chitosan molecular chain	heavy metal rejection, atifouling activity	heavy metal purification	32
Bueno et al.	2014	chitosan and alginate based	biomedical application	containing ratio of surfactant (Pluronic F68)	porosity, mechanical strength	wound dressing	6
Ma et al.	2014	chitosan	biomedical application	cross-linking with tripolyphosphate	mechanical strength, biocompatibility	guided bone regeneration	17

PES: polyethersulfone, PDMCHEA:poly (2-methacryloyloxy ethyl trimethylammonium chloride-co-2-hydroxyethyl acrylate), PPEES: poly (1, 4-phenylene ether ether sulfone),
PVP: polyvinylpyrrolidone

Kashima, Nomoto and Imai 2015.

Table 2. Recent investigations of chitosan membrane (2012–2014).

http://dx.doi.org/10.5772/61998

Chitosan is dissolved in an acid aqueous solution, such as acetic acid. To form a water-insoluble chitosan membrane, the acetic acid has to be neutralized by basic components. The authors have reported preparation of chitosan membrane using casting methods [16]. For example, 4 g of chitosan was dissolved in 198 mL of 1.7 mol·L^{-1} acetic acid aqueous solution. The solution was mixed magnetically for 12 h at room temperature to obtain a mature solution. Some insoluble matter was removed by vacuum filtration using a filter paper (Grade No. 1, ADVANTEC, Japan). The concentration of the chitosan casting solution was 0.02 $g_{\text{-chitosan}}\cdot mL_{\text{-chitosan casting solution}}^{-1}$. Ten grams of the casting solution was dispensed in a glass Petri dish (inner diameter 7.75 cm) and dried in a thermostatic chamber at 333 K for 24 h. A dried chitosan membrane formed on the glass Petri dish. A 20 mL of sodium hydroxide (NaOH) aqueous solution was directly introduced onto the dried chitosan membrane. The chitosan membrane was continuously immersed for 3 h in NaOH solution for neutralization. The concentration of the supplied NaOH ranged from 0.1 to 5 mol·L^{-1}. In this case, the stoichiometric equivalent concentration of NaOH was estimated to be 0.83 mol·L^{-1}. After neutralization, the swollen membrane was easily separated from the glass Petri dish and was washed fully with pure water to remove excess NaOH.

2.2. Dominant role of the acid–base neutralization in chitosan membrane characteristics

The authors previously reported the dominant role of the acid–base neutralization process in forming chitosan membranes for controlling some membrane properties [16].

2.2.1. Mechanical strength

The mechanical strength, maximum tensile stress (Figure 2a), and maximum strain (Figure 2b) at membrane rupture presented bell-shaped curves with peaks when the NaOH concentration increased. For an 81% *DD* chitosan membrane, the maximum stress and the maximum strain reached maximum values of 8.6 MPa and 80.5%, respectively, in a membrane prepared with 1 mol·L^{-1} NaOH. This concentration was almost the same as the stoichiometric equivalent concentration (0.83 mol·L^{-1}). In contrast, in a 98% *DD* chitosan membrane, the maximum stress and the maximum strain increase to 17.9 MPa and 134% when the membrane was prepared with 2 mol·L^{-1} NaOH.

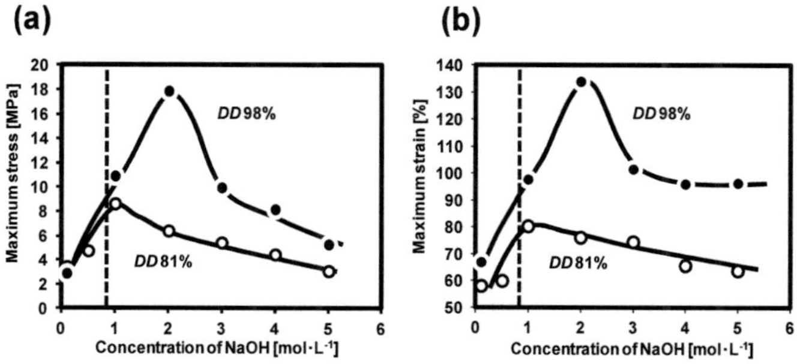

Figure 2. Effect of NaOH concentration in neutralization on the maximum stress (a) and maximum strain (b) at membrane rupture in reference [16]. The dashed line represents the stoichiometric neutralization concentration of NaOH (0.83 mol·L^{-1}).

Authors	Year	Membrane materials	Application	Raguration factors	Controled properties	Coments	Ref.
Sajjan et al.	2015	GTMAC grafted chitosan	membrane separation	mixing ration of chitosan	hydrophilicity, water/isopropanaol selectivity	pervaporative dehydration	33
Bibi et al.	2015	chitosan containing carbon nanotube	membrane separation	with or without carbon nanotube	pore size, swelling ratio, adsorption of naphthalene	water treatment	34
Ji et al.	2015	chitosan blended PDMCHEA	membrane separation	mixing ratio of PDMCHEA	water flux, mechanical strength, hydrophilicity	nanofiltration membrane	35
Uragami et al.	2015	chitosan	membrane separation	molecular weight and degree of deacetylation	permeation flux, water/ethanol selectivity	pervaporative dehydration	18
Premakshi et al.	2015	chitosan	membrane separation	additive concentration of zeolite	mechanical strength, permeation flux, water/isopropanol selectivity	pervaporative dehydration	36
Santamaria et al.	2015	chitosan	fuel cell	concentration of chitosan, ratio of chitosan and phosphotungstic acid	membrane thickness, conductivity	hydrogen-oxygen fuel cell	37
Zhu et al.	2015	PES and PVP blended chitosan and montmorillonite	membrane separation	additive concentration	water flux, salts rejection	dyes purification	38
Li et al.	2015	chitosan	biomedical application	concentration of genipin as a cross-linker	swelling ratio, mechanical strength	cultivation medium of epithelial cells	39
Alias et al.	2015	chitosan blended SiO_2	membrane separation	mixing ration of chitosan and SiO_2	pore size, water uptake	proton battery	19
Hegab et al.	2015	chitosan	membrane separation	surface modification by using graphen oxide	hydrophilicity, NaCl rejection, water flux, antifouling activity	reverse osmosis	40
Zhang et al.	2015	poly (vinyl alcohol) blended chitosan	biomedical application	blending ratio of chitosan	mechanical strength, water vapor, oxygen permeability, antibacterial activity	wound dressing	41
Panda et al.	2015	chitosan coated iron-oxide-polyacrylonitrile	membrane separation	chitosan coating, concentration of Fe_3O_4	permeability, MWCO, pore size, hydrophilicity, mechanical strength	remove of humic acid	42
Antunes et al.	2015	chitosan and arginine	biomedical application	-	-	wound regeneration	43
Song et al.,	2015	chitosan containing carbon nanotube	antibacterial medium	blending ratio among chitosan, cationic chitosan, carbon nanotubes and silicon cuouple agent	hydrophobicity, mechanical strength	surface modification	44
Wang and He	2015	chitosan blended copolymer of polyacrylamide/polystyrene	fuel cell	blending ratio of chitosan and polyacrylamide/polystyrene	methanol permeability, wataer uptake, mecanical strength	methanol alkaline fuel cells	45
Malini et al.	2015	chitosan contained ZnO	antibacterial medium	combined with ZnO nano particules	antibacterial activity, antifouling activity	nanocomposite membrane	46

GTMAC: Glycidyltrimethylammonium chloride, PES: polyethersulfone, PDMCHEA:poly (2-methacryloyloxy ethyl trimethylammonium chloride-co-2-hydroxyethyl acrylate), PVP: polyvinylpyrrolidone

Kashima, Nomoto and Imai 2015.

Table 3. Most recent investigations of chitosan membrane (2015).

http://dx.doi.org/10.5772/61998

2.2.2. *Water permeation flux*

Figure 3 demonstrates that the water permeation flux and the volumetric void fraction of the swollen membrane were almost linearly correlated. The NaOH concentration in the neutralization process enhanced both the water permeation flux and the void fraction in the membrane. Neutralization using a high NaOH concentration weakens hydrogen bonding between chitosan polymer chains. The void fraction in the swollen membrane can assumed to be the volume of a water permeation channel occupying the membrane. This result suggests that the membrane structure involving water permeation was dominantly regulated by the NaOH concentration during neutralization.

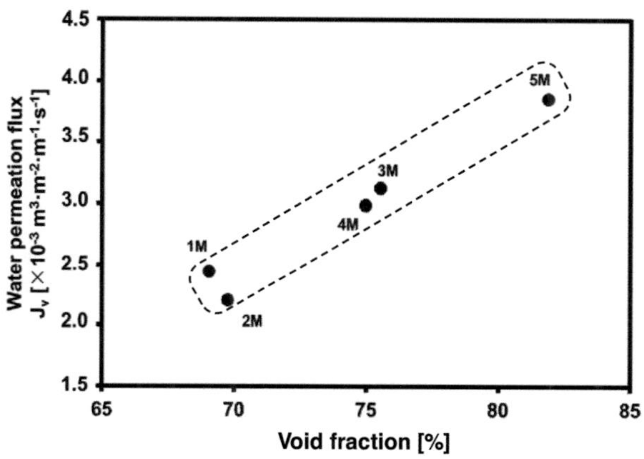

Figure 3. Correlation between the water permeation flux and the void fraction of 81% deacetylation degree chitosan membrane in reference [16]. The concentrations of NaOH used in the neutralization process are indicated at adjacent to keys, respectively.

2.2.3. *Mass transfer characteristics*

The effective diffusion coefficient of the model components was evaluated based on the mass transfer flux. The membrane was sandwiched between twin mass transfer cells. The feed solution contained the model components (urea 60 Da, Methyl Orange 327 Da, Rose Bengal 1017 Da, and Sirius Red 1373 Da) at the desired concentrations, and the stripping solution was deionized water. The overall mass transfer coefficient K_{OL} was evaluated from the mass transfer flux according to Eq. (2).

$$\ln\left(1-\frac{2C_s}{C_{fi}}\right)=-\frac{2AK_{OL}}{V}t \tag{2}$$

Here, C_s and C_{fi} are the concentrations of the stripping and initial feed solutions, A is the mass transfer area of the membrane, V is the volume of aqueous phase in each transfer cell, and t is the operation time.

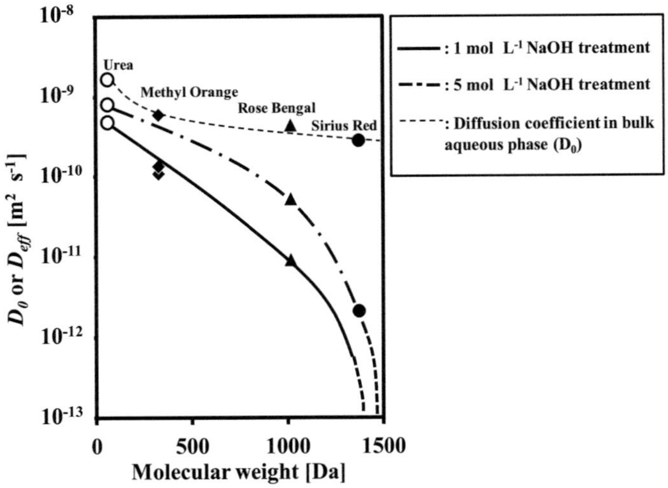

Figure 4. Change of effective diffusion coefficient (D_{eff}) versus molecular weight of tested components (81% deacetylation) [16].

The overall mass transfer coefficient K_{OL} included the film mass transfer resistances (k_{L1}^{-1} and k_{L2}^{-1}) and the membrane mass transfer coefficient (k_m) as seen in Eq. (3).

$$K_{OL}^{-1} = k_{L1}^{-1} + k_m^{-1} + k_{L2}^{-1} \tag{3}$$

The aqueous phases in the mass transfer cells were sufficiently stirred to attain a fully developed turbulent condition ($Re > 10^4$). The film mass transfer resistances (k_{L1}^{-1} andk_{L2}^{-1}) within the overall mass transfer resistance (K_{OL}^{-1}) were ignored under fully turbulent conditions. This directly indicated the membrane mass transfer coefficient (k_m). The effective diffusion coefficient was evaluated from k_m using Eq. (4):

$$k_m = \frac{D_{eff}}{\ell} \tag{4}$$

The initial membrane thickness in the swollen state (ℓ) was measured with a micrometer (Mitutoyo Corporation, Kanagawa, Japan).

Figure 4 depicts the change in the effective diffusion coefficient in the chitosan membrane with the molecular weight of the tested components for 81% *DD*. The effective diffusion coefficient of a chitosan membrane prepared from NaOH of 1mol·L^{-1} changed dramatically at a molecular size corresponding to Rose Bengal (1017 Da). This result suggested that the size distribution of the mass transfer channel in the membrane was monodisperse and similar to the molecular size of Rose Bengal.

The effective diffusion coefficient of a membrane prepared from 5 mol·L^{-1} NaOH greatly increased relative to a membrane prepared from 1 mol·L^{-1}. The molecular diffusion channel

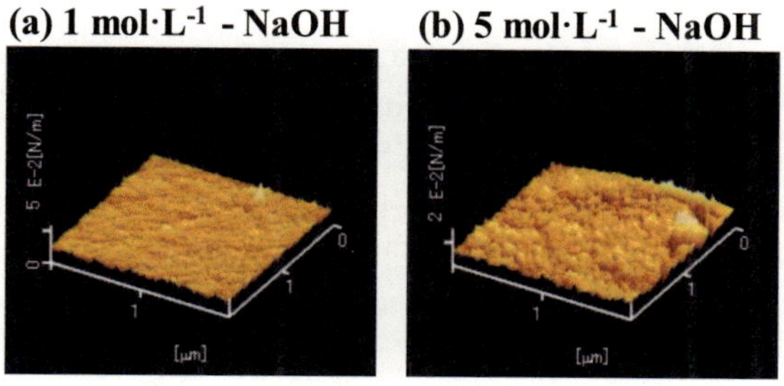

Figure 5. Scanning probe microscope photographs of chitosan membrane surface. (a) 1 mol L^{-1} NaOH, (b) 5 mol L^{-1} NaOH [16].

formed by the polymer networks became enlarged due to the higher NaOH concentration employed.

The dominant role of the acid–base neutralization process in chitosan membranes was revealed to control membrane properties involving mechanical strength, the water permeation flux, and the effective diffusion coefficient.

2.2.4. Morphology of chitosan membrane

Figure 5 presents scanning probe microscope (SPM) photographs of chitosan membrane surfaces. The chitosan membrane made from 1 mol·L^{-1} NaOH had a very smooth surface. In contrast, the membrane made from 5 mol·L^{-1} NaOH had a somewhat rough surface.

2.3. Other regulation factors

2.3.1. Deacetylation degree

The deacetylation degree (*DD*) of a chitosan membrane can be stoichiometrically controlled by an acetic anhydride additive. Effects of the deacetylation degree on membrane properties, such as water permeation flux, have been found [50]. The water permeation flux of a membrane prepared from 99.2% *DD* chitosan was remarkably enhanced about 100-fold compared with a 76.5% *DD* chitosan membrane. The water permeation flux increased exponentially with increasing deacetylation degree from 81.8% to 92.2% [51].

Increasing the deacetylation degree enhances the surface hydrophilicity of the membrane surface because the acetamido groups on the chitosan membrane are converted into amino groups. This advances the effective formation of hydrogen bonds between the amino and hydroxyl groups and between the amino groups in the chitosan molecules, thus resulting in a dense molecular chain in the membrane [18].

2.3.2. Molecular weights of chitosan

Chitosan membranes prepared by Uragami and coworkers from different molecular weights (13–201 kDa) were tested during pervaporation dehydration of an ethanol aqueous solution.

The permeation flux decreased with increasing molecular weight from 13 kDa to 90 kDa, then increased from 90 kDa to 201 kDa. In contrast, the water permeation selectivity from the ethanol aqueous solution increased remarkably with increasing molecular weight up to 90 kDa. Over 90 kDa, the selectivity decreased. These results were understood to indicate that the 90 kDa molecular weight effectively formed entanglements and thus produced the strongest hydrogen bonds between the molecular chains of chitosan [18], although the mechanical strength increased with increasing average molecular weight of chitosan [52].

2.3.3. *Addition of a cross-linker*

Basically, chitosan can be formed into an insoluble membrane by acid–base neutralization in the casting solution without using a cross-linker. However, the addition of a cross-linker to improve membrane characteristics was carefully investigated.

The most prominent effect of an additive cross-linker is improved mechanical strength. Figure 6 depicts the change of mechanical strength of chitosan membrane (maximum stress vs. maximum strain at membrane rupture) with increasing amount of cross-linker. In a common trend, increasing the amount of cross-linker forms a rigid membrane. The addition of genipin as a cross-linker increased the maximum stress and reduced the maximum strain at membrane rupture [39]. Increasing added amounts of glutaraldehyde, methanol with ethylene glycol diglycidyl ether, and gallic acid also increased the mechanical strength. However, excess addition of cross-linker formed rigid and fragile structures on the chitosan membrane [25–26, 53].

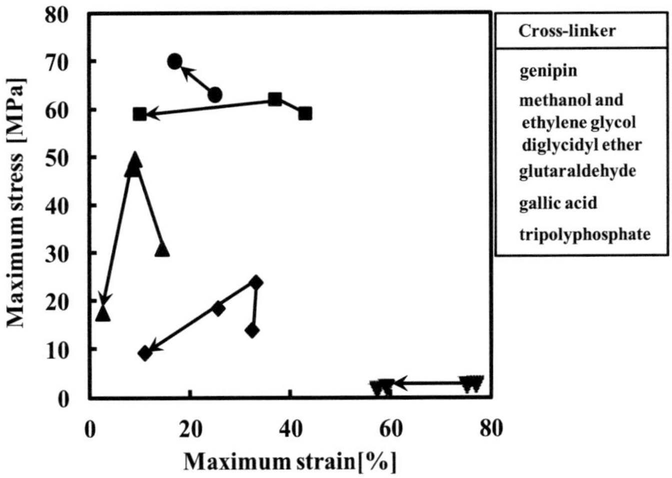

Figure 6. Change of mechanical strength of chitosan membranes with increasing amount of cross-linker. ●: chitosan membrane cross-linked with genipin [54]. ■: chitosan membrane cross-linked with methanol and ethylene glycol diglycidyl ether [53]. ▲: chitosan membrane cross-linked with glutaraldehyde [25]. ◆: chitosan membrane cross-linked with gallic acid [26], ▼: chitosan cross-linked with tripolyphosphate [55]. Arrows indicate increasing additive amount of each cross-linker.

http://dx.doi.org/10.5772/61998

Other researchers have reported that cross-linking with sodium tripolyphosphate enhanced the mechanical strength of chitosan membrane [17]. From another viewpoint, Shenvi and coworkers determined the effect of the pH value of the cross-linking solution on the membrane properties. A chitosan membrane cross-linked by tripolyphosphate solution at pH 5 realized a higher rejection rate of NaCl and $MgSO_4$ than a membrane cross-linked at pH 9 [28].

The enhancement of pervaporative dehydration ability of water–alcohol mixtures by adding cross-linker has been frequently reported [56]. A chitosan membrane cross-linked with toluene-2,4-diisocyanate exhibited high pervaporative dehydration ability with an isopropanol aqueous solution in which the selectivity of water α was performed as 472 [57]. The selectivity of water α was calculated using the following equation:

$$\alpha = \frac{y\,(1-x)}{x\,(1-y)} \tag{5}$$

Here, y is the permeate weight fraction of water and x is the feed weight fraction of water.

2.3.4. *Hybrid approaches with other materials*

Some hybrid approaches using chitosan and other materials have been proposed to overcome problems in practical use [35]. To make a porous structure on a chitosan membrane, SiO_2, a porogen agent, was mixed with chitosan cast solution and removed from the membrane. The average pore size reached a maximum of 8.5 μm when the mixing ratio of chitosan and SiO_2 was 1:2 [19].

Chitosan membranes are frequently combined with nanomaterial to form a composite medium. A chitosan membrane composited with a multi-walled carbon nanotube was modified by adding perfluorooctanesulfonyl fluoride (PFOSF). The added PFOSF enhanced antibacterial activity due to its remarkable hydrophobicity. The mechanical strength increased with increasing multi-walled carbon nanotubes [44]. Chitosan composited with ZnO nanoparticles was also investigated for antibacterial activity [46]. Chitosan membrane has a good potential as a composited nanomaterial medium for biologically based nanotechnology.

2.3.5. *Chemical modification of the chitosan molecular chain*

Chitosan has great potential for chemical modification. Functionalized chitosan is often used as the main membrane body, coating agent, or additive agent for functionalization of surface modifications. Graphene oxide functionalizing a chitosan membrane as a surface modification agent on a commercial polyamide membrane has been investigated. The water flux and NaCl rejection of the chitosan membrane were increased by the modification, due to the formation of a dense, thin layer of graphene-oxide-functionalized chitosan, which is better than natural chitosan membrane [40]. Kumar and coworkers prepared various chitosan membranes: a natural chitosan blended with polysulfone membrane, an *N*-succinylchitosan blended with polysulfone membrane, and an *N*-propylphosphonyl chitosan blended with polysulfone membrane. These membranes were applied to a heavy metal purification process. *N*-succinylchitosan blended with polysulfone membrane sufficiently purified Cu, Ni, and Cd [32].

3. Alginate membrane

Alginate is abundantly and sustainably produced by marine biological resources, especially brown seaweed. It has been widely applied in the food industry [58] and as a thickener [59], a suspending agent [60], an emulsion stabilizer [61], a gelling agent [62], and a film-forming agent [63]. In addition, alginate was continuously developed as useful materials for biomedical applications, especially for controlled delivery of drugs and other biologically active compounds and for the encapsulation of cells [64]. In recent pioneering works, alginate has developed as membrane material with excellent molecular selectivity for water-soluble components [65]. Recent studies of various alginate membranes are listed in Table 4 and Table 5 referred to previous literatures [10-11, 15, 22, 66-95].

3.1. Chemical composition and membrane formation of alginate

The molecular chain of alginate is constructed of a block copolymer of β-D-mannuronate (Figure 7a) and α-L-guluronate (Figure 7b) [96]. These two uronates construct a polymeric block in an alginate polymer chain with the following three types of block: homopolymeric blocks of α-L-guluronate (GG blocks), blocks with an alternating sequence in varying proportions of guluronate and mannuronate (MG blocks), and homopolymeric blocks of β-D-mannuronate (MM blocks) [97]. GG blocks chelate alkaline earth-metal ions because of the spatial arrangement of the pyranose ring and the hydroxyl oxygen atoms, and thus create a much stronger interaction than MM blocks and MG blocks [98–99].

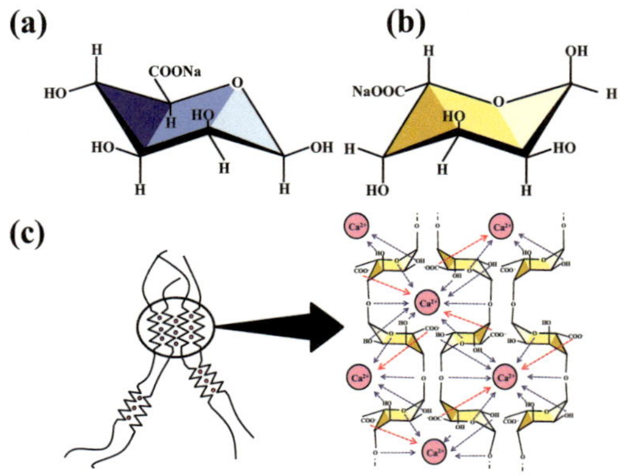

Figure 7. Residues of alginate molecular chain. (a) β-D-mannuronate. (b) α-L-guluronate. (c) Chelation of homopolymeric blocks of α-L-guluronate junction with calcium ions. Alginate was cross-linked by divalent cations according to the "Egg-box" model.

http://dx.doi.org/10.5772/61998

Authors	Year	Membrane material	Application	Raguration factor	Controled properties of membrane	Coments	Ref.
Toti and Aminabhavi	2004	sodium alginate	membrane separation	containing ratio of PVA and PEG	swelling ratio, water permeability, water/isopropanol selectivity	pervaporative dehydration	66
Kalyani et al.	2008	sodium alginate	membrane separation	with or without phosphoric acid as a cross-linker	swelling ratio, water permeation flux, water/ethanol selectivity	pervaporative dehydration	67
Kashima and Imai	2010	calcium alginate	membrane separation	concentration of $CaCl_2$ as a cross-linker	mechanical strength, swelling ratio, water permeation flux, effective diffusion coefficient	mass transfer characteristics	10
Saraswathi et al.	2011	calcium alginate blended dextrin	membrane separation	blending ratio of alginte and dextrin	swelling ratio, water permeability, water/isopropanol selectivity	pervaporative dehydration	68
Li et al.	2011	sodium alginate blended gelatin	membrane separation	blending ratio of gelatin	crystallinity, water uptake, water permeability, water/ C_3H_6 selectivity	hollow fiber	69
Taskin et al.	2011	sodium alginate photografted with itaconic acid	material science	grafting time and temperture, concentration of benzophenone	swelling ratio, hydrophilicity	hydrophilic material	70
Chen et al.	2011	sodium alginate containing humic acid	membrane separation	containing humic acid	adsorption ability of Cu(II)	adsorption of Cu(II)	71
Kashima and Imai	2011	calcium alginate	membrane separation	homopolymeric blocks of α-L-guluronate	mechanical strength void fraction, mass transfer characteristics	mass transfer characteristics	11
Li et al.	2012	sodium alginate blended PVA and PSf	membrane separation	blending ratio of PVA	crystallinity, water uptake, water vapor flux	dehumidification	72
Nigiz et al.	2012	sodium alginate containing zeolite	membrane separation	containing ratio of zeolite	water permeation flux, water/ethanol selectivity	pervaporative dehydration	73
Sajjan et al.	2013	sodium alginate blended chitosan-wrapped MWCNTs	membrane separation	blending ratio of chitosan-wrapped MWCNTs	swelling ratio, permation flux, water/isopropanol selectivity	pervaporative dehydration	74
Flynn et al.	2012	sodium alginate blended PVA	membrane separation	membrane thickness	permeation flux, water/ethanol selectivity	pervaporative dehydration	75
Shi et al.	2013	calcium alginate containing $CaCO_3$, PAA and PUA	drug release	containing with or without $CaCO_3$, PAA and PUA	swelling ratio, drug release ability	biomineralization	76
Chen et al.	2013	sodium algiante containing APTEOS	membrane separation	containing APTEOS	adsorption ability of Cr(III)	adsorption of Cr(III)	77
Zhu et al.	2013	sodium alginate	membrane separation	concentration of sodium tartrate as a cross-linker	CO_2 permeability, CO_2/N_2 selectivity	CO_2 separation	78
Adoor et al.	2013	sodium alginate containing phosphotungstic acid	membrane separation	containing ratio of phosphotungstic acid	water permeation flux, water/ethanol and water/isopropanaol selectivity	pervaporative dehydration	79
Moraes et al.	2013	alginate blended chitosan	membrane separation	blended chitosan	swelling ratio, thickness, adsorption of herbicide	water treatment	15

APTEOS: 3-aminopropyltriethoxysilane, MWCNTs: multiwalled cabon nanotubes, PAA: polyacrylic acid, PAN: polyacrylonitrile, PEG: polyethylene glycol, PSf: polysulfone, PUA: poly (urethane-amine), PVA: poly (vinyl alcohol),

Kashima, Nomoto and Imai 2015.

Table 4. Recent investigations of alginate membrane (2004–2013).

Authors	Year	Membrane material	Application	Raguration factor	Controled properties of membrane	Coments	Ref.
Zhang et al.	2014	sodium alginate supported on polypropylene	membrane separation	kind of cross-linking metal ion	mechanical strength, hydrophilicity, permeation flux, water/acetic acid selectivity	pervaporative dehydration	80
Gao et al.	2014	sodium alginate and hyaluronic acid supported on PAN	membrane separation	concentration of Ca^{2+} as a cross-linker, coating sequence of alginate and hyaluronic acid	swelling ratio, hydrophilicity, permeation flux, water/ethanol selectivity	pervaporative dehydration	81
Kuila and Ray	2014	sodium alginate blended PVA	membrane separation	blending ratio of alginate and PVA	mechanical strength, permeation flux, water/dioxane selectivity	pervaporative dehydration	82
Kuila and Ray	2014	sodium alginate blended CMC	membrane separation	blending ratio of alginate and CMC	permeation flux, benzene/cyclohexan selectivity	pervaporative separation	83
Yang et al.	2014	sodium alginate blended PVA	fuel cell	blending ratio of sodium alginate and PVA, time of crosslinking with glutaraldehyde	swelling ratio, ionic conductivity, methanol permeability	direct methanol fuel cell	84
Cabello et al.	2014	alginate blended carrageenan	fuel cell	blending ratio of alginate and carrageenan	mechanical strength, water uptake, proton conductivity, methanol permeation flux	direct methanol fuel cell	85
Cao et al.	2014	sodium alginate containing graphene oxides	membrane separation	containing ratio of graphene oxides	hydrophilicity, crystallinity, swelling ratio, permeation flux, water/ethanol selectivity	pervaporative dehydration	86
Yoo and Ghosh	2014	calcium alginate	biomedical application	concentration of alginate and $CaCl_2$, cross-linking time	membrane thickness	microporous hollow fiber	87
Kamoun et al.	2015	sodium alginate blended PVA	biomedical application	blending ratio of sodium alginate and PVA	water uptake, mechanical strength, release property, adsorption of BSA	wound dressing	88
Li et al.	2015	calcium alginate	biomedical application	concentration of ethanol as solvent of $CaCl_2$	swelling ratio, thickness, mechanical strength	pharmaceutical products	89
Zhang et al.	2015	calcium alginate blended polyacrylamide	membrane separation	with or without polyacrylamide	water permeation flux, adsorption of BSA (antifouling activity), rejection of Brilliant Blue	nanofiltration	22
Saozo et al.	2015	calcium alginate	edible film	concentration of alginate and calcium gluconolactate, heat treatment	thickness, mechanical strength, color	membrane preparation	90
Zhao et al.	2015	calcium alginate	membrane separation	concentration of sodium alginate, modification by PEG	water permeation flux, rejection of PEG and methylene blue, antifouling activity	ultrafiltration, nanofiltration	91
Jie et al.	2015	calcium alginate containing carbon nanotubes	membrane separation	containing ratio of carbon nanotubes	mechanical strength, antifouling activity, water permeation flux, dye rejection	nanofiltration	92
Kirdponpattara et al	2015	alginate blended cellulose	biomedical application	blending alginate and bacterial cellulose	pore size, void fraction, mechanical strength, biocompatibility	tissue enginering	93
Shao et al.	2015	sodium alginate blended cellulose loaded with AgSD	biomedical application	loaded ratio of AgSD	swelling ratio, antibacterial ability	wound dressing	94
Uragami et al.	2015	alginate blended DNA	membrane separation	kind of cross-linking metal ion	hydrophilicity, swelling ratio, density, water/ethanol selectivity	pervaporative dehydration	95

AgSD: silver sulfadiazine, CMC: carboxymethyl cellulose, DNA: deoxyribonucleic acid, PEG: polyethylene glycol, PVA: poly (vinyl alcohol),

Kashima, Nomoto and Imai 2015

Table 5. Most recent investigations of alginate membrane (2014–2015).

Sodium alginate easily forms a cross-link with the presence of divalent cations such as Ca^{2+}, resulting in a highly compacted and dense gel network. GG blocks are constructed mainly of a cross-linked zone with Ca^{2+}. This is called an "Egg-box junction" (Figure 7c), where the ions were assimilated to "Eggs" [100].

Sodium alginate forms a cross-link in the presence of divalent cations such as Ca^{2+}, resulting in a highly compacted gel network. Homopolymeric blocks of α-L-guluronate are constructed mainly of a cross-linked zone with divalent cations. This section describes the impact of this kind of divalent cations and the mass fraction of homopolymeric blocks of α-L-guluronate in an alginate polymer chain (F_{GG}).

The authors previously reported the preparation of alginate membrane described below [10–11]. A 20 mL sodium alginate aqueous solution (10 $g \cdot L^{-1}$) was placed in a Petri dish. The solution was gradually dried at a mild temperature in a desiccator (298 K) or an electrical dryer (303 K) for 24 h to prevent heat degradation. A dried, thin film of sodium alginate was obtained on the Petri dish.

An electrolyte aqueous solution ($CaCl_2$, $SrCl_2$, and $BaCl_2$) was directly introduced onto the dried, thin sodium alginate film as a source of divalent cation for cross-linking. The concentration range of the electrolyte aqueous solutions was 0.1–1.0 $mol \cdot L^{-1}$. A stable alginate membrane was quickly formed in the Petri dish at room temperature. After 20 min, the prepared swollen membrane was spontaneously separated from the surface of the Petri dish. The membrane remained in the electrolyte aqueous solution for further 20 min. The membrane was totally immersed in the electrolyte aqueous solution for 40 min. The prepared membrane was repeatedly washed with pure water to remove excess electrolyte, and then stored in pure water.

3.2. Effect of homopolymeric block of α-L-guluronate (F_{GG})

The authors previously reported that membrane properties are evidently controlled by the mass fraction (F_{GG}) [11, 65, 101]. At first, the mass of GG blocks in the actual alginate polymer chain was determined by acid hydrolysis combined with Bitter–Muir's carbazole sulfuric acid method [102]. The actual alginate was separated into three fractions (GG block, MM block, and MG block) by the acid-hydrolysis method. Acid-hydrolysis protocols were employed according to a previously published method [11, 103]. Their obtained fractions were then determined by Bitter–Muir's carbazole sulfuric acid method using an optical density of 530 nm (UV-1200, Shimadzu, Kyoto, Japan). The calibration curve for mannuronic acid lactone was obtained and used as the standard component of uronic acid. These methods produced good intensity and accuracy of coloration [104].

The mass of the GG block (W_{GG}) was obtained from a fractional solution of the GG block by acid hydrolysis. The masses of the MM block (W_{MM}) and MG block (W_{MG}) were determined in the same way. The mass fraction of the homopolymeric blocks of α-L-guluronate (F_{GG}) was then calculated using the following formula:

$$F_{GG} = \frac{W_{GG}}{W_{GG} + W_{MM} + W_{MG}} \quad (6)$$

F_{GG} was therefore shown to be a key factor in regulating membrane properties. In this study, two sodium alginates were examined, F_{GG} 0.18 and F_{GG} 0.56. Calcium alginate membrane prepared from five types of sodium alginate (F_{GG} 0.18, 026, 0.35, 0.45, and 0.56) were regulated by mixing with two types of sodium alginate (F_{GG} 0.18 and 0.56).

3.2.1. Mechanical strength

The effect of F_{GG} on mechanical strength is depicted in Figure 8. The mechanical strength was determined based on the membrane deformation. The maximum stress was defined at the rupture of the membrane. This value increased with increasing F_{GG}. In contrast, the maximum strain decreased with increasing F_{GG}. The mechanical properties of the calcium alginate membrane changed from elastic to plastic transfer, depending on F_{GG}.

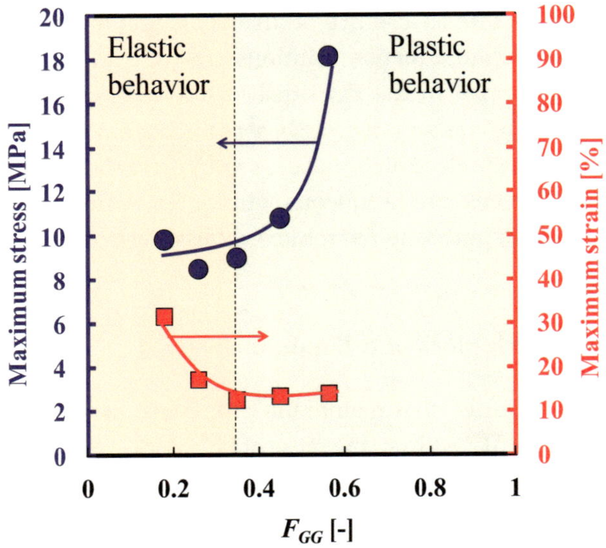

Figure 8. Effect of F_{GG} on mechanical strength of calcium alginate membrane [11].

3.2.2. Mass transfer characteristics

An alginate membrane performs superior molecular size recognition on low-molecular-weight components from 60 Da to 600Da [10]. Figure 9 demonstrates that the effective diffusion coefficient in the membrane (D_{eff}) of saccharides (glucose, G, 180Da; maltose, M, 324 Da; raffinose, R, 504 Da) can be efficiently changed by regulating F_{GG}. The effective diffusion coefficient in the membrane prepared from F_{GG} 0.56 alginate was proportional to the –4.6th power of the molecular volume. The diffusion coefficient in bulk aqueous phase (D_0) was inversely proportional to the 0.6th power of the molecular volume [105]. In contrast, the effective diffusion coefficient in the membrane prepared from F_{GG} 0.18 alginate was proportional to the –2.9th power of the molecular volume. The higher F_{GG} membrane exhibited more sensitive molecular size recognition. "Egg-box junction" zones constructed from calcium ions and homopolymeric blocks of α-L-guluronate dominantly regulated the mass transfer mechanism of the alginate membrane.

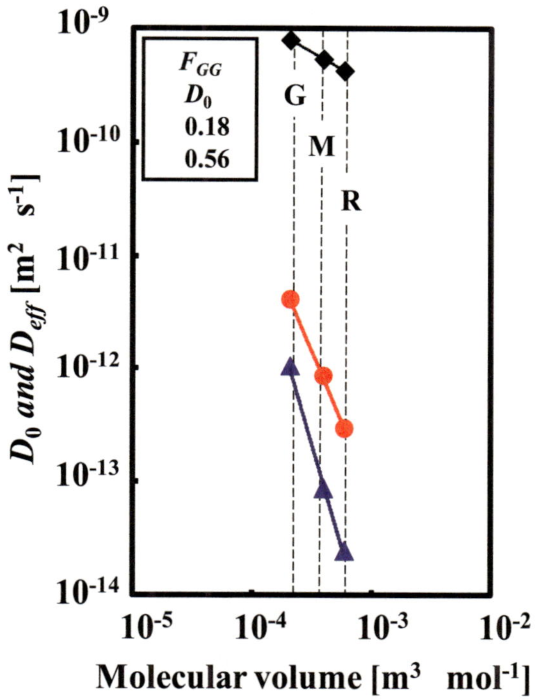

Figure 9. The effective diffusion coefficient changed remarkably in our experimental range from 180 Da to 504 Da. The symbol ◆ indicates the diffusion coefficient in bulk aqueous solution. Temperature: 303 K. Agitation rate: 14.2 s^{-1} [65].

The effect of F_{GG} on the mass transfer characteristics was evaluated using a permeation test with urea aqueous solution as a typical molecule model. The effective diffusion coefficient (D_{eff}) of urea in the membrane decreased with increasing F_{GG}. Figure 10 illustrates the correlation of the volumetric void fraction of the swollen membrane with the ratio between the effective diffusion coefficient of urea in the membrane and the diffusion coefficient in bulk aqueous phase (D_0). Both the effective diffusion coefficient and the volumetric water fraction were restrained by increasing F_{GG}. The strong dependency of the effective diffusion coefficient of urea on the void fraction contributes to good understanding of the formation of a mass transfer channels in the alginate membrane. These channels are speculated as being monodisperse in a similar size with urea molecule. The mass transfer channel was governed by the mass fraction of homopolymeric blocks of α-L-guluronate (F_{GG}).

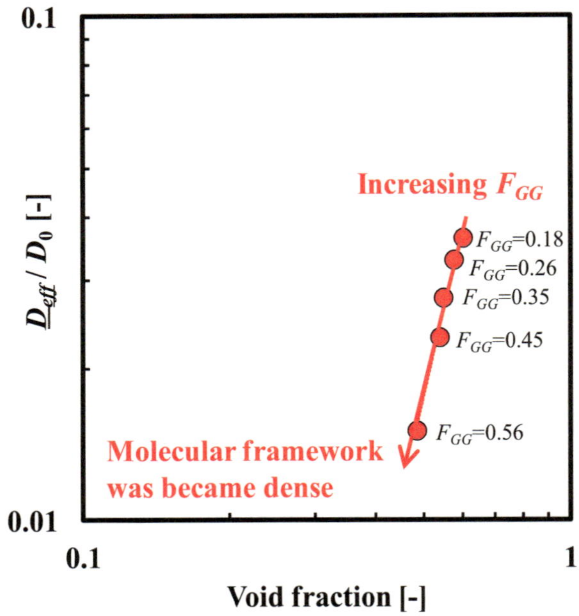

Figure 10. Correlation of the volumetric void fraction of the swollen membrane with the effective diffusion coefficient of urea [11].

3.2.3. *Morphology of the alginate membrane*

Figure 11 presents SPM photographs of the calcium alginate membrane surface. The distribution of membrane asperity clearly decays with increasing F_{GG}. Especially in higher F_{GG} conditions, many of the GG blocks formed a higher population density of Egg-box junction chelating with Ca^{2+}, and the atomic force of the polymer networks was uniform.

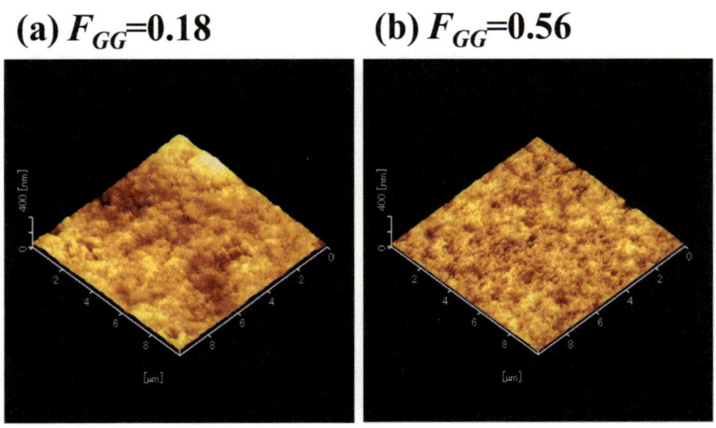

Figure 11. Scanning probe microscope photographs of calcium alginate membrane surface. (a) F_{GG} = 0.18, (b) F_{GG} = 0.56.

http://dx.doi.org/10.5772/61998

3.3. Effect of cross-linking divalent cations

The authors prepared a stable alginate membrane cross-linked with $CaCl_2$, $SrCl_2$, and $BaCl_2$. These membranes are sufficiently stable in aqueous phase to apply for practical use.

3.3.1. *Water permeability*

The water permeation coefficient was evaluated based on the pure water permeation flux as presented in Figure 12. It was decreased logarithmically with increasing concentration of cross-linker. This suggests that the number of permeation channels and/or the size of the permeation channels decreased with increasing concentration of cross-linking ions. The effect of cross-linker concentration on water permeability appeared strongly in the $BaCl_2$ used, suggesting that Ba^{2+} structures have a much stronger interaction in α-L-guluronate block. Binding of divalent cation to the three kind of copolymer blocks have been investigated [106–107]. The following orders of binding strength indicated by sign of equality were reported:

GG block: Ba > Sr> Ca >> Mg
MM block: Ba > Sr ≈ Ca ≈ Mg
MG block: Ba ≈ Sr ≈ Ca ≈ Mg

Hence, the alginate membrane cross-linked with $BaCl_2$ constructed a dense polymer network due to strong bonding between alginate molecular chain and Ba^{2+}.

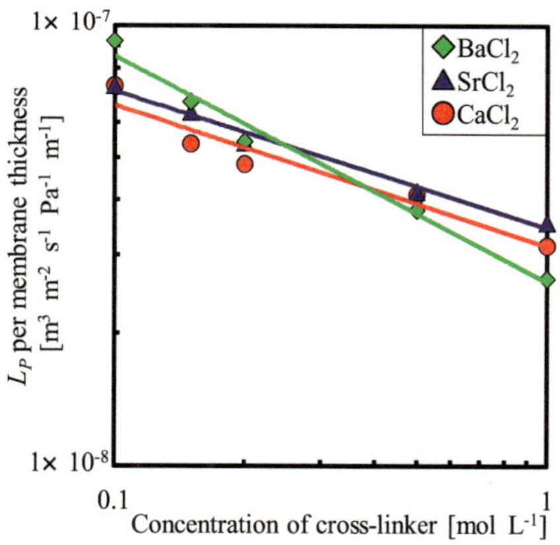

Figure 12. Effect of concentration of cross-linking electrolyte solution on water permeation coefficient.

3.3.2. *Volumetric water fraction*

The volumetric water content of a swollen membrane can be assumed to an indicator of the void fraction of the membrane structure [108]. Figure 13 illustrates the effect of the concentration of electrolyte solution on the void fraction of the membrane. The void fraction decreased with increasing concentration of the cross-linking electrolyte solution. An alginate membrane cross-linked with $BaCl_2$ was highly densified by the provided Ba^{2+}.

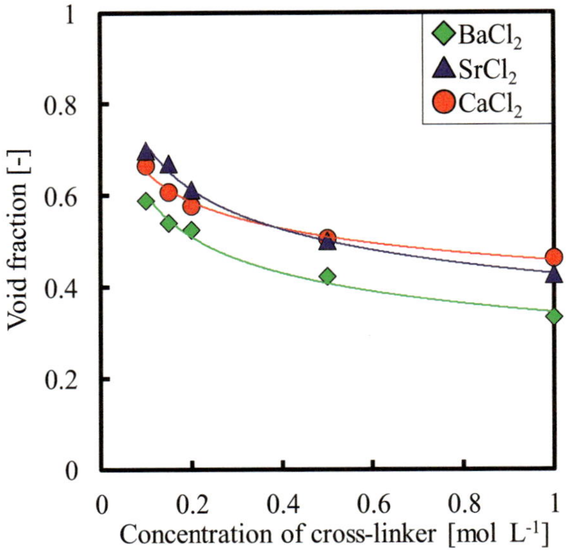

Figure 13. Effect of concentration of cross-linking electrolyte solution on void fraction of swollen alginate membrane.

3.4. Other regulation factors for controlling membrane properties

3.4.1. Addition of cross-linker other than metal ions

Conventional cross-linkers forming a polymer membrane were also examined to realize the alginate membrane. Glutaraldehyde is most commonly employed as a cross-linker to form sodium alginate membranes [83–84]. These membranes cross-linked with glutaraldehyde were developed for use in organic dehydration by pervaporation [66, 79, 82]. A sodium alginate membrane cross-linked with phosphoric acid was prepared for pervaporative dehydration of ethanol aqueous solution [67]. This resulted in 3.5×10^{-2} kg·m^{-2}·h^{-1} of permeation flux and 2182 of selectivity as defined by Eq. (4). A sodium alginate membrane cross-linked by sodium tartrate was characterized by CO_2 capture from CO_2/N_2 [78].

3.4.2. Hybrids with other polymers or materials

Many efforts have been made to enhance the performance of the alginate membrane by blending it with different hydrophilic polymers. An alginate-blended DNA membrane cross-linked with Mg^{2+} or Ca^{2+} was investigated with regard to the permeation flux of ethanol aqueous solution for pervaporation [95]. The permeation flux and selectivity of a calcium alginate membrane with DNA in pervaporative dehydration for an ethanol aqueous solution were measured as 1.2×10^{-2} kg·m^{-2}·h^{-1} and 5500, respectively. In contrast, a magnesium alginate membrane with DNA exhibited a permeation flux and selectivity of 1.2×10^{-2} kg·m^{-2}·h^{-1} and 6500.

Hybrid membranes of sodium alginate and dextrin cross-linked with glutaraldehyde had a permeation flux of 9.65×10^{-2} kg·m^{-2}·h^{-1} permeation flux and a selectivity of 8991 in pervaporative dehydration for an isopropanol aqueous solution [68].

Alginate membranes are frequently combined with nanomaterials to form a composite medium. A calcium alginate membrane containing multi-walled carbon nanotubes was prepared as a new nanofiltration membrane. It had high mechanical strength, antifouling ability, and high rejection of small organic molecules (Congo Red, 697Da) [92].

4. Conclusion

Strategic regulation of an oceanic biopolymer membrane to control its characteristics in membrane separation technology was demonstrated. An oceanic biopolymer chitosan membrane can be easily prepared by casting chitosan in dilute aqueous organic acids and neutralizing it with an alkaline aqueous solution. The dominant role of neutralization for chitosan membrane involves the mechanical strength, the permeation flux, and the mass transfer characteristics. Other regulating factors, such as the deacetylation degree, the average molecular weight, and the addition of a cross-linker were presented.

A calcium alginate membrane performs superior molecular size recognition on low-molecular-weight components from 60 Da to 600 Da due to a dense polymer network consisting of homopolymeric blocks of α-L-guluronate and calcium ions. The kind of cross-linking ion used is also able to control the membrane properties. Barium ion performed a stronger cross-linker in the alginate membrane rather than Ca^{2+} and Sr^{2+}, and a highly complex structure was formed. The mass fraction of homopolymeric blocks of α-L-guluronate and cross-linking metal ions were impact factors in regulating the mass transfer characteristics, water permeability, and mechanical strength.

These oceanic polymers are biodegradable, biocompatible, environmentally friendly, stable, and easily available from renewable agricultural resources. In the future, oceanic biopolymer membranes should be developed as an alternative to artificial polymer membranes.

Nomenclature

A Area of membrane surface [m^2]

C_{fi} Initial concentration in the feed solution [mol·L^{-1}]

C_s Concentration in the stripping solution [mol·L^{-1}]

D_0 Diffusion coefficient in a bulk aqueous phase estimated from Wilke–Chang's empirical formula [m^2·s^{-1}]

DD Deacetylation degree [%]

D_{eff} Effective diffusion coefficient [$m^2 \cdot s^{-1}$]

F_{GG} Mass fraction of GG block inalginate determined by Eq. (5) [–]

K_{OL} Overall mass transfer coefficient [$m \cdot s^{-1}$]

K_{OL}^{-1} Overall mass transfer resistance [$(m \cdot s^{-1})^{-1}$]

k_m Membrane mass transfer coefficient [$m \cdot s^{-1}$]

k_{L1} Film mass transfer coefficient in the feed phase [$m \cdot s^{-1}$]

k_{L2} Film mass transfer coefficient in the stripping phase [$m \cdot s^{-1}$]

ℓ Membrane thickness in the initial state [m]

n_{AGS} Molar number of acetylglucosamine residues in polymer chain of chitin or chitosan [mol]

n_{GS} Molar number of glucosamine residues in polymer chain of chitin or chitosan [mol]

Re Reynolds number [–]

t Time [s]

V Volume of aqueous solution in the transfer cell [m^3]

W_{GG} Mass of GG block in sodium alginate [kg]

W_{MG} Mass of MG block in sodium alginate [kg]

W_{MM} Mass of MM block in sodium alginate [kg]

x Feed weight fraction of water in pervaporation [–]

y Permeate weight fraction of water in pervaporation [–]

α separation factor on pervaporative dehydration defined by Eq. (4) [–]

Author details

Keita Kashima[1], Ryuhei Nomoto[2] and Masanao Imai[2*]

*Address all correspondence to: XLT05104@nifty.com

1 Department of Materials Chemistry and Bioengineering,National Institute of Technology, Oyama College, Oyama, Tochigi, Japan

2 Course in Bioresource Utilization Sciences,Graduate School of Bioresource Sciences, Nihon University, Fujisawa, Kanagawa, Japan

References

[1] Xiong B, Richard T L, Kumar M. Integrated acidogenic digestion and carboxylic acid separation by nanofiltration membranes for the lignocellulosic carboxylate platform. J. Membrane Science. 2015;489: 275–283. DOI: 10.1016/j.memsci.2015.04.022

[2] Xiao T, Wang P, Yang X, Cai X, Lu J. Fabrication and characterization of novel asymmetric polyvinylidene fluoride (PVDF) membranes by the nonsolvent thermally induced phase separation (NTIPS) method for membrane distillation applications. J. Membrane Science. 2015;489: 160–174.DOI: 10.1016/j.memsci.2015.03.081

[3] Sawada S, Ursino C, Galiano F, Simone S, Drioli E, Figoli A. Effect of citrate-based non-toxic solvents on poly (vinylidene fluoride) membrane preparation via thermally induced phase separation. J. Membrane Science. 2015;493: 232–242. DOI: 10.1016/j.memsci.2015.07.003

[4] Baker R W. Membrane Technology and Applications. 3rd ed. Chichester: John Wiley & Sons Ltd; 2012. 575p. DOI: 10.1002/9781118359686

[5] Ma B, Qin A, Li X, Zhao X, He C. Structure and properties of chitin whisker reinforced chitosan membranes. International J. Biological Macromolecules. 2014;64: 341–346. DOI:10.1016/j.ijbiomac.2013.12.015

[6] Bueno C Z., Dias A M A, de Sousa H J C, Braga M E M, Moraes A M. Control of the properties of porous chitosan-alginate membranes through the addition of different proportions of Pluronic F68. Materials Science and Engineering C 2014;44: 117–125. DOI: 10.1016/j.msec.2014.08.014

[7] Loeb S, Sourirajan S. Sea water demineralization by means of an osmotic membrane. In: Advances in Chemistry. Vol. 38.America: ACS Publications; 1963. p. 117–132. DOI: 10.1021/ba-1963-0038.ch009

[8] Wu P, Imai M. Novel biopolymer composite membrane involved with selective mass transfer and excellent water permeability. In: Ning R Y, editor. Advancing Desalination. Croatia: InTech; 2012. p. 57–81. DOI: 10.5772/50697

[9] Torres F G, Troncoso O P, Piaggio F, Hijar A. Structure–property relationships of a biopolymer network: The eggshell membrane. Acta Biomaterialia. 2010;6: 3687–3693. DOI: 10.1016/j.actbio.2010.03.014

[10] Kashima K, Imai M, Suzuki I. Superior molecular size screening and mass-transfer characterization of calcium alginate membrane. Desalination and Water Treatment. 2010;17: 143–149. DOI: 10.5004/dwt.2010.1710

[11] Kashima K, Imai M. Dominant impact of the α-L-guluronic acid chain on regulation of the mass transfer character of calcium alginate membranes. Desalination and Water Treatment. 2011;34: 257–265. DOI: 10.5004/dwt.2011.2892

[12] Michalak I, Mucha M. The release of active substances from selected carbohydrate biopolymer membranes. Carbohydrate Polymers. 2012;87: 2432–2438. DOI: 10.1016/j.carbpol.2011.11.013

[13] Wu P, Imai M. Excellent dyes removal and remarkable molecular size rejection of novel biopolymer composite membrane. Desalination and Water Treatment. 2013;51: 5237–5247. DOI: 10.1080/19443994.2013.768836

[14] Lakra R, Saranya R, Lukka Thuyavan Y, Sugashini S, Meera K M, Begum S, Arthanareeswaran G. Separation of acetic acid and reducing sugars from biomass derived hydrosylate using biopolymer blend polyethersulfone membrane. Separation and Purification Technology. 2013;118: 853–861. DOI: 10.1016/j.seppur.2013.08.023

[15] Moraes M A, Cocenza D S, Vasconcellos F da C, Fraceto L F, Beppu M M. Chitosan and alginate biopolymer membranes for remediation of contaminated water with herbicides. J. Environmental Management. 2013;131: 222–227. DOI: 10.1016/j.jenvman.2013.09.028

[16] Nomoto R, Imai M. Dominant role of acid-base neutralization process in forming chitosan membrane for regulating mechanical strength and mass transfer characteristics. J. Chitin and Chitosan Science. 2014;2: 197–204. DOI: 10.1166/jcc.2014.1055

[17] Ma S, Chen Z, Qiao F, Sun Y, Yang X, Deng X, Cen L, Cai Q, Wu M, Zhang X, Gao P. Guided bone regeneration with tripolyphosphate cross-linked asymmetric chitosan membrane. J. Dentistry. 2014;42: 1603–1612. DOI: 10.1016/j.jdent.2014.08.015

[18] Uragami T, Saito T, Miyata T. Pervaporative dehydration characteristics of an ethanol/water azeotrope through various chitosan membranes. Carbohydrate Polymers. 2015;120: 1–6. DOI: 10.1016/j.carbpol.2014.11.032

[19] Alias S S, Ariff Z M, Mohamad A A. Porous membrane based on chitosan–SiO_2 for coin cell proton battery. Ceramics International. 2015;41: 5484–5491. DOI: 10.1016/j.ceramint.2014.12.119

[20] Puspasari T, Pradeep N, Peinemann K-V. Crosslinked cellulose thin film composite nanofiltration membranes with zero salt rejection. J. Membrane Science 2015;491: 132–137. DOI: 10.1016/j.memsci.2015.05.002

[21] Livazovic S, Li Z, Behzad A R, Peinemann K-V, Nunes S P. Cellulose multilayer membranes manufacture with ionic liquid. J. Membrane Science. 2015;490: 282–293. DOI: 10.1016/j.memsci.2015.05.009

[22] Zhang X, Lin B, Zhao K, Wei J, Guo J., Cui W, Jiang S, Liu D, Li J. A free-standing calcium alginate/polyacrylamide hydrogel nanofiltration membrane with high antifouling performance: Preparation and characterization. Desalination. 2015:365: 234–241. DOI: 10.1016/j.desal.2015.03.015

[23] Patel A K. Chitosan: Emergence as potent candidate for green adhesive market. Biochemical Engineering J..2015; 102: 74–81. DOI: 10.1016/j.bej.2015.01.005

[24] Bai H, Zhang H, He Y, Liu J, Zhang B, Wang J. Enhanced proton conduction of chitosan membrane enabled by halloysite nanotubes bearing sulfonate polyelectrolyte brushes. J. Membrane Science. 2014;454: 220–232. DOI: 10.1016/j.memsci.2013.12.005

[25] Tasselli F, Mirmohseni A, Seyed Dorraji M S, Figoli A. Mechanical, swelling and adsorptive properties of dry–wet spun chitosan hollow fibers crosslinked with glutaraldehyde. Reactive and Functional Polymers. 2013;73: 218–223. DOI: 10.1016/j.reactfunctpolym.2012.08.007

[26] Sun X, Wang Z, Kadouh H, Zhou K, The antimicrobial, mechanical, physical and structural properties of chitosan–gallic acid films. LWT - Food Science and Technology. 2014;57: 83–89. DOI: 10.1016/j.lwt.2013.11.037

[27] Wanichapichart P, Bootluck W, Thopan P, Yu L D. Influence of nitrogen ion implantation on filtration of fluoride and cadmium using polysulfone/chitosan blend membranes. Nuclear Instruments and Methods in Physics Research Section B: Beam Interactions with Materials and Atoms. 2014;326: 195–199. DOI: 10.1016/j.nimb.2013.10.091

[28] Shenvi S, Ismail A F, Isloor A M. Preparation and characterization study of PPEES/chitosan composite membrane crosslinked with tripolyphosphate. Desalination. 2014;344: 90–96. DOI: 10.1016/j.desal.2014.02.026

[29] Han F, Dong Y, Song A, Yin R, Li S. Alginate/chitosan based bi-layer composite membrane as potential sustained-release wound dressing containing ciprofloxacin hydrochloride. Applied Surface Science. 2014;311: 626–634. DOI: 10.1016/j.apsusc.2014.05.125

[30] Karim Z, Mathew A P, Grahn M, Mouzon J, Oksman K. Nanoporous membranes with cellulose nanocrystals as functional entity in chitosan: Removal of dyes from water. Carbohydrate Polymers. 2014;112: 668–676. DOI: 10.1016/j.carbpol.2014.06.048

[31] Dudek G, Gnus M, Turczyn R, Strzelewicz A, Krasowska M. Pervaporation with chitosan membranes containing iron oxide nanoparticles. Separation and Purification Technology. 2014;133: 8–15. DOI: 10.1016/j.seppur.2014.06.032

[32] Kumar R, Isloor A M, Ismail A F. Preparation and evaluation of heavy metal rejection properties of polysulfone/chitosan, polysulfone/N-succinyl chitosan and polysulfone/N-propylphosphonyl chitosan blend ultrafiltration membranes. Desalination. 2014;350: 102–108. DOI: 10.1016/j.desal.2014.07.010

[33] Sajjan A M, Premakshi H G, Kariduraganavar M Y. Synthesis and characterization of GTMAC grafted chitosan membranes for the dehydration of low water content isopropanol by pervaporation. J. Industrial and Engineering Chemistry. 2015;25: 151–161. DOI: 10.1016/j.jiec.2014.10.027

[34] Bibi S, Yasin T, Hassan S, Riaz M, Nawaz M. Chitosan/CNTs green nanocomposite membrane: Synthesis, swelling and polyaromatic hydrocarbons removal. Materials Science and Engineering C. 2015;46: 359–365. DOI: 10.1016/j.msec.2014.10.057

[35] Ji Y-L, An Q-F, Zhao F-Y, Gao C-J. Fabrication of chitosan/PDMCHEA blend positively charged membranes with improved mechanical properties and high nanofiltration performances. Desalination. 2015;357: 8–15. DOI: 10.1016/j.desal.2014.11.005

[36] Premakshi H G, Ramesh K, Kariduraganavar M Y. Modification of crosslinked chitosan membrane using NaY zeolite for pervaporation separation of water–isopropanol mixtures. Chemical Engineering Research and Design. 2015;94: 32–43. DOI: 10.1016/j.cherd.2014.11.014

[37] Santamaria M, Pecoraro C M, Quarto F D, Bocchetta P. Chitosan–phosphotungstic acid complex as membranes for low temperature H_2–O_2 fuel cell. J. Power Sources. 2015;276: 189–194. DOI: 10.1016/j.jpowsour.2014.11.147

[38] Zhu J, Tian M, Zhang Y, Zhang H, Liu J. Fabrication of a novel "loose" nanofiltration membrane by facile blending with Chitosan–Montmorillonite nanosheets for dyes purification. Chemical Engineering J.. 2015;265: 184–193. DOI: 10.1016/j.cej.2014.12.054

[39] Li Y-H, Cheng C-Y, Wang N-K, Tan H-Y, Tsai Y-J, Hsiao C-H, Ma D H-K, Yeh L-K. Characterization of the modified chitosan membrane cross-linked with genipin for the cultured corneal epithelial cells. Colloids and Surfaces B: Biointerfaces. 2015;126: 237–244. DOI: 10.1016/j.colsurfb.2014.12.029

[40] Hegab H M, Wimalasiri Y, Ginic-Markovic M, Zou L. Improving the fouling resistance of brackish water membranes via surface modification with graphene oxide functionalized chitosan. Desalination. 2015;365: 99–107. DOI: 10.1016/j.desal.2015.02.029

[41] Zhang D, Zhou W, Wei B, Wang X, Tang R, Nie J, Wang J. Carboxyl-modified poly (vinyl alcohol)-crosslinked chitosan hydrogel films for potential wound dressing. Carbohydrate Polymers. 2015;125: 189–199. DOI: 10.1016/j.carbpol.2015.02.034

[42] Panda S R, Mukherjee M, De S. Preparation, characterization and humic acid removal capacity of chitosan coated iron-oxide-polyacrylonitrile mixed matrix membrane. J. Water Process Engineering. 2015;6: 93–104. DOI: 10.1016/j.jwpe.2015.03.007

[43] Antunes B P, Moreira A F, Gaspar V M, Correia I J. Chitosan/arginine–chitosan polymer blends for assembly of nanofibrous membranes for wound regeneration. Carbohydrate Polymers. 2015;130: 104–112. DOI: 10.1016/j.carbpol.2015.04.072

[44] Song K, Gao A, Cheng X, Xie K. Preparation of the superhydrophobic nano-hybrid membrane containing carbon nanotube based on chitosan and its antibacterial activity. Carbohydrate Polymers. 2015;130: 381–387. DOI:10.1016/j.carbpol.2015.05.023

[45] Wang J, He R. Formation and evaluation of interpenetrating networks of anion exchange membranes based on quaternized chitosan and copolymer poly (acrylamide)/polystyrene. Solid State Ionics. 2015;278: 49–57. DOI: 10.1016/j.ssi.2015.05.017

[46] Malini M, Thirumavalavan M, Yang W-Y, Lee J-F, Annadurai G. A versatile chitosan/ZnO nanocomposite with enhanced antimicrobial properties. International Journal of Biological Macromolecules. 2015;80: 121–129. DOI: 10.1016/j.ijbiomac.2015.06.036

[47] Deuchi K, Kanauchi O, Imasato Y, Kobayashi E. Decreasing effect of chitosan on the apparent fat digestibility by rats fed on high-fat diet. Bioscience, Biotechnology, and biochemistry. 1994;58: 1613–1616. DOI: 10.1271/bbb.58.1613

[48] Yao H-T, Huang S-Y, Chiang M-T. A comparative study on hypoglycemic and hypocholesterolemic effects of high and low molecular weight chitosan in streptozotocin-induced diabetic rats. Food and Chemical Toxicology. 2008;46: 1525–1534. DOI: 10.1016/j.fct.2007.12.012

[49] Lu G, Kong L, Sheng B, Wang G, Gong Y, Zhang X. Degradation of covalently cross-linked carboxymethyl chitosan and its potential application for peripheral nerve regeneration. European Polymer J.. 2007;43: 3807–3818. DOI: 10.1016/j.eurpolymj.2007.06.016

[50] Takahashi T, Imai M, Suzuki I. Water permeability of chitosan membrane involved in deacetylation degree control. Biochemical Engineering J.. 2007;36: 43–48. DOI: 10.1016/j.bej.2006.06.014

[51] Takahashi T, Imai M, Suzuki I. Cellular structure in an N-acetyl-chitosan membrane regulate water permeability. Biochemical Engineering J.. 2008;42: 20–27. DOI: 10.1016/j.bej.2008.05.013

[52] Santos C, Seabra P, Veleirinho B, Delgadillo I, Lopes da Silva J A. Acetylation and molecular mass effects on barrier and mechanical properties of shortfin squid chitosan membranes. European Polymer J.. 2006;42: 3277–3285. DOI: 10.1016/j.eurpolymj.2006.09.001

[53] Ghosh A, Ali M A, Walls R. Modification of microstructural morphology and physical performance of chitosan films. International J. Biological Macromolecules. 2010;46: 179–186. DOI: 10.1016/j.ijbiomac.2009.11.006

[54] Jin J, Song M, Hourston D J. Novel chitosan-based films cross-linked by genipin with improved physical properties. Biomacromolecules. 2004;5: 162–168. DOI: 10.1021/bm034286m

[55] Remuñán-López C, Bodmeier R. Mechanical, water uptake and permeability properties of crosslinked chitosan glutamate and alginate films. J. Controlled Release. 1997;44: 215–225. DOI: 10.1016/S0168-3659 (96)01525-8

[56] Lee Y M, Nam S Y, Woo D J. Pervaporation of ionically surface crosslinked chitosan composite membranes for water-alcohol mixtures. J. Membrane Science. 1997;133: 103–110. DOI: 10.1016/S0376-7388 (97)00089-6

[57] Devi D A, Smitha B, Sridhar S, Aminabhavi T M. Pervaporation separation of isopropanol/water mixtures through crosslinked chitosan membranes. J. Membrane Science. 2005;262: 91–99. DOI: 10.1016/j.memsci.2005.03.051

[58] Cottrell I W, Kovacs P. Alginate. In: Davidson R L, editor.Handbook of water-soluble gums and resins. New York: McGraw-HillInc.;1980. P. 2-1–2-43.

[59] Baffoun A, Viallier P, Dupuis D, Haidara H. Drying morphologies and related wetting and impregnation behaviours of 'sodium-alginate/urea' inkjet printing thickeners. Carbohydrate Polymers. 2005; 61: 103–110. DOI: 10.1016/j.carbpol.2005.03.018

[60] Richardson J C, Dettmar P W, Hampson F C, Melia C D. Oesophageal bioadhesion of sodium alginate suspensions: 2. Suspension behaviour on oesophageal mucosa.European J. Pharmaceutical Sciences. 2005; 24: 107–114. DOI: 10.1016/j.ejps.2004.10.001

[61] Rescignano N, Fortunati E, Armentano I, Hernandez R, Mijangos C, Pasquino R, Kenny J M. Use of alginate, chitosan and cellulose nanocrystals as emulsion stabilizers in the synthesis of biodegradable polymeric nanoparticles. J. Colloid Interface Science. 2015; 445: 31–39. DOI: 10.1016/j.jcis.2014.12.032

[62] Yang Y, Campanella O H, Hamaker B R, Zhang G, Gu Z. Rheological investigation of alginate chain interactions induced by concentrating calcium cations. Food Hydrocolloids. 2013; 30: 26–32. DOI: 10.1016/j.foodhyd.2012.04.006

[63] Li J, He J, Huang Y, Li D, Chen X. Improving surface and mechanical properties of alginate films by using ethanol as a co-solvent during external gelation. Carbohydrate Polymers. 2015; 123: 208–216. DOI: 10.1016/j.carbpol.2015.01.040

[64] Ha T L B, Quan T M, Vu D N, Si D M. Naturally Derived Biomaterials: Preparation and Application. In: Andrades J A, editor. Regenerative Medicine and Tissue Engineering. Croatia: InTech; 2013. p. 247–274. DOI: 10.5772/55668

[65] Kashima K, Imai M. Impact factors to regulate mass transfer characteristics of stable alginate membrane performed superior sensitivity on various organic chemicals. Procedia Engineering. 2012; 42: 964–977. DOI: 10.1016/j.proeng.2012.07.490

[66] Toti U S, Aminabhavi T M. Different viscosity grade sodium alginate and modified sodium alginate membranes in pervaporation separation of water + acetic acid and water + isopropanol mixtures. J. Membrane Science. 2004;228: 199–208. DOI: 10.1016/j.memsci.2003.10.008

[67] Kalyani S, Smitha B, Sridhar S, Krishnaiah A. Pervaporation separation of ethanol–water mixtures through sodium alginate membranes. Desalination. 2008;229: 68–81. DOI: 10.1016/j.desal.2007.07.027

[68] Saraswathi M, Madhusudhan Rao K, Prabhakar M N, Prasad C V, Sudakar K, Naveen Kumar H M P, Prasad M, Chowdoji Rao K, Subha M C S. Pervaporation studies of sodium alginate (SA)/dextrin blend membranes for separation of water and isopropanol mixture. Desalination. 2011;269: 177–183. DOI: 10.1016/j.desal.2010.10.059

[69] Li Y, Jia H, Cheng Q, Pan F, Jiang Z. Sodium alginate–gelatin polyelectrolyte complex membranes with both high water vapor permeance and high permselectivity. J. Membrane Science. 2011;375: 304–312. DOI: 10.1016/j.memsci.2011.03.058

[70] Taşkın G, Şanlı O, Asman G. Swelling assisted photografting of itaconic acid onto sodium alginate membranes. Applied Surface Science. 2011;257: 9444–9450. DOI: 10.1016/j.apsusc.2011.06.030

[71] Chen J H, Liu Q L, Hu S R, Ni J C, He Y S. Adsorption mechanism of Cu (II) ions from aqueous solution by glutaraldehyde crosslinked humic acid-immobilized sodium alginate porous membrane adsorbent. Chemical Engineering J.. 2011;173: 511–519. DOI: 10.1016/j.cej.2011.08.023

[72] Li Y, Jia H, Pan F, Jiang Z, Cheng Q. Enhanced anti-swelling property and dehumidification performance by sodium alginate–poly (vinyl alcohol)/polysulfone composite hollow fiber membranes. J. Membrane Science. 2012;407–408: 211–220. DOI: 10.1016/j.memsci.2012.03.049

[73] Nigiz F U, Dogan H, Hilmioglu N D. Pervaporation of ethanol/water mixtures using clinoptilolite and 4A filled sodium alginate membranes. Desalination. 2012;300: 24–31. DOI: 10.1016/j.desal.2012.05.036

[74] Sajjan A M, Jeevan Kumar B K, Kittur A A, Kariduraganavar M Y. Novel approach for the development of pervaporation membranes using sodium alginate and chitosan-wrapped multiwalled carbon nanotubes for the dehydration of isopropanol. J. Membrane Science. 2013;425–426: 77–88. DOI: 10.1016/j.memsci.2012.08.042

[75] Flynn E J, Keane D, Holmes J D, Morris M A. Unusual trend of increasing selectivity and decreasing flux with decreasing thickness in pervaporation separation of ethanol/water mixtures using sodium alginate blend membranes. J. Colloid and Interface Science. 2012;370: 176–182. DOI: 10.1016/j.jcis.2011.12.022

[76] Shi J, Shi J, Du C, Chen Q, Cao S. Thermal and pH dual responsive alginate/$CaCO_3$ hybrid membrane prepared under compressed CO_2. J. Membrane Science. 2013;433: 39–48. DOI: 10.1016/j.memsci.2013.01.021

[77] Chen J H, Xing H T, Guo H X, Li G P, Weng W, Hu S R. Preparation, characterization and adsorption properties of a novel 3-aminopropyltriethoxysilane functionalized sodium alginate porous membrane adsorbent for Cr (III) ions. J. Hazardous Materials. 2013;248–249: 285–294. DOI: 10.1016/j.jhazmat.2013.01.042

[78] Zhu Y, Wang Z, Zhang C, Wang J, Wang S. A novel membrane prepared from sodium alginate cross-linked with sodium tartrate for CO_2 capture. Chinese J. Chemical Engineering. 2013;21: 1098–1105. DOI: 10.1016/S1004-9541 (13)60574-1

[79] Adoor S G, Rajineekanth V, Nadagouda M N, Rao K C, Dionysiou D D, Aminabhavi T M. Exploration of nanocomposite membranes composed of phosphotungstic acid in sodium alginate for separation of aqueous–organic mixtures by pervaporation. Separation and Purification Technology. 2013;113: 64–74. DOI: 10.1016/j.seppur.2013.03.051

[80] Zhang W, Xu Y, Yu Z, Lu S, Wang X. Separation of acetic acid/water mixtures by pervaporation with composite membranes of sodium alginate active layer and microporous polypropylene substrate. J. Membrane Science. 2014;451: 135–147. DOI: 10.1016/j.memsci.2013.09.027

[81] Gao C, Zhang M, Ding J, Pan F, Jiang Z, Li Y, Zhao J. Pervaporation dehydration of ethanol by hyaluronic acid/sodium alginate two-active-layer composite membranes. Carbohydrate Polymers. 2014;99: 158–165. DOI: 10.1016/j.carbpol.2013.08.057

[82] Kuila S B, Ray S K. Dehydration of dioxane by pervaporation using filled blend membranes of polyvinyl alcohol and sodium alginate. Carbohydrate Polymers. 2014;101: 1154–1165. DOI: 10.1016/j.carbpol.2013.09.086

[83] Kuila S B, Ray S K. Separation of benzene–cyclohexane mixtures by filled blend membranes of carboxymethyl cellulose and sodium alginate. Separation and Purification Technology. 2014;123: 45–52. DOI: 10.1016/j.seppur.2013.12.017

[84] Yang J-M, Wang N-C, Chiu H-C. Preparation and characterization of poly (vinyl alcohol)/sodium alginate blended membrane for alkaline solid polymer electrolytes membrane. J. Membrane Science. 2014;457: 139–148. DOI: 10.1016/j.memsci.2014.01.034

[85] Pasini Cabello S D, Mollá S, Ochoa N A, Marchese J, Giménez E, Compañ V. New bio-polymeric membranes composed of alginate-carrageenan to be applied as polymer electrolyte membranes for DMFC. J. Power Sources. 2014;265: 345–355. DOI: 10.1016/j.jpowsour.2014.04.093

[86] Cao K, Jiang Z, Zhao J, Zhao C, Gao C, Pan F, Wang B, Cao X, Yang J. Enhanced water permeation through sodium alginate membranes by incorporating graphene oxides. J. Membrane Science. 2014;469: 272–283. DOI: 10.1016/j.memsci.2014.06.053

[87] Yoo S M, Ghosh R. Fabrication of alginate fibers using a microporous membrane based molding technique. Biochemical Engineering J.. 2014;91: 58–65. DOI: 10.1016/j.bej.2014.07.006

[88] Kamoun E A, Kenawy E-R S, Tamer T M, El-Meligy M A, Eldin M S M. Poly (vinyl alcohol)-alginate physically crosslinked hydrogel membranes for wound dressing

applications: Characterization and bio-evaluation. Arabian J. Chemistry. 2015;8: 38–47. DOI: 10.1016/j.arabjc.2013.12.003

[89] Li J, He J, Huang Y, Li D, Chen X. Improving surface and mechanical properties of alginate films by using ethanol as a co-solvent during external gelation. Carbohydrate Polymers. 2015;123: 208–216. DOI: 10.1016/j.carbpol.2015.01.040

[90] Soazo M, Báez G, Barboza A, Busti P A, Rubiolo A, Verdini R, Delorenzi N J. Heat treatment of calcium alginate films obtained by ultrasonic atomizing: Physicochemical characterization. Food Hydrocolloids. 2015;51: 193–199. DOI: 10.1016/j.foodhyd.2015.04.037

[91] Zhao K, Zhang X, Wei J, Li J, Zhou X, Liu D, Liu Z, Li J. Calcium alginate hydrogel filtration membrane with excellent anti-fouling property and controlled separation performance. J. Membrane Science. 2015;492: 536–546. DOI: 10.1016/j.memsci.2015.05.075

[92] Jie G, Kongyin Z, Xinxin Z, Zhijiang C, Min C, Tian C, Junfu W. Preparation and characterization of carboxyl multi-walled carbon nanotubes/calcium alginate composite hydrogel nano-filtration membrane. Materials Letters. 2015;157: 112–115. DOI: 10.1016/j.matlet.2015.05.080

[93] Kirdponpattara S, Khamkeaw A, Sanchavanakit N, Pavasant P, Phisalaphong M. Structural modification and characterization of bacterial cellulose–alginate composite scaffolds for tissue engineering. Carbohydrate Polymers. 2015;132: 146–155. DOI: 10.1016/j.carbpol.2015.06.059

[94] Shao W, Liu H, Liu X, Wang S, Wu J, Zhang R, Min H, Huang M. Development of silver sulfadiazine loaded bacterial cellulose/sodium alginate composite films with enhanced antibacterial property. Carbohydrate Polymers. 2015;132: 351–358. DOI: 10.1016/j.carbpol.2015.06.057

[95] Uragami T, Banno M, Miyata T. Dehydration of an ethanol/water azeotrope through alginate-DNA membranes cross-linked with metal ions by pervaporation. Carbohydrate Polymers. 2015;134: 38–45. DOI: 10.1016/j.carbpol.2015.07.054

[96] Haug A, Larsen B, Smidsrød O. A study of the constitution of alginic acid by partial acid hydrolysis. Acta Chemica Scandinavica. 1966;20: 183–190. DOI: 10.3891/acta.chem.scand.20-0183

[97] Haug A, Larsen B, Smidsrød O. Uronic acid sequence in alginate from different sources. Carbohydrate Research. 1974;32: 217–225. DOI: 10.1016/S0008-6215 (00)82100-X

[98] Gacesa P. Alginates. Carbohydr. Polym. 1988;8: 161–182. DOI: 10.1016/0144-8617(88)90001-X

[99] Draget K I, Smidsrød O, Skjåk-Bræk G. Alginates from algae. In: Steinbuchel A, Rhee S K, editors. Polysaccharides and Polyamides in the Food Industry, Properties, Pro-

duction and Patents. Weinheim: Wiley–VCH; 2005. P. 1-30. DOI: 10.1002/3527600035.bpol6008

[100] Grant G T, Morris E R, Rees D A, Smith P J C, Thom D. Biological interactions between polysaccharides and divalent cations: The egg-box model. FEBS Letters. 1973;32: 195–198. DOI: 10.1016/0014-5793 (73)80770-7

[101] Kashima K, Imai M. Advanced membrane material from marine biological polymer and sensitive molecular-size recognition for promising separation technology. In: Ning R Y, editor. Advancing Desalination. Croatia: InTech; 2012. p. 3–36. DOI: 10.5772/50734

[102] Bitter T, Muir H M. A modified uronic acid carbazole reaction. Analytical Biochemistry. 1962;4: 330–334. DOI: 10.1016/0003-2697 (62)90095-7

[103] Anzai H, Uchida N, Nishide E. Determination of D-mannuronic to L-guluronic acids ratio in acid hydrolysis of alginate under improved conditions. Nippon Suisan Gakkaishi. 1990;56: 73–81. DOI: 10.2331/suisan.56.73

[104] Anzai H, Uchida N, Nishide E. Comparative studies of colorimetric analysis for uronic acids. Bulletin of the College of Agriculture and Veterinary Medicine Nihon University. 1986;43: 53–56 (in Japanese).

[105] Wilke C R, Chang P. Correlation of diffusion coefficients in dilute solutions. A. I. Ch. E. J. 1955;1: 264–270. DOI: 10.1002/aic.690010222

[106] Smidsrød O. Molecular basis for some physical properties of alginates in the gel state. Faraday Discussions of the Chemical Society.1974; 57: 263–274. DOI: 10.1039/DC9745700263

[107] Donati I, Paoletti S. Material Properties of Alginates, In: B H. A. Rehm editor, Alginates: Biology and Applications. Berlin: Springer; 2009. p. 1–53. DOI: 10.1007/978-3-540-92679-5_1

[108] So M T, Eirich F R, Strathmann H, Baker R W. Preparation of Asymmetric Loeb-Sourirajan Membranes. J. Polymer Science: Polymer Letters Edition. 1973;11: 201–205. DOI: 10.1002/pol.1973.130110311

Chapter 9

Bio-Interfaces Engineering Using Laser-Based Methods for Controlled Regulation of Mesenchymal Stem Cell Response *In Vitro*

Valentina Dinca, Livia Elena Sima, Laurentiu Rusen, Anca Bonciu, Thomas Lippert, Maria Dinescu and Maria Farsari

Additional information is available at the end of the chapter

http://dx.doi.org/10.5772/61516

Abstract

The controlled interfacial properties of materials and modulated behaviours of cells and biomolecules on their surface are the requirements in the development of a new generation of high-performance biomaterials for regenerative medicine applications. Roughness, chemistry and mechanics of biomaterials are all sensed by cells. Organization of the environment at the nano- and the microscale, as well as chemical signals, triggers specific responses with further impact on cell fate. Particularly, human mesenchymal stem cells (hMSCs) hold a great promise in both basic developmental biology studies and regenerative medicine, as progenitors of bone cells. Their fate can be affected by various key regulatory factors (e.g. soluble growth factors, intrinsic, extrinsic environmental factors) that can be delivered by a fabricated scaffold. For example, when cultured on engineered environments that reproduce the physical features of the bone, hMSCs express tissue-specific transcription factors and consequently undergo an osteogenic fate. Therefore, producing smart bio-interfaces with targeted functionalities represents the key point in effective use of hierarchically topographical and chemical bioplatforms. In this chapter, we review laser-based approaches (e.g. Matrix-Assisted Pulsed Laser Evaporation (MAPLE), Laser-Induced Forward Transfer (LIFT), laser texturing and laser direct writing) used for the design of bio-interfaces aimed at controlling stem cell behaviour *in vitro*.

Keywords: Bio-interfaces, laser processing, topography, protein-based coatings, stem cells

1. Introduction

Currently, there is an increased interest within regenerative medicine applications in materials and methods for controlling the interface characteristics of materials in report with the modulated behaviour of cells and biomolecules on their surface. Tissue engineering approaches rely on guided tissue regeneration using materials that serve as templates for ingrowths of host cells and tissue, or on cells that have been implanted as part of an engineered device [1-3]. The factors considered within this context refer to understanding the extracellular and intracellular factors modulating cell functions, optimizing or replicating complex scaffold architecture and arrangement *in vitro* resembling to tissue, developing new materials and processing techniques compatible with bio-interfaces and enhanced strategies for low inflammatory response [4-14]. Although great progress has been made in materials engineering and fabrication of scaffolds and implant devices, there are still important issues to be clarified depending on the type of application, target tissue, materials and techniques available for biomaterial production [9-15].

Stem cells hold a great promise for a wide range of applications in regenerative medicine as well as key elements in model systems aimed at understanding fundamental processes in the field of developmental biology. Therefore, any progress in controlling their behaviour remains an important challenge [16-18]. Stem cells, from multipotent stem cells, pluripotent embryonic stem cells to induced pluripotent stem cells, have been the subject of numerous investigations related to identifying specific intrinsic and extrinsic regulatory cues, defining niches and characterizing their potential of self-renewal and differentiation [19, 20]. Consequently, a wide variety of key regulatory factors were discovered, from soluble growth factors, extracellular matrix (ECM) interactions and cell–cell interactions to micro- and nano-engineered environments or combinations thereof.

The domains of preventive and therapeutic healthcare and tissue engineering converge into an interdisciplinary field, where the interaction between materials interfaces and proteins or cells is dictated by materials bulk but more importantly by interface characteristics. Biomaterials are used widely for studying cell–substrate interactions, scaffolding or implant functionality [15-20]. The requirements for the use of natural or synthetic biomaterials are related to the degradation or stability (mechanical integrity) within fluids and fluid transport, the presence of cell-recognizable surface chemistries, the ability to deliver active biomolecules and to induce signal transduction.

The current trend is producing either 'smart' biomimicking synthetic or natural composites or enhancing bulk materials (e.g. metals, ceramics and polymers) by either chemical, physical and/or topographic surface modification or surface functionalization [15-17].

Natural biomaterials include collagen, fibrinogen, hyaluronic acid, glycosaminoglycans (GAGs), laminin, heparan sulphate proteoglycans, hydroxyapatite (HA) (being bioactive, biocompatible and of similar mechanical properties as native tissue), fibrinogen and fibrin cellulose, chitosan and silk fibroin. These were and are successfully used as 2D and 3D micro- and nano-engineered environments for stem cell culture. For example, Matrigel™, which contains a variety of ECM components, was used in tissue culture as such or in combination with various growth factors for improving neovasculature formation in an ischaemic mouse model [21], with a view to produce tube-like vascular structures [22]. When designing a new biomaterial, there are several aspects to take into account, such as micro- and nanoscale manipulation of scaffold composition, mechanical strength, control over porosity and 3D architecture, adequate surface area and adsorption kinetics and chemistry of bulk degradation of the scaffold, material-processing abilities, logistical issues of cost, compatibility with sterilization techniques and shelf life [14-16].

An enhanced biological response (surface–cell interactions, osseointegration, bone-to-implant interface strength, resistance for long-term functional loading) requires different surface treatments to be applied, from hydroxyapatite aggregation to Mg ion implantation, sandblasting or acid-etching. However, contamination of the surface with chemical compounds occurs often, which leads to modified biological response; so more *debris- or contaminant-free* techniques are necessary.

Therefore, besides material composition importance in addressing the current challenges in tissue engineering, the micro- and nano-fabrication methods provide the input for development of precise topographical architecture for 2D or 3D substrates. Various methods have been used, from self-assembly, lithography, photolithography, soft lithography, polymer demixing, phase separation and electrospinning [23-27] to surface roughness modification, with sandblasting, anodic oxidation and acid-etching [28, 29].

Laser-based method for surface modification is a promising alternative as it can be automated and is reproducible, does not generate contamination and, moreover, it can confer a variety of nano- and microstructures with increased roughness and stable characteristics for long-term bio-interaction assays. Laser deposition (i.e. Matrix-Assisted Pulsed Laser Evaporation) can create complex coatings onto 2D and 3D substrates, while direct writing and/or texturing of polymers represent two different approaches used to create 2D and 3D topographical features as physical guidance structures. This chapter will provide information about how bio-interfaces can influence cell fate, with a focus on their effect on stem cells and on bio-interfaces engineering using laser-based methods. In addition, it will cover the controlled regulation of stem cell response *in vitro* using biomaterials with defined features.

2. Surface topographical key factors for triggering stem cell response

Various factors are responsible for the behaviour of cells in response to their own niche. Considering that surface topography and chemistry represent major factors at cell–substrate interface dictating biological reactions, it becomes crucial to elucidate the complex interplay between cells, molecular signal pathways and the effect of external factors (e.g. an implant surface, functionalization and topographic features) within well-controlled cell culture systems. Nevertheless, since, in the last years, mesenchymal stem cells (MSCs) have been ideal targets in regenerative therapies, understanding of the pathways and the cues affecting stem cell differentiation to the needed lineage is critical. The ideal tissue substrate will therefore have to deliver the optimal combination of physical and biochemical signals for the spatio-temporal control of stem cell commitment and differentiation [30]. This section of the chapter focuses on describing current state of the art in topography controlling MSC fate.

2.1. Mechanical factors that influence stem cell differentiation

MSCs are multipotent cells able to differentiate to at least three lineages: osteoblast, chondrocyte and adipocyte. With a view to recreate specific niche microenvironments to direct MSC differentiation to a required phenotype, specific control elements have to be defined. Recent high-throughput approaches have aimed at defining specific chemical [31, 32] and/or biophysical [33, 34] factors controlling stem cell fate.

The formation, evolution and breakdown of the cellular components are highly influenced by the presence of topographical cues on the surface. The stress applied through focal adhesions *via* external forces and stiffness sensing induces the biological response of cells within their environment [35]. Matrix stiffness is the principal driver of stem cell specification [36], followed by soluble factors. Cultivating MSCs on extracellular matrix (ECM) derived from either osteoblasts or chondrocytes has revealed that cell-specific ECMs are capable of modulating the BMP-2-induced osteogenic versus chondrogenic differentiation [37]. A study regarding the matrix stiffness modulation of the effect of TGFβ on MSC fate concluded that stiffness is an important determinant for differentiation while requiring soluble input for unique lineage-specific outcome [38]. It has been shown that a pattern of mechanical stress is induced by the presence of textures in a nano- or micron-size range, which is directly correlated to cell spreading or arrangement in mESCs, but not in mESC-derived differentiated cells [39].

Stressing the cytoskeletal filaments networks has impact on alteration of nuclear protein assembly, gene transcription, DNA replication or RNA processing, due to nuclear shape changes, which generates modifications in chromatin organization [39], with further impact on gene expression programmes. Physical interactions between stem cells and ECM that govern cell fate have been reviewed in Ref. [40]. Several lessons have been more recently learned by studying polymer scaffolds that are important for understanding how the ECM proteins regulate MSC behaviour and for the future development of customized culture systems as already available for embryonic stem cells [41, 42]. Recent studies have revealed that it is the extracellular matrix tethering that regulates stem cell fate rather than substrate/ECM stiffness *per se* [43]. The authors showed that MSC differentiation is influenced by the stiffness of PAAm hydrogels but not of PDMS substrates. They concluded that differentiation on soft hydrogels is independent of substrate stiffness and is regulated by the ability of the cells to remodel the collagen fibres covalently linked onto the polymer structures. Controlled substrate geometries were designed using microcontact printing and fibronectin coating of patterned islands. Geometric features that increased actomyosin contractility promoted osteogenesis with an efficiency of 60–70% [44]. Using tunable stiffness, hydrogel researchers explored the lineage specification outcome when MSCs pre-cultured on soft or stiff substrates were transferred to gels of opposite stiffness [45]. They surprisingly found that rewiring of MSC lineage is possible by switching the biophysical microenvironment. Advanced tissue-mimetic matrices have been created using poly(octadecene-alt-maleic anhydride) (POMA) coated with fibronectin to tightly anchor MSC-derived ECM. [46]. This research paves the way for fabrication of native-like niches to control stem cell fate.

http://dx.doi.org/10.5772/61516

2.2. Engineering biosurface topographical cues at the micro- and nanoscale to control cell function

It has been shown that texturing of biomaterials increases cell adhesion in comparison to flat, smooth surfaces. Topography of a material can be characterized by different roughness parameters (Ra, Sa, Sm – measured in nanometres) determined by AFM, confocal microscopy, optical profilometry, SEM or tactile profilometry [47]. Knowing that the *in vivo* microenvironment is not flat, it is important to understand the function of cells grown on topographies that mimic natural conditions. Since the early definition of 'contact guidance' effect on cellular behaviour, a variety of different substrate parameters have been designed in micro and nano ranges, implying organization of the nanofeatures with different organization, chemistry and rigidity of the natural or synthetic biosurfaces, to meet the desired criteria and complexity of the native tissue architecture [48]. While *nanostructure* topography and roughness are important for cell adhesion, orientation, motility, antigen presentation, cytoskeleton polymerization, activation of tyrosine kinases and modulation of intracellular signalling pathways, *microstructure* texturing of biomaterials has impact on cell morphology, migration and tissue organization. Microfabrication technologies have been widely used in the development of substrates, scaffolds or biomaterials comprising precise topographical features.

A variety of nano- or microtopographies within hundreds of nanometres up to tens of microns range (e.g. nanoposts and nanogratings: 150, 400 and 600 nm in diameter and width; microposts and pits: 300 nm–10 μm in diameter, 3–50 μm in height) were obtained by various techniques (photolithography, electron beam lithography, reactive imprint lithography, etching, replications, hot embossing, injection moulding) [10-16] and used to stimulate hMSC-stimulated osteogenic differentiation. Although most of the studies present interesting and promising results, only a single or a small selection of parameters related to cell response were measured. Proposed platforms designed to study MSC fate regulation by both substrate mechanics and dynamic loading in 2D and 3D are reviewed in Ref. [49]. Noteworthy, few bio-interfaces obtained by polymer-processing approaches were able to initiate osseoinduction *per se*, without the need of differentiation factors. Seminal studies of Matthew Dalby *et al.* on electron beam lithography processed polycaprolactone (PCL) structures have proved that spatial organization of 120 nm pits in a square arrangement with centre-to-centre spacing of 300 nm and 50 nm offset in pit position had osteoinductive properties [50]. More recently, the same group produced substrates that are able to maintain MSC phenotype and multipotency by reducing the level of offset to almost zero [51].

2.3. Micro- and nanotopography of biosurfaces engineered by laser methods

Within the above-discussed context, by providing a control over the scale and patterning (both chemical and topographic), development in specific cell-regulating cues is implied, with the application ranging from basic cell biology to tissue engineering. Specific control over patterning involves microfabrication approaches: microcontact printing, abrasion, photolithography, hot embossing, electrospinning and laser ablation [52-57]. As most of the bioapplications require sterile conditions, laser-based techniques (i.e. laser direct irradiation or texturing, Laser-Induced Forward Transfer (LIFT), Matrix-Assisted Pulsed Laser Evaporation, photopolymerization) are contact-free techniques and could be integrated with required sterile processes. Tailoring surface textures and their features on multiple scales can be controlled by directly and precisely processing by rapid scanning of focused laser beams (nanometre range up to the millimetre range on flat and curved surfaces). Moreover, by using ultrashort pulse lasers, the mechanical properties of the materials remain unchanged after the laser processing. In the cases where controlled biocoating is necessary, by using MAPLE, the thickness, porosity, architecture of mono- or multilayer can be easily tuned by controlling laser and target parameters. Among the main advantages, these are included: no limitations in the use of materials to be deposited or structured, no difficulty in controlling the thickness of the deposited layers, the ability to deposit multilayers without interlayer blending and compatibility in processing a wide range of nanoparticles, polymeric and biologic materials [52-57].

2.3.1. Engineering microstructured thin-film biosurfaces: gradients and porous surfaces by excimer laser direct texturing

The surface texturing in the context of specific surface roughness and architecture is also required for influencing cell directionality, proliferation and differentiation as the specific surface textures can be used to influence the functional properties. Using laser texturing, the surface topography of the material is altered, thus increasing the adhesion of the cells to the substrate and promoting the growth of these in a desired direction.

The most used approach was to modify surface roughness either by mechanical methods (high-pressure blasting with metal oxide particles, high-pressure liquid jet) or by chemical means (acid treatment), presenting the disadvantage of leaving the surface contaminated or undermining the structural integrity of implants. Similar to laser-based methods like direct laser metal sintering (DLMS), laser (i.e. CO_2) surface treatment modification has substantially broadened laser application for treating Ti alloys, allowing implants to be produced more economically than by traditional techniques. Laser irradiation using pulses with duration in the nanosecond (ns) to the femtosecond range can be used for direct surface texturing of a wide variety of materials: metals, ceramics and polymers.

In the recent works by Dinca *et al.* [58], texturing by excimer lasers (193 nm) and characterization of chitosan–collagen-based structures were performed with the goal of determining the optimal morpho-chemical characteristics of these structures for *in vitro* tailoring protein adsorption and cell and degradability behaviour. The chitosan–collagen surface processing by excimer lasers with 193 nm irradiation, combined with tilting of the sample, roughness gradients (from 5 nm to 213 nm) were obtained due to the change in fluence value over the exposed area. By monitoring four different types of mammalian cells (i.e. L 929 Fibroblasts, HEP G2 hepatocytes, OLN 93 oligodendrocytes and M63 osteoblasts) adhered onto structured chitosan–collagen surfaces, an indication on how cell growth is conditioned by the substrate topography was given. For example, if fibroblast cells had a preferential adhesion on the smooth regions (below 30 nm) of the structured films compared to the rough regions (up to 210 nm), osteoblast cells behaved completely opposite. Different behaviour was observed for the other two cell lines studied. The hepatocytes were round and separate on smooth surfaces, while spheroids were formed on rough surfaces. OLN exhibited bipolar elongated morphology to tri-polar branched cells within the areas with roughness in the range of 160–213 nm, while no cell integration could be seen on the areas characterized by the roughness below 120 nm.

Excimer laser processing can also provide other types of surface geometries (foams, bubbles) by varying wavelength, pulse number and sample positioning. Figure 1 presents such examples of chitosan–collagen surfaces generated by two different applied fluences using single pulse irradiation (248 nm KrF laser) onto horizontally placed samples. It can be observed that the laser irradiation of the polymer films caused different modifications of their morphological characteristics, and due to the local heat and pressure generated by the laser radiation, 2D and 3D polymer 'bubble' (Figure 1a–c), or 'sponge-like' (Figure 1d–f) structures on the surface were obtained.

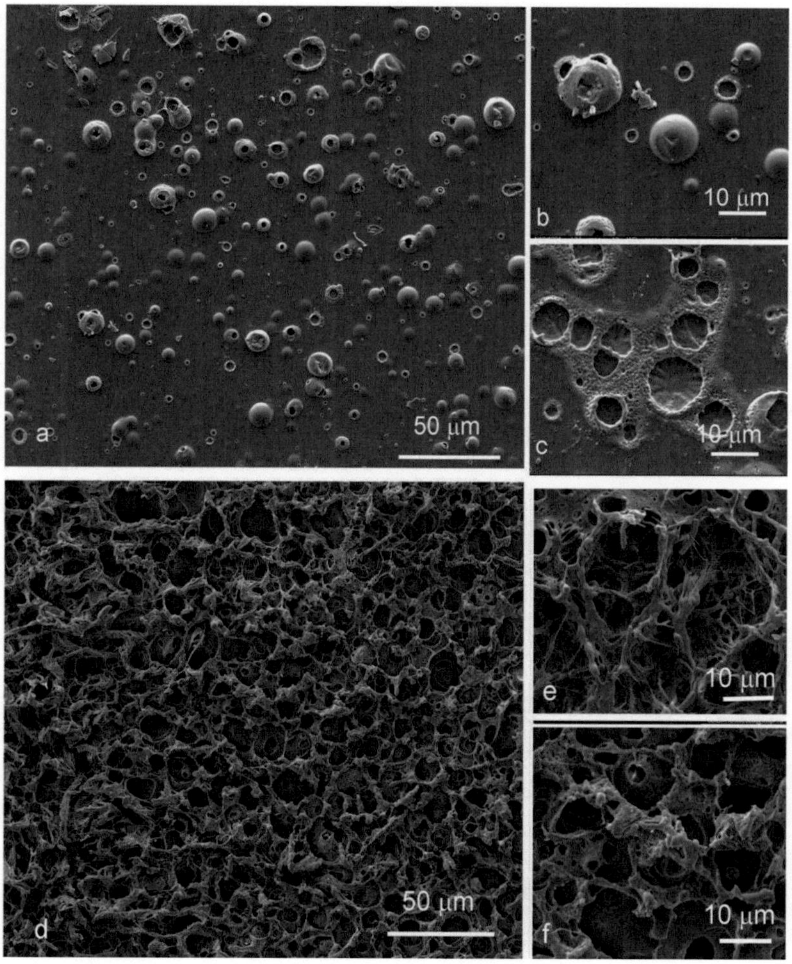

Figure 1. SEM images of various microbubble structures induced by KrF excimer single laser pulse (0.35 J/cm²) (a–c), and microfoam (d–f) (0.7 J/cm²).

In vitro studies showed an increased proliferation of mouse NIH/3T3 fibroblasts on chitosan foams obtained by excimer laser (KrF laser) irradiation as compared with untreated surfaces [59].

Obtaining microstructured thin-film surfaces and porous surfaces from naturally derived polymers (such as chitosan, collagen) could be used as artificial membranes for skin engineering and/or in cell directional growth and tissue regeneration.

2.3.2. Periodic nanostructured surfaces induced by femtosecond laser texturing

Nevertheless, machining and surface patterning of polymers and biopolymers using ultrashort pulse lasers take place with reduced mechanical and thermal deformation as compared with processing performed with longer ns laser pulses, and particular structures such as self-organizing textures can be fabricated on different materials (e.g. on metals, ceramics, semiconductors and glass). Self-organizing effects are caused by the laser material interaction and can be used to create patterns with dimensions independent of the focused laser spot size. Recent work by Rusen *et al.* [60] combined the advantages of natural biopolymer characteristics (chitosan) with the flexibility in surface texturing by ultrafast laser for creating functional microstructured surfaces for cell–substrate *in vitro* studies. A Ti: Sapphire femtosecond laser irradiation (λ = 775 nm and 387 nm) was used for tailoring surface morphological characteristics of chitosan-based films (i.e. polymer 'bubbles', 'fingertips' and 'sponge-like' structures). In the case of 'bubble' or 'fingertips' types of structures, the heights of the irradiated area were from several hundred nanometres up to few micrometres, and along with increasing fluences (beginning from the fluence of 1500 mJ/cm^2), the characteristic 'sponge-like' and even folded filaments at the irradiated area edges were observed. One example of 'fingertips' type of structure is shown in Figure 2, where regular ripple structures with a typical period of a few hundred nanometres were observed inside the created textures.

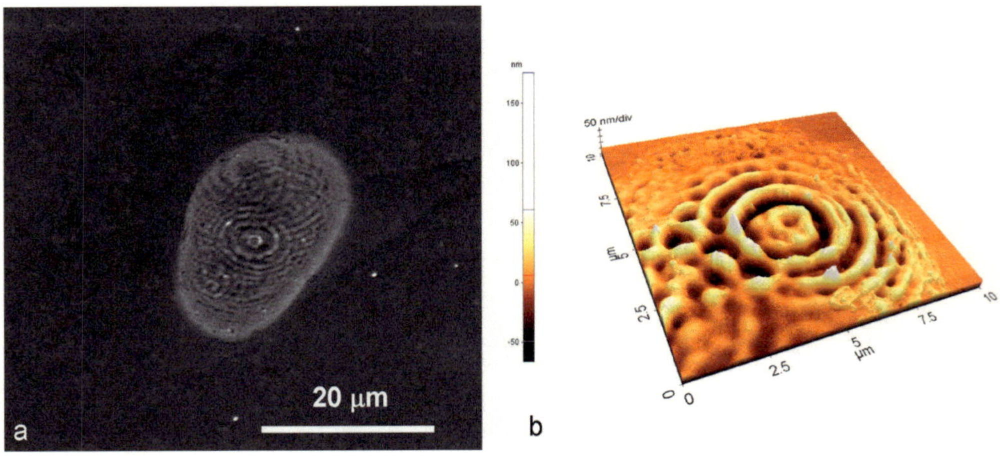

Figure 2. SEM image (a) and AFM image (b) of fingerprint-like structures obtained by Ti: Sapphire femtosecond laser irradiation (λ = 387 nm) in chitosan films.

OLN cells cultured on the patterned surface showed that early cell growth was conditioned by specific microtopography and indicate possible uses for the structures as cellular platform applications.

A similar effect was observed in the case of ceramics. Zirconia sheet cut with the dimension of 10 × 10 × 5 mm^3 from Zirkonzahn (Zirkon Translucent-ZRAB0490, Lot ZB 0070A) was used as received before the laser irradiation (1.9 Jcm^{-2} fluence), while the traverse speed was set at 0.1 mm/sec. Droplet-shaped microcavity (DμC) arrays were obtained by setting 20 pulses per cavity and translating the samples in X, respectively Y directions with a fixed step of 35 μm (Figure 3). Control planar Zirconia substrates were used as a comparison throughout our study. It was shown that the increase in surface area induced increased spreading of cell on all directions, which triggered cell morphology modification towards polygonal shape (Figure 4a) and consequently increased circularity of stem cell nuclei (Figure 4b).

Cells grown onto DμC pattern (Figure 4, DμC, left, 2000× magnification) spread efficiently over the microcavities. Cytoskeleton filaments extended closely over their tubule-granular topography (Figure 4, DμC, right, 10000× magnification, arrows).

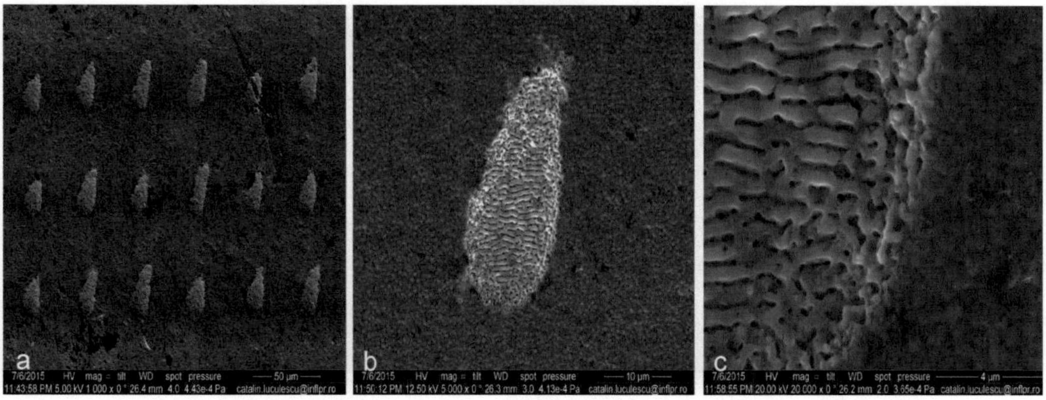

Figure 3. SEM images of droplet-shaped microcavity (DμC) arrays obtained in zirconia (a) and the corresponding close-up images of one DμC (b) and ripple-like structures inside each DμC (c).

2.3.3. *Laser engineering contact guidance structures*

'Contact guidance' mechanism was proposed by Harrison in 1911 [61], and since then, micro- and nanoridges were used to control not only cellular morphology and orientation but direct cell migration as well. Intriguingly, researchers proved by time-lapse microscopy that opposite to cells on the flat surface, mouse MSCs extended forward the long, thin process and left the wider edge trailing while migrating in both directions along the tracks, when seeded onto a groove pattern (10-μm pitch and 1600-nm step height) [62]. Consequently, cells occupy the bottom of the grooves while projecting long, thin extensions along the silicon gratings.

In order to screen for conditions to control MSC orientation and morphology, we have produced different microtopographies using excimer (Exitech KrF excimer laser, PPM-601E

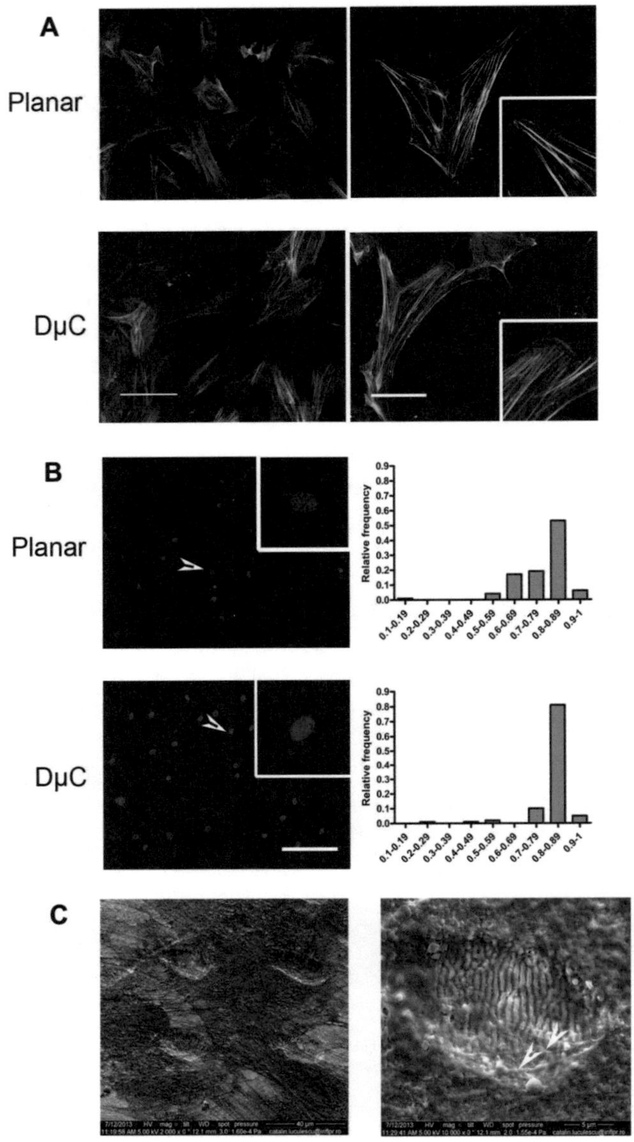

Figure 4. Modifications in cell and nuclei morphology upon growth on DμC arrays. (a) Fluorescence microscopy of hMSCs labelled for actin filaments at 8 days after seeding on femtosecond laser ablated droplet-like cavities as compared with planar zirconia. *Left panel* (10×) shows adhesion of osteoprogenitor cells onto irradiated micropatterns (scale bar = 200 μm); *right panel* (20×) evidences stress fibre organization onto the irradiated areas (sce bar = 100 μm) with emphasis on cell membrane protrusions (insets). (b) Nuclei orientation and shape. *Laleft panel* – fluorescence microscopy of DAPI-stained hMSCs nuclei 8 days after seeding onto DμC topographies or planar control. Insets provide high magnification details of a characteristic nucleus (pointed with white arrowheads onto the enlarged images). *Right panel* – frequency distribution of nuclei with specific circularity index. Note that the nuclei develop more rounded shapes on DμC arrays. Scale bar = 200 μm. (c) SEM analysis of hMSCs–zirconia interface emphasizing cytoskeleton organization (see white arrowheads on high magnification image).

Gen 6 Tool), laser micromachining by mask projection with half-tone masks. By scanning the mask under the beam and the polycarbonate (PC) sample, micropatterns of different shapes (dots, cones, pyramids, pits, funnels, inverted pyramids) with a depth/height of 2.5 to 10 μm and a width of 5 to 25 μm were generated on a polycarbonate (PC) substrate.

Cells grown on microlens-like topography (Figure 5a) lay their body over the upper part of the structures and project long dendrites along these tracks (Figure 5b,c). When grown onto reversed U-shaped lines separated by deep grooves (2 μm) (Figure 5d), cells elongate and align to surface geometry. When grooves are less deep (Figure 5e) and more frequent (Figure 5f), cells are able to diverge from their aligned profile and respectively spread over a number of lines recuperating their fibroblast-like shape, while maintaining a controlled direction. Steep edges of the V-shaped lines' topography did not have the same restrictive effect (Figure 5g), as cells are able to protrude laterally *via* long filopodia that contact the peaks. When deep grooves are ablated into PC surface, cells tend to bridge the highest areas of the substrate, with membrane domains remaining suspended over the gaps. It is the case of V-shaped lines (Figure 5g,h), pyramids (reversed or standing) (Figure 5j,k) and reversed microlenses (Figure 5m) with increased heights, from 5 to 10 μm. On milder topographies of the same structures (Figure 5i,l,n,o), cells are able to access even the deeper parts of the substrate.

The surface morphological characteristics of chitosan films were tailored (i.e. ridges, grooves structures) by using multiple pulses from Ti: Sapphire femtosecond laser irradiation (λ = 775 nm and 387 nm) and varying the scanning speed (Figure 6),. *In vitro* tests on the patterned surface showed that early growth for both OLN and fibroblast cells (Figure 6 b,d). was conditioned by the microtopography and indicate possible uses of the structures in biomedical applications.

Substrates bearing anisotropic microscale [63] or sub-micron [64] geometries have been shown to control hMSC alignment and elongation and consequently cell differentiation [65]. MSC elongation was correlated with osteogenic commitment [66] while stem cell circularity inclines the differentiation balance towards adipogenesis [67]. Combinations of sub-microscale texture and microgroove patterns, as well as distinct direction of the texture in relation to that of the microgrooves, have been also subject to primary cell adhesion assays [68, 69].

Studies *in vivo* have confirmed that aligned topographies induce differentiation of progenitor adult stem cells, with remarkable impact for tissue engineering [70, 71]. Aligned topographies enhance osseointegration and orient the local microstructure of attached tissue. Moreover, textured interfaces prevent the colonization by fibroblastic and macrophage cells that arrive early during wound healing and initiate the encapsulation of smooth substrates [72].

2.4. 2D and 3D pattern formation by Laser-Induced Forward Transfer for steering cell adhesion

LIFT is a laser-assisted direct-write process in which the materials to be transferred are in the form of a rheological fluid, polymer-based composite or fine powder, placed on a transparent support and transferred by a single pulse onto a receiver. The main key feature that separates LIFT from other direct-write techniques is that it offers the advantages of controlled and

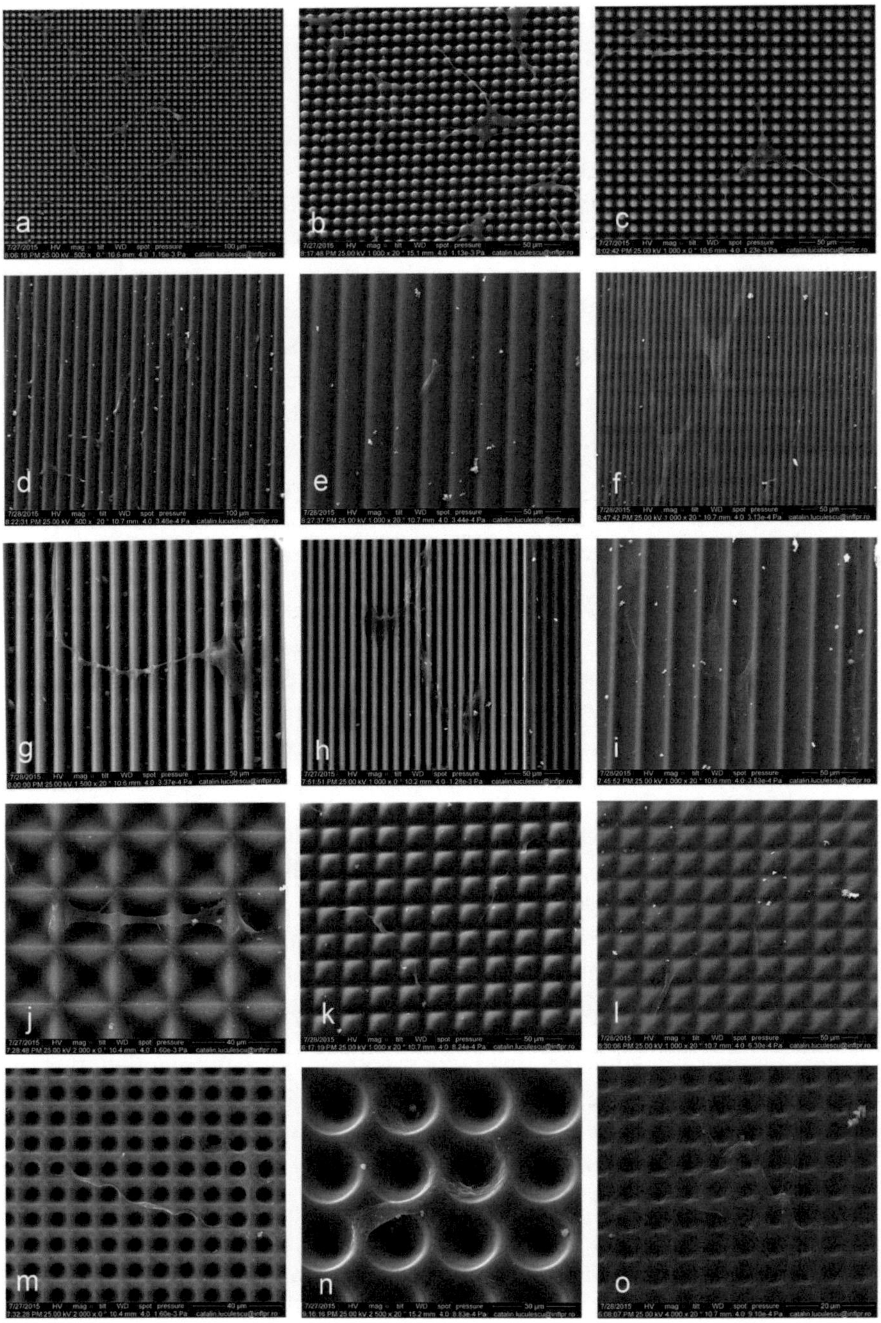

Figure 5. Interaction of hMSCs with polycarbonate microtopographies obtained by KrF excimer laser irradiation. SEM analyses were performed 72 hours post-seeding of cells onto structured microlenses (a–c), reversed U-shaped lines (d–f), V-shaped lines (g), U-shaped grooves (h,i), reversed pyramids (j), pyramids (k,l) and reversed microlenses (m–o).

http://dx.doi.org/10.5772/61516

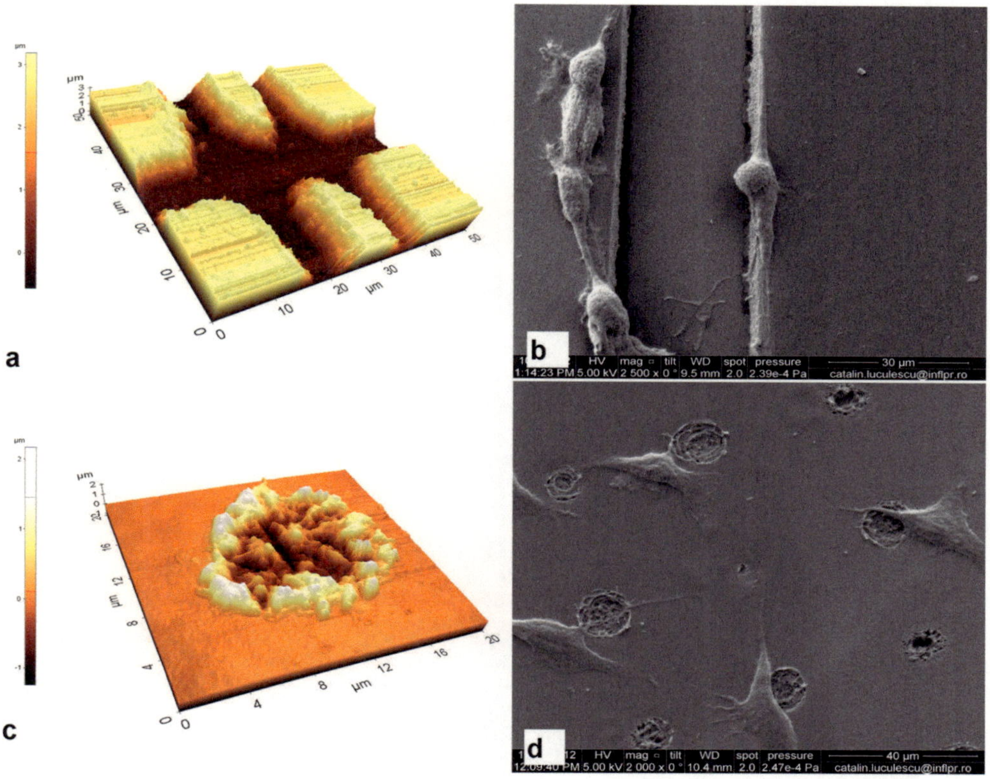

Figure 6. Architecture of chitosan ridges (a) and pits (c) scaffold made by fs multiple pulses irradiation. 3D AFM images (a,c) and SEM images (b,d) of the corresponding OLN and fibroblast cells response and organization onto the different surface textures.

localized micrometre-size pixels/clusters, similar to inkjet printing, but faster (up to 200.000 pixels/sec) and without the limitation in the nature of the material to be transferred [73-81]. The technique does not depend on the donor material properties, which therefore allows the use of non-soluble organic compounds, and the realization of complex 2D and 3D structures onto any type of surface *via* a single laser shot. Although LIFT was successfully used for transfer of biological compounds (peptides, enzymes, proteins, cells), its ability to form patterns to be used as a biological platform for studying cells behaviour is still under progress. Two-dimensional spatially controlled polyethyleneimine (PEI) micropatterns were obtained by Dinca *et al.* [82] using Dynamic Release Layer-assisted LIFT (DRL-LIFT) onto PEG-coated glass substrate to study the adhesion and cellular behaviour of SH-SY5Y human neuroblastoma cell confinement and behaviour within the created patterns. We demonstrated that the separation distances between the polymer pixels and on the surface chemistry influence the positioning of neuron aggregates inside the polymer pattern and the formation of interconnecting neurite fascicles. In this context, new directions were related to combining the advantages of DRL-LIFT with a soft substrate (i.e. Thermanox) for obtaining surface functionalization concomitantly with micro- and nano-'porous' polymeric structures. The structures obtained with different topographical properties were used as LIFT-printed platforms for fibroblast and oligodendrocyte cellular behaviour study [83]. In all cases, the transfer was achieved using a single pulse from a XeCl excimer laser (Compex, Lambda Physik, 308 nm, 30 ns). Not only 2D polymer pattern can be obtained by DRL-LIFT, but also 3D patterns [84]. For example, polystyrene microbeads arrays were printed by LIFT for selective attachment of OLN cell adhesion and the study of changes in cell shape and spreading induced by the beads arrangement (Figure 7).

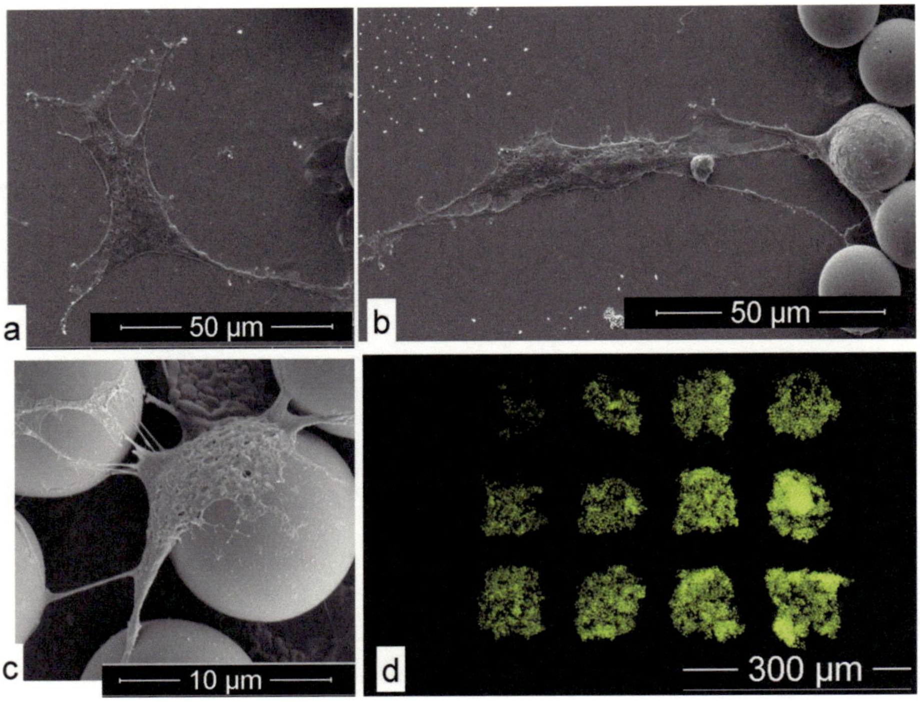

Figure 7. Interaction of OLN cells with polystyrene microbead microtopographies obtained by LIFT. SEM analyses were performed 24 hours post-seeding of cells onto control surfaces (a) and transferred microbeads (b–c). Fluorescence micrographs (d) of OLN cells cultured for 24 hours on PS microbead array.

Cells grown on microbead arrays (Figure 7a) spread onto flat surfaces and start projecting long dendrites along the bead-patterned array area (Figure 7b), their body adhering over the upper part of the structures (Figure 7c). Interestingly, when washed gently to check the adhesion strength, the remaining cells were grouped strictly over the microbead pattern, as seen in Figure 7d.

2.5. Biofunctional coatings onto 2D and 3D surfaces obtained by laser evaporation: Matrix-Assisted Pulsed Laser Evaporation

Although Pulsed Laser Deposition (PLD) was successfully used for deposition of few polymers (polyepichlorhydrin [85], poly(methyl methacrylate) [86], polyhydroxybutyrate [87] polyisobutylene (PIB)), proteins (silk protein, bovine serum albumin protein [88]) and crystalline HA on Ti implants [89] onto various substrates, when biological molecules are implied, another technique, namely MAPLE, has been used for deposition of various composites or hybrid biofunctional materials [52-57, 90, 91]. Both methods are performed under vacuum, comprising a rotating target and substrate holder, with the difference in target preparation. If for PLD the solid target is made of alloys, metals, ceramics, etc., in the case of MAPLE, the material (1–5% in weight) to be deposited is suspended into a solvent and frozen. The target is irradiated and the resultant laser plasma plume transports the molecules on the substrate placed in parallel and at a short distance (3–5 cm) [53, 90].

The coating characteristics are related to wavelength, pulse duration, repetition rate and, especially in the case of MAPLE, of solvent absorption and target composition and percentage (preferable under 5–10% in mass). For example, using the same laser parameters (Nd: YAG, 266 nm, 150 kpulses), HA coating with different aspects, from highly porous (Figure 8, left) to compact (Figure 8, right) agglomeration of HA nanoparticles, was obtained when target composition was changed by different solvents.

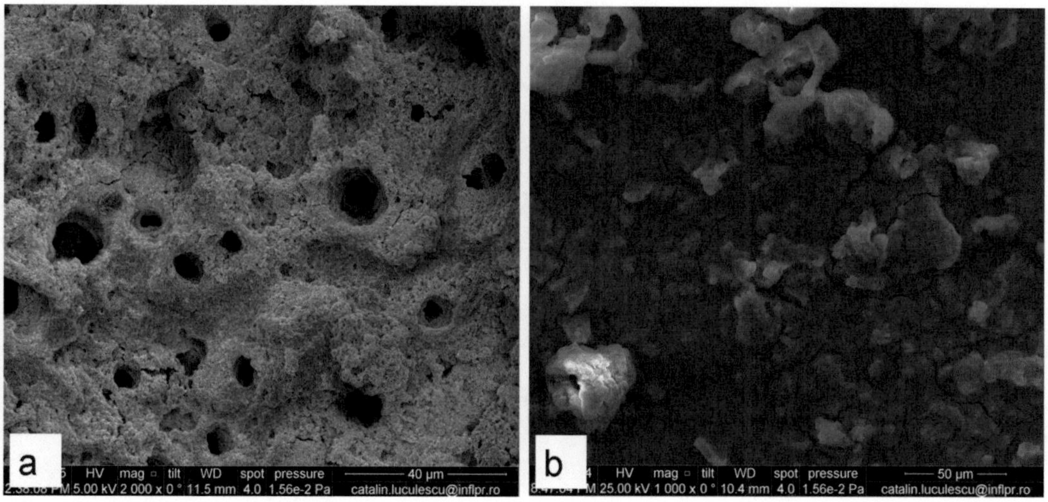

Figure 8. SEM images of HA coatings deposited on glass by MAPLE (266 nm, 150 kpulses, 500 mJ/cm2, and 2% HA in chloroform and DMSO target composition) when different solvents led to the formation of different topographies. Porous structures were obtained when chloroform was used (a) while the DMSO led to more compact arrangement of the film (b).

It is known that adherent cells may sense and respond to micro/nanoscale ECM topographical cues through cell–ECM adhesive interactions. To note, the characteristic size dimension of topographical features at the cell–ECM interface as fibrillary collagens or fibronectin fibrils in the ECM of connective tissues is in the broad size range between about 10 nm and 10 μm. The MAPLE process was shown to produce protein and protein–copolymer coatings with a wide range of morphologies and without changing the functionalities of the proteins embedded in the polymeric matrix [90, 92]. MAPLE-deposited fibronectin layers were shown to induce superior osteoprogenitor cell attachment *via* dense and extended stress fibres [93]. Levan and oxidized levan nanostructured thin films were similarly synthetized and shown to support osteoblast cell proliferation [94]. Moreover, biopolymer thin-film gradients were successfully obtained by combinatorial MAPLE (C-MAPLE) in a single step using simultaneous irradiation of two targets containing levan and oxidized levan [95]. Distinct areas of the gradient were shown to impact differently the MAPK cell signalling pathway, which indicates a potential application of these assemblies for cell fate regulation [96]. Control can be exerted over the morphology and thickness of MAPLE coatings by varying four process parameters: solution concentration, applied fluence, deposition distance and deposition time. Combinations of these parameters can generate thin films with complex morphologies. For example, low concentration protein (0.5–1%) and high fluence (500–700 mJ/cm^2) would create fibrillary structures, similar to the ECM structure and topography in the case of collagen (Figure 9) and chitosan (Figure 10). AFM images of chitosan and collagen reveal fibrillary, porous or conglomerate types of structures. Conglomerates can be seen especially at chitosan at low and medium fluences, while higher fluence can influence the morphology in periodic ways, creating a variety of new shapes on the surface.

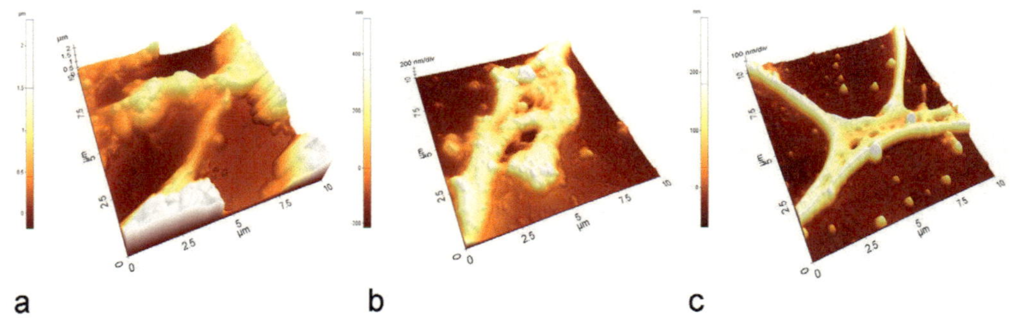

Figure 9. Examples of AFM images (10 × 10 μm^2) of fibrous collagen structures obtained by MAPLE (Nd:YAG laser, 266 nm), using different fluences: (a) 700 mJ/cm^2, (b) 600 mJ/cm^2 and (c) 500 mJ/cm^2)

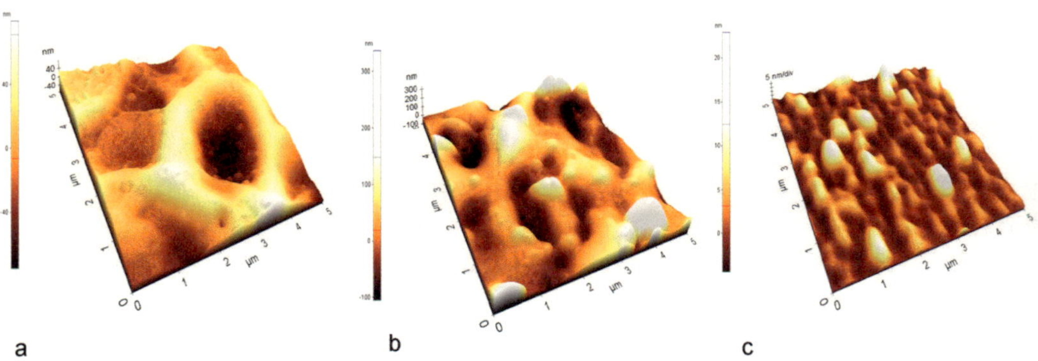

Figure 10. Examples of AFM images (5 × 5 μm^2) of porous and fibrous chitosan structures obtained by MAPLE (Nd:YAG laser, 266 nm), using different fluences: (a) 700 mJ/cm^2, (b) 600 mJ/cm^2 and (c) 500 mJ/cm^2)

AFM images of chitosan and collagen reveal fibrillary, porous or conglomerate types of structures. Conglomerates can be seen especially at chitosan at low and medium fluences, while higher fluence can influence the morphology in periodic ways, creating a variety of new shapes on the surface. More importantly, MAPLE can be easily used to coat 2D and 3D structures [92]. For example, PMMA structured substrates obtained by photolithography (Figure 11a) were coated with laminin (Figure 11b) and the effect on OLN cells was studied (Figure 11c–e). SEM images of OLN-93 cell after 2-day culture are shown in Figure 11. The cells cultured onto laminin-coated structured PMMA substrates span from bipolar elongated morphology, with preferential alignment along the line arrays with line dimensions below 10 μm.

We also used laminin (Figure 11g–j) to coat structures that were previously generated by laser irradiation of polycarbonate (PC) (Figure 11f) and studied the combined effect of topography and chemistry on MSCs. We observed that cells aligned to the 1.5-μm separated lines both on PC (Figure 11f) and on laminin-coated structure (Figure 11g–j) as compared to the random orientation they have on the laminin-coated non-irradiated PC areas (Figure 11h). A more detailed analysis showed that cells that extended partially to smooth areas between the lines (delimited by white dots in Figures 11i1 and j1) developed small protrusions from their lamellipodia (Figure 11i1,j1 – white arrows), regardless of cells that encountered the flat surface laterally (Figure 11 i–i1) or along the lines (Figure 11 j–j1). However, when cells adhered with the whole body onto the laser-processed microtopography, they 'scanned' the environment with long, thin filopodia (Figure 11i,j – red arrows).

2.5.1. *From 2D to complex 3D biosurfaces: Multi-Photon Lithography*

Multi-Photon Lithography (MPL) is a 3D printing technology that allows the construction of readily assembled structures with sub-100-nm resolution [97, 98]. It is based on the phenomenon of non-linear photon absorption and polymerization; the beam of a sub-picosecond-length pulse laser is focused inside the volume of a transparent photosensitive material, causing it to absorb two or more photons and polymerize locally. Moving the beam according to a CAD model, one can fabricate a realistic micromodel of this design.

Despite it being fairly new technology, MPL very quickly found application in tissue engineering and cell growth investigations, and, to date, several biomaterials have been employed to this purpose [99]. These are mostly negative photoresists, such as hydrogels [100-104], acrylate materials [105, 106], the epoxy-based photoresist SU-8 [107] and organic–inorganic hybrid materials (Figure 12)[108-113]. Special mention should be made of the 3D structuring of natural polymers and proteins, pioneered by Campagnola *et al.*, who worked on cross-linked proteins, such as bovine serum albumin (BSA), fibrogen and fibronectin, and collagen [114-120]. Seidlits *et al.* have crosslinked protein inside a hydrogel for neural cell guidance [121]. BSA and also avidin using flavin mononucleotide as a photoinitiator were crosslinked by Turunen *et al.* [122], while Torgesen *et al.* photopolymerized BSA encapsulating a live *C. elegans* worm [123]. The same team mixed BSA with a modified gelatin, to make stable and well-defined MPL-made scaffolds [124].

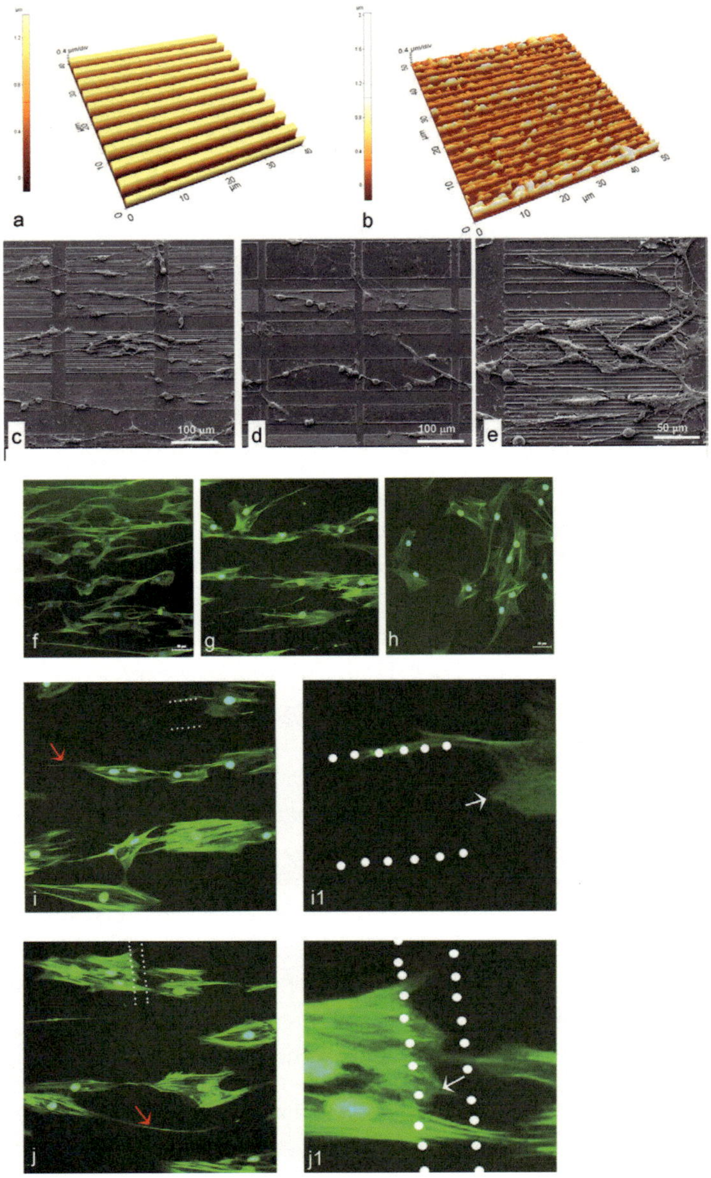

Figure 11. Cell distribution onto laminin-coated line structured PMMA and PC substrates. AFM analysis of PMMA microgroove profile before (a) and after (b) laminin coating. SEM images of OLN-93 cell alignment after 2-day culture on laminin-coated PMMA topographies (c–e). Fluorescence images of MSCs cells aligned to the 1.5 μm separated lines on PC (f) and laminin coated structure (g-j). Random orientation of MSCs on the laminin-coated non-irradiated PC areas (h). Close up images of the small protrusions from MSCs lamellipodia (i1,j1 – white arrows) onto treated areas.

There is also a lot of work on modifying natural polymers for MPL, such as polycaprolactone (Figure 13) [125], polylactide (Figure 14) [126, 127], gelatin[128] and hyaluronic acid [129]. This avenue of research has provided the most promising results, as far as biodegradable materials are concerned.

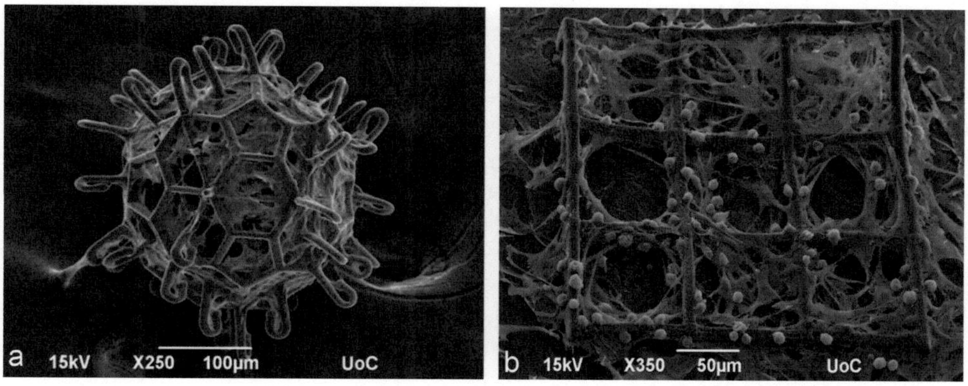

Figure 12. A 3D free-standing scaffold made of an organic–inorganic hybrid material loaded with pre-osteoblastic cells, coloured green for easy-viewing [113] (a) and a scaffold decorated with mineralized amyloid peptides and loaded with pre-osteoblastic cells [110] (b).

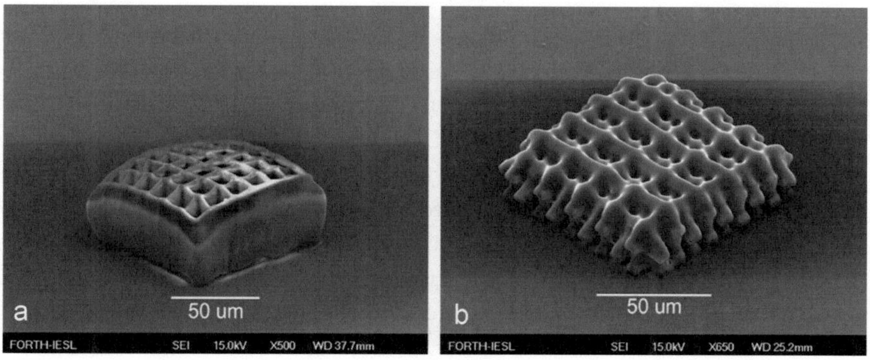

Figure 13. scaffolds made using chemically modified poly-e-caprolactone [125].

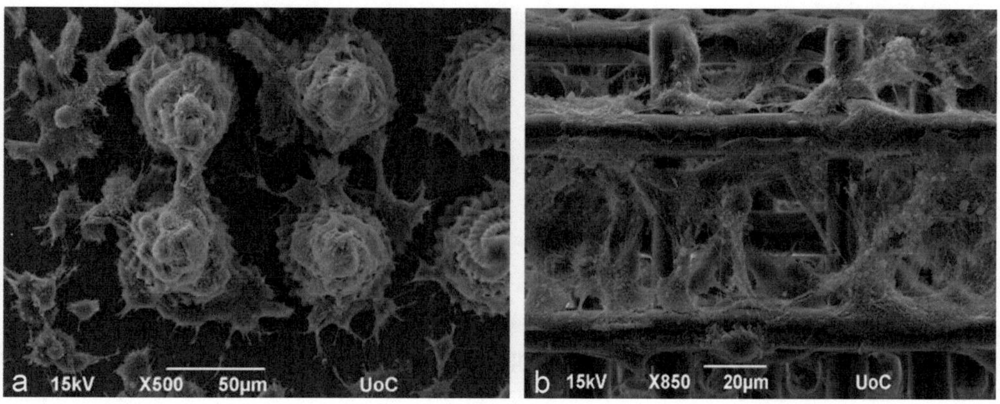

Figure 14. Cells growing on polylactide structures [126].

3. Conclusions and future prospects

Combining topographical features for mimicking the natural architecture of the ECM environment with chemical factors, such as biologically active molecules for controlling interfacial interaction between biological entities and materials, provides a challenging approach for identifying important factors that determine stem cell fate, such as extrinsic signals from the niche that impact gene expression.

This chapter briefly reported the recent progress in our research and others over topographical and chemical cues of biosurfaces on cells with accent on the hMSCs.

Although significant progress has been achieved in this field, the materials discussed above still have obvious limitations in practical applications that need to be overcome in the future. Our results indicate that substrate micropattern features play a key role in hMSCs' spreading response and more importantly that even smoother patterns than grooves and ridges are able to change cell shape. Whether the substrate conformations presented are able to induce stem cell commitment to a specific differentiation pathway is a topic of future investigation.

Although the majority of the previous works in this field have used simple topographical features such as grooves, future attempts should be more focused on realistic substrates with a higher degree of biomimetic relevance to impose multidirectional cues within the cellular microenvironment. Such novel substrates will enable addressing questions on how cells globally integrate biophysical signals from their surrounding microenvironment. Furthermore, topographical features could be integrated with chemical stimuli, such as soluble factors, to enhance cellular process, such as stem cell differentiation. The continuous advancements in the field of cell–substrate topography interactions will not only benefit fundamental biological studies but also have significant implications in the field of tissue engineering through fabrication of engineering synthetic and implantable substrates with controlled features. Therefore, the design of topographical and chemical features for engineering smart biointerfaces with multiple and synergetic functionalities represents the key point in effective use of hierarchically topographical and chemical bioplatforms targeting controlled regulation of stem cell fate.
Conflict of interest

The authors declare no competing interests.

Acknowledgements

The research leading to these results has received funding from the Romanian Ministry of National Education PN-II-PT-PCCA 239/2014. V.D. acknowledges the financial support from Sciex Grant project no.12.313 and from a grant of the Romanian National Authority for Scientific Research, CNCS – UEFISCDI, project number PN-II-RU-TE-2011-3-0289. L.E.S. is grateful for the financial support from the European Social Fund POSDRU 2007–2013 through the contract POSDRU/89/1.5/S/60746 (postdoctoral fellowship). Partial funding was obtained

http://dx.doi.org/10.5772/61516

from the Romanian Ministry of Research and Education, through the Romanian Academy Project 1/2011 of the Institute of Biochemistry.

Author details

Valentina Dinca[1*], Livia Elena Sima[2], Laurentiu Rusen[1], Anca Bonciu[1], Thomas Lippert[3], Maria Dinescu[1] and Maria Farsari[4]

*Address all correspondence to: valentina.dinca@inflpr.ro

1 National Institute for Lasers, Plasma and Radiation Physics, Magurele RO, Bucharest, Romania

2 Institute of Biochemistry of the Romanian Academy, Bucharest, Romania

3 Paul Scherrer Institute, General Energy Research Department, Villigen PSI, Switzerland

4 Institute of Electronic Structure and Laser (IESL), Foundation for Research and Technology Hellas (FORTH), Heraklion, Greece

Valentina Dinca and Livia Elena Sima contributed equally to this work

References

[1] Anselme K, Bigerelle M. 2011.Role of materials surface topography on mammalian cell response Int Mater Rev 56, 243-66.

[2] Axente E, Sima F, Sima LE, Erginer M, Eroglu MS, Serban N, et al. 2014.Combinatorial MAPLE gradient thin film assemblies signalling to human osteoblasts Biofabrication 6.

[3] Bae H, Chu HH, Edalat F, Cha JM, Sant S, Kashyap A, et al. 2014.Development of functional biomaterials with micro- and nanoscale technologies for tissue engineering and drug delivery applications Journal of tissue engineering and regenerative medicine 8, 1-14.

[4] Basu S, Campagnola PJ. 2004.Properties of crosslinked protein matrices for tissue engineering applications synthesized by multiphoton excitation J Biomed Mater Res A 71A, 359-68.

[5] Basu S, Campagnola PJ. 2004.Enzymatic activity of alkaline phosphatase inside protein and polymer structures fabricated via multiphoton excitation Biomacromolecules 5, 572-9.

[6] Basu S, Cunningham LP, Pins GD, Bush KA, Taboada R, Howell AR, et al. 2005.Multiphoton excited fabrication of collagen matrixes cross-linked by a modified benzophenone dimer: Bioactivity and enzymatic degradation Biomacromolecules 6, 1465-74.

[7] Bellayr IH, Walters TJ, Li Y. 2010.Scarless wound healing The journal of the American College of Certified Wound Specialists 2, 40-3.

[8] Blind O, Klein LH, Dailey B, Jordan L. 2005.Characterization of hydroxyapatite films obtained by pulsed-laser deposition on Ti and Ti-6AL-4v substrates Dent Mater 21, 1017-24.

[9] Campbell JJ, Hume RD, Watson CJ. 2014.Engineering Mammary Gland in Vitro Models for Cancer Diagnostics and Therapy Mol Pharmaceut 11, 1971-81.

[10] Castillejo M, Rebollar E, Oujja M, Sanz M, Selimis A, Sigletou M, et al. 2012.Fabrication of porous biopolymer substrates for cell growth by UV laser: The role of pulse duration Appl Surf Sci 258, 8919-27.

[11] Chang JC, Fujita S, Tonami H, Kato K, Iwata H, Hsu SH. 2013.Cell orientation and regulation of cell-cell communication in human mesenchymal stem cells on different patterns of electrospun fibers Biomed Mater 8.

[12] Chatzinikolaidou M, Rekstyte S, Danilevicius P, Pontikoglou C, Papadaki H, Farsari M, et al. 2015.Adhesion and growth of human bone marrow mesenchymal stem cells on precise-geometry 3D organic-inorganic composite scaffolds for bone repair Mat Sci Eng C-Mater 48, 301-9.

[13] Chu H, Wang Y. 2012.Therapeutic angiogenesis: controlled delivery of angiogenic factors Therapeutic delivery 3, 693-714.

[14] Claeyssens F, Hasan EA, Gaidukeviciute A, Achilleos DS, Ranella A, Reinhardt C, et al. 2009.Three-Dimensional Biodegradable Structures Fabricated by Two-Photon Polymerization Langmuir 25, 3219-23.

[15] Cunningham LP, Veilleux MP, Campagnola PJ. 2006.Freeform multiphoton excited microfabrication for biological applications using a rapid prototyping CAD-based approach Opt Express 14, 8613-21.

[16] Dalby MJ, Gadegaard N, Tare R, Andar A, Riehle MO, Herzyk P, et al. 2007.The control of human mesenchymal cell differentiation using nanoscale symmetry and disorder Nat Mater 6, 997-1003.

[17] Danilevicius P, Georgiadi L, Pateman CJ, Claeyssens F, Chatzinikolaidou M, Farsari M. 2015.The effect of porosity on cell ingrowth into accurately defined, laser-made, polylactide-based 3D scaffolds Appl Surf Sci 336, 2-10.

[18] Danilevicius P, Rezende RA, Pereira FDAS, Selimis A, Kasyanov V, Noritomi PY, et al. 2015.Burr-like, laser-made 3D microscaffolds for tissue spheroid encagement Biointerphases 10.

[19] Dhandayuthapani B, Yoshida Y, Maekawa T, Kumar DS. 2011.Polymeric Scaffolds in Tissue Engineering Application: A Review Int J Polym Sci.

[20] Dinca V, Allonde P, Delaporte P, Ion V, Rusen L, Filipescu M, et al.Excimer laser texturing of natural composite polymer surfaces for studying cell-to-substrate specific response Appl Surf Sci.

[21] Dinca V, Florian PE, Sima LE, Rusen L, Constantinescu C, Evans RW, et al. 2014.MAPLE-based method to obtain biodegradable hybrid polymeric thin films with embedded antitumoral agents Biomed Microdevices 16, 11-21.

[22] Dinca V, Kasotakis E, Catherine J, Mourka A, Mitraki A, Popescu A, et al. 2007.Development of peptide-based patterns by laser transfer Appl Surf Sci 254, 1160-3.

[23] Dinca V, Kasotakis E, Catherine J, Mourka A, Ranella A, Ovsianikov A, et al. 2008.Directed three-dimensional patterning of self-assembled peptide fibrils Nano Lett 8, 538-43.

[24] Dinca V, Mattle T, Papavlu AP, Rusen L, Luculescu C, Lippert T, et al. 2013.Polyethyleneimine patterns obtained by laser-transfer assisted by a Dynamic Release Layer onto Themanox soft substrates for cell adhesion study Appl Surf Sci 278, 190-7.

[25] Dinca V, Palla-Papavlu A, Paraico I, Lippert T, Wokaun A, Dinescu M. 2011.2D spatially controlled polymer micro patterning for cellular behavior studies Appl Surf Sci 257, 5250-4.

[26] Dinca V, Ranella A, Farsari M, Kafetzopoulos D, Dinescu M, Popescu A, et al. 2008.Quantification of the activity of biomolecules in microarrays obtained by direct laser transfer Biomed Microdevices 10, 719-25.

[27] Dinescu M, Matei A, Dinca V, Papavlu AP, Di Pietrantonio F, Cannata D, et al. 2013.Laser Processing of Organic Materials: Applications in Tissue Engineering and Chemical Sensing Rom Rep Phys 65, 1019-31.

[28] Dolatshahi-Pirouz A, Nikkhah M, Gaharwar AK, Hashmi B, Guermani E, Aliabadi H, et al. 2014.A combinatorial cell-laden gel microarray for inducing osteogenic differentiation of human mesenchymal stem cells Sci Rep 4, 3896.

[29] Doraiswamy A, Patz T, Narayan RJ, Dinescu M, Modi R, Auyeung RCY, et al. 2006.Two-dimensional differential adherence of neuroblasts in laser micromachined CAD/CAM agarose channels Appl Surf Sci 252, 4748-53.

[30] Duocastella M, Fernandez-Pradas JM, Dominguez J, Serra P, Morenza JL. 2008.Printing biological solutions through laser-induced forward transfer Appl Phys a-Mater 93, 941-5.

[31] Ehninger A, Trumpp A. 2011.The bone marrow stem cell niche grows up: mesenchymal stem cells and macrophages move in J Exp Med 208, 421-8.

[32] Engler AJ, Sen S, Sweeney HL, Discher DE. 2006.Matrix elasticity directs stem cell lineage specification Cell 126, 677-89.

[33] Farsari M, Filippidis G, Sambani K, Drakakis TS, Fotakis C. 2006.Two-photon polymerization of an Eosin Y-sensitized acrylate composite J Photoch Photobio A 181, 132-5.

[34] Galli C, Piemontese M, Lumetti S, Ravanetti F, Macaluso GM, Passeri G. 2012.Actin cytoskeleton controls activation of Wnt/beta-catenin signaling in mesenchymal cells on implant surfaces with different topographies Acta biomaterialia 8, 2963-8.

[35] Giljean S, Bigerelle M, Anselme K, Haidara H. 2011.New insights on contact angle/roughness dependence on high surface energy materials Appl Surf Sci 257, 9631-8.

[36] Gobaa S, Hoehnel S, Roccio M, Negro A, Kobel S, Lutolf MP. 2011.Artificial niche microarrays for probing single stem cell fate in high throughput Nature methods 8, 949-55.

[37] Guarino V, Ambrosio L. 2010.Temperature-driven processing techniques for manufacturing fully interconnected porous scaffolds in bone tissue engineering P I Mech Eng H 224, 1389-400.

[38] Guarino V, Urciuolo F, Alvarez-Perez MA, Mele B, Netti PA, Ambrosio L. 2012.Osteogenic differentiation and mineralization in fibre-reinforced tubular scaffolds: theoretical study and experimental evidences J R Soc Interface 9, 2201-12.

[39] Guilak F, Cohen DM, Estes BT, Gimble JM, Liedtke W, Chen CS. 2009.Control of stem cell fate by physical interactions with the extracellular matrix Cell stem cell 5, 17-26.

[40] Harrison RG. 1911.On the Stereotropism of Embryonic Cells Science 34, 279-81.

[41] Hernandez-Perez MA, Garapon C, Champeaux C, Shahgaldian P, Coleman A, Mugnier J. 2003.Pulsed laser deposition of bovine serum albumin protein thin films Appl Surf Sci 208, 658-62.

[42] Honda Y, Ding XT, Mussano F, Wiberg A, Ho CM, Nishimura I. 2013.Guiding the osteogenic fate of mouse and human mesenchymal stem cells through feedback system control Sci Rep-Uk 3.

[43] Honegger T, Elmberg T, Berton K, Peyrade D. 2011.Visible microlaser two-photon polymerization in a microfludic cell: A resist study Microelectron Eng 88, 2725-8.

[44] Hosseinkhani M, Mehrabani D, Karimfar MH, Bakhtiyari S, Manafi A, Shirazi R. 2014.Tissue engineered scaffolds in regenerative medicine World journal of plastic surgery 3, 3-7.

[45] Hubbell JA. 1995.Biomaterials in Tissue Engineering Bio-Technol 13, 565-76.

[46] Ikada Y. 2006.Challenges in tissue engineering J R Soc Interface 3, 589-601.

[47] Im BJ, Lee SW, Oh N, Lee MH, Kang JH, Leesungbok R, et al. 2012.Texture direction of combined microgrooves and submicroscale topographies of titanium substrata influence adhesion, proliferation, and differentiation in human primary cells Archives of oral biology 57, 898-905.

[48] Ji Y, Ghosh K, Shu XZ, Li BQ, Sokolov JC, Prestwich GD, et al. 2006.Electrospun three-dimensional hyaluronic acid nanofibrous scaffolds Biomaterials 27, 3782-92.

[49] Juodkazis S, Mizeikis V, Misawa H. 2009.Three-dimensional microfabrication of materials by femtosecond lasers for photonics applications J Appl Phys 106.

[50] Kecskemeti G, Smausz T, Kresz N, Toth Z, Hopp B, Chrisey D, et al. 2006.Pulsed laser deposition of polyhydroxybutyrate biodegradable polymer thin films using ArF excimer laser Appl Surf Sci 253, 1185-9.

[51] Khademhosseini A, Langer R, Borenstein J, Vacanti JP. 2006.Microscale technologies for tissue engineering and biology P Natl Acad Sci USA 103, 2480-7.

[52] Kilian KA, Bugarija B, Lahn BT, Mrksich M. 2010.Geometric cues for directing the differentiation of mesenchymal stem cells P Natl Acad Sci USA 107, 4872-7.

[53] Kinney MA, McDevitt TC. 2013.Emerging strategies for spatiotemporal control of stem cell fate and morphogenesis Trends in biotechnology 31, 78-84.

[54] Kufelt O, El-Tamer A, Sehring C, Schlie-Wolter S, Chichkov BN. 2014.Hyaluronic Acid Based Materials for Scaffolding via Two-Photon Polymerization Biomacromolecules 15, 650-9.

[55] Kwon SH, Lee TJ, Park J, Hwang JE, Jin M, Jang HK, et al. 2013.Modulation of BMP-2-Induced Chondrogenic Versus Osteogenic Differentiation of Human Mesenchymal Stem Cells by Cell-Specific Extracellular Matrices Tissue Eng Pt A 19, 49-58.

[56] Ladoux B, Nicolas A. 2012.Physically based principles of cell adhesion mechanosensitivity in tissues Rep Prog Phys 75.

[57] LaFratta CN, Fourkas JT, Baldacchini T, Farrer RA. 2007.Multiphoton fabrication Angew Chem Int Edit 46, 6238-58.

[58] Lee J, Abdeen AA, Kilian KA. 2014.Rewiring mesenchymal stem cell lineage specification by switching the biophysical microenvironment Sci Rep-Uk 4.

[59] Li D, Zhou JX, Chowdhury F, Cheng JJ, Wang N, Wang F. 2011.Role of mechanical factors in fate decisions of stem cells Regen Med 6, 229-40.

[60] Luches A, Caricato AP. 2011.Matrix assisted pulsed laser evaporation: the surface cluster problem Appl Phys B-Lasers O 105, 503-8.

[61] MacQueen L, Sun Y, Simmons CA. 2013.Mesenchymal stem cell mechanobiology and emerging experimental platforms J R Soc Interface 10, 20130179.

[62] Malinauskas M, Danilevicius P, Baltriukiene D, Rutkauskas M, Zukauskas A, Kairyte Z, et al. 2010.3d Artificial Polymeric Scaffolds for Stem Cell Growth Fabricated by Femtosecond Laser Lith J Phys 50, 75-82.

[63] Malinauskas M, Farsari M, Piskarskas A, Juodkazis S. 2013.Ultrafast laser nanostructuring of photopolymers: A decade of advances Phys Rep 533, 1-31.

[64] McMurray RJ, Dalby MJ, Tsimbouri PM. 2015.Using biomaterials to study stem cell mechanotransduction, growth and differentiation Journal of tissue engineering and regenerative medicine 9, 528-39.

[65] McMurray RJ, Gadegaard N, Tsimbouri PM, Burgess KV, McNamara LE, Tare R, et al. 2011.Nanoscale surfaces for the long-term maintenance of mesenchymal stem cell phenotype and multipotency Nat Mater 10, 637-44.

[66] McMurray RJ, Wann AKT, Thompson CL, Connelly JT, Knight MM. 2013.Surface topography regulates wnt signaling through control of primary cilia structure in mesenchymal stem cells Sci Rep-Uk 3.

[67] Melissinaki V, Gill AA, Ortega I, Vamvakaki M, Ranella A, Haycock JW, et al. 2011.Direct laser writing of 3D scaffolds for neural tissue engineering applications Biofabrication 3.

[68] Melkoumian Z, Weber JL, Weber DM, Fadeev AG, Zhou YE, Dolley-Sonneville P, et al. 2010.Synthetic peptide-acrylate surfaces for long-term self-renewal and cardiomyocyte differentiation of human embryonic stem cells Nat Biotechnol 28, 606-U95.

[69] Nagel M, Hany R, Lippert T, Molberg M, Nüesch FA, Rentsch D. 2007.Aryltriazene Photopolymers for UV-Laser Applications: Improved Synthesis and Photodecomposition Study Macromolecular Chemistry and Physics 208, 277-86.

[70] Nair LS, Bhattacharyya S, Laurencin CT. Nanotechnology and Tissue Engineering: The Scaffold Based Approach. Nanotechnologies for the Life Sciences: Wiley-VCH Verlag GmbH & Co. KGaA; 2007.

[71] Nelson CM, Raghavan S, Tan JL, Chen CS. 2003.Degradation of Micropatterned Surfaces by Cell-Dependent and -Independent Processes† Langmuir 19, 1493-9.

[72] Nie ZH, Kumacheva E. 2008.Patterning surfaces with functional polymers Nat Mater 7, 277-90.

[73] Norman J, Desai T. 2006.Methods for fabrication of nanoscale topography for tissue engineering scaffolds Ann Biomed Eng 34, 89-101.

[74] Nune M, Kumaraswamy P, Krishnan UM, Sethuraman S. 2013.Self-Assembling Peptide Nanofibrous Scaffolds for Tissue Engineering: Novel Approaches and Strategies for Effective Functional Regeneration Curr Protein Pept Sc 14, 70-84.

[75] Nuutinen T, Silvennoinen M, Paivasaari K, Vahimaa P. 2013.Control of cultured human cells with femtosecond laser ablated patterns on steel and plastic surfaces Biomed Microdevices 15, 279-88.

[76] Oh S, Brammer KS, Li YSJ, Teng D, Engler AJ, Chien S, et al. 2009.Stem cell fate dictated solely by altered nanotube dimension P Natl Acad Sci USA 106, 2130-5.

[77] O'Neal R, Wang HL, MacNeil RL, Somerman MJ. 1994.Cells and materials involved in guided tissue regeneration Current opinion in periodontology 141-56.

[78] Ovsianikov A, Deiwick A, Van Vlierberghe S, Dubruel P, Moller L, Drager G, et al. 2011.Laser Fabrication of Three-Dimensional CAD Scaffolds from Photosensitive Gelatin for Applications in Tissue Engineering Biomacromolecules 12, 851-8.

[79] Ovsianikov A, Malinauskas M, Schlie S, Chichkov B, Gittard S, Narayan R, et al. 2011.Three-dimensional laser micro- and nano-structuring of acrylated poly(ethylene glycol) materials and evaluation of their cytoxicity for tissue engineering applications Acta biomaterialia 7, 967-74.

[80] Ovsianikov A, Mironov V, Stampf J, Liska R. 2012.Engineering 3D cell-culture matrices: multiphoton processing technologies for biological and tissue engineering applications Expert review of medical devices 9, 613-33.

[81] Palla-Papavlu A, Dinca V, Ion V, Moldovan A, Mitu B, Luculescu C, et al. 2011.Characterization of polymer thin films obtained by pulsed laser deposition Appl Surf Sci 257, 5303-7.

[82] Palla-Papavlu A, Dinca V, Paraico I, Moldovan A, Shaw-Stewart J, Schneider CW, et al. 2010.Microfabrication of polystyrene microbead arrays by laser induced forward transfer J Appl Phys 108.

[83] Park JS, Chu JS, Tsou AD, Diop R, Tang Z, Wang A, et al. 2011.The effect of matrix stiffness on the differentiation of mesenchymal stem cells in response to TGF-beta Biomaterials 32, 3921-30.

[84] Pique A, Chrisey DB, Fitz-Gerald JM, McGill RA, Auyeung RCY, Wu HD, et al. 2000.Direct writing of electronic and sensor materials using a laser transfer technique J Mater Res 15, 1872-5.

[85] Pique A, McGill RA, Chrisey DB, Leonhardt D, Mslna TE, Spargo BJ, et al. 1999.Growth of organic thin films by the matrix assisted pulsed laser evaporation (MAPLE) technique Thin Solid Films 355, 536-41.

[86] Piskin E. 1995.Biodegradable Polymers as Biomaterials J Biomat Sci-Polym E 6, 775-95.

[87] Pitts JD, Campagnola PJ, Epling GA, Goodman SL. 2000.Submicron multiphoton free-form fabrication of proteins and polymers: Studies of reaction efficiencies and applications in sustained release Macromolecules 33, 1514-23.

[88] Pitts JD, Howell AR, Taboada R, Banerjee I, Wang J, Goodman SL, et al. 2002.New photoactivators for multiphoton excited three-dimensional submicron cross-linking of proteins: Bovine serum albumin and type 1 collagen Photochem Photobiol 76, 135-44.

[89] Popat KC, Leoni L, Grimes CA, Desai TA. 2007.Influence of engineered titania nanotubular surfaces on bone cells Biomaterials 28, 3188-97.

[90] Prewitz MC, Seib FP, von Bonin M, Friedrichs J, Stissel A, Niehage C, et al. 2013.Tightly anchored tissue-mimetic matrices as instructive stem cell microenvironments Nature methods 10, 788-94.

[91] Qin XH, Torgersen J, Saf R, Muhleder S, Pucher N, Ligon SC, et al. 2013.Three-Dimensional Microfabrication of Protein Hydrogels via Two-Photon-Excited Thiol-Vinyl Ester Photopolymerization J Polym Sci Pol Chem 51, 4799-810.

[92] Raimondi MT, Eaton SM, Lagana M, Aprile V, Nava MM, Cerullo G, et al. 2013.Three-dimensional structural niches engineered via two-photon laser polymerization promote stem cell homing Acta biomaterialia 9, 4579-84.

[93] Ranga A, Gobaa S, Okawa Y, Mosiewicz K, Negro A, Lutolf MP. 2014.3D niche microarrays for systems-level analyses of cell fate Nature communications 5, 4324.

[94] Rashidi H, Yang J, Shakesheff KM. 2014.Surface engineering of synthetic polymer materials for tissue engineering and regenerative medicine applications Biomater Sci-Uk 2, 1318-31.

[95] Rebollar E, Villavieja N, Gaspard S, Oujja M, Corrales T, Georgiou S, et al. 2007.Pulsed laser deposition of polymers doped with fluorescent probes. Application to environmental sensors J Phys Conf Ser 59, 305-9.

[96] Rodin S, Domogatskaya A, Strom S, Hansson EM, Chien KR, Inzunza J, et al. 2010.Long-term self-renewal of human pluripotent stem cells on human recombinant laminin-511 Nat Biotechnol 28, 611-U102.

[97] Rusen L, Cazan M, Mustaciosu C, Filipescu M, Simion S, Zamfirescu M, et al. 2014.Tailored topography control of biopolymer surfaces by ultrafast lasers for cell-substrate studies Appl Surf Sci 302, 256-61.

[98] Rusen L, Dinca V, Mitu B, Mustaciosu C, Dinescu M. 2014.Temperature responsive functional polymeric thin films obtained by matrix assisted pulsed laser evaporation for cells attachment-detachment study Appl Surf Sci 302, 134-40.

[99] Rusen L, Mustaciosu C, Mitu B, Filipescu M, Dinescu M, Dinca V. 2013.Protein-resistant polymer coatings obtained by matrix assisted pulsed laser evaporation Appl Surf Sci 278, 198-202.

[100] Seidlits SK, Schmidt CE, Shear JB. 2009.High-Resolution Patterning of Hydrogels in Three Dimensions using Direct-Write Photofabrication for Cell Guidance Adv Funct Mater 19, 3543-51.

[101] Selimis A, Mironov V, Farsari M. 2015.Direct laser writing: Principles and materials for scaffold 3D printing Microelectron Eng 132, 83-9.

[102] Sellinger A, Leveugle E, Fitz-Gerald JM, Zhigilei LV. 2008.Generation of surface features in films deposited by matrix-assisted pulsed laser evaporation: the effects of the stress confinement and droplet landing velocity Appl Phys a-Mater 92, 821-9.

[103] Serra P, Colina M, Fernandez-Pradas JM, Sevilla L, Morenza JL. 2004.Preparation of functional DNA microarrays through laser-induced forward transfer Appl Phys Lett 85, 1639-41.

[104] Sill TJ, von Recum HA. 2008.Electrospinning: applications in drug delivery and tissue engineering Biomaterials 29, 1989-2006.

[105] Sima F, Axente E, Sima LE, Tuyel U, Eroglu MS, Serban N, et al. 2012.Combinatorial matrix-assisted pulsed laser evaporation: Single-step synthesis of biopolymer compositional gradient thin film assemblies Appl Phys Lett 101.

[106] Sima F, Davidson P, Pauthe E, Sima LE, Gallet O, Mihailescu IN, et al. 2011.Fibronectin layers by matrix-assisted pulsed laser evaporation from saline buffer-based cryogenic targets Acta biomaterialia 7, 3780-8.

[107] Sima F, Mutlu EC, Eroglu MS, Sima LE, Serban N, Ristoscu C, et al. 2011.Levan Nanostructured Thin Films by MAPLE Assembling Biomacromolecules 12, 2251-6.

[108] Sima LE, Buruiana EC, Buruiana T, Matei A, Epurescu G, Zamfirescu M, et al. 2013.Dermal cells distribution on laser-structured ormosils Journal of tissue engineering and regenerative medicine 7, 129-38.

[109] Sima LE, Filimon A, Piticescu RM, Chitanu GC, Suflet DM, Miroiu M, et al. 2009.Specific biofunctional performances of the hydroxyapatite-sodium maleate copolymer hybrid coating nanostructures evaluated by in vitro studies J Mater Sci-Mater M 20, 2305-16.

[110] Slaughter BV, Khurshid SS, Fisher OZ, Khademhosseini A, Peppas NA. 2009.Hydrogels in regenerative medicine Advanced materials 21, 3307-29.

[111] Sorrell JM, Baber MA, Caplan AI. 2009.Influence of Adult Mesenchymal Stem Cells on In Vitro Vascular Formation Tissue Eng Pt A 15, 1751-61.

[112] Su PJ, Tran QA, Fong JJ, Eliceiri KW, Ogle BM, Campagnola PJ. 2012.Mesenchymal Stem Cell Interactions with 3D ECM Modules Fabricated via Multiphoton Excited Photochemistry Biomacromolecules 13, 2917-25.

[113] Subramani C, Cengiz N, Saha K, Gevrek TN, Yu X, Jeong Y, et al. 2011.Direct Fabrication of Functional and Biofunctional Nanostructures Through Reactive Imprinting Advanced materials 23, 3165-+.

[114] Sugita Y, Ishizaki K, Iwasa F, Ueno T, Minamikawa H, Yamada M, et al. 2011.Effects of pico-to-nanometer-thin TiO2 coating on the biological properties of microroughened titanium Biomaterials 32, 8374-84.

[115] Suske E, Scharf T, Krebs HU, Junkers T, Buback M. 2006.Mechanism of poly(methyl methacrylate) film formation by pulsed laser deposition J Appl Phys 100.

[116] Tan J, Shen H, Saltzman WM. 2001.Micron-scale positioning of features influences the rate of polymorphonuclear leukocyte migration Biophysical journal 81, 2569-79.

[117] Terzaki K, Kalloudi E, Mossou E, Mitchell EP, Forsyth VT, Rosseeva E, et al. 2013.Mineralized self-assembled peptides on 3D laser-made scaffolds: a new route toward 'scaffold on scaffold' hard tissue engineering Biofabrication 5.

[118] Terzaki K, Kissamitaki M, Skarmoutsou A, Fotakis C, Charitidis CA, Farsari M, et al. 2013.Pre-osteoblastic cell response on three-dimensional, organic-inorganic hybrid material scaffolds for bone tissue engineering J Biomed Mater Res A 101, 2283-94.

[119] Torgersen J, Ovsianikov A, Mironov V, Pucher N, Qin XH, Li ZQ, et al. 2012.Photosensitive hydrogels for three-dimensional laser microfabrication in the presence of whole organisms Journal of biomedical optics 17.

[120] Trappmann B, Gautrot JE, Connelly JT, Strange DGT, Li Y, Oyen ML, et al. 2012.Extracellular-matrix tethering regulates stem-cell fate Nat Mater 11, 642-9.

[121] Turunen S, Kapyla E, Terzaki K, Viitanen J, Fotakis C, Kellomaki M, et al. 2011.Pico- and femtosecond laser-induced crosslinking of protein microstructures: evaluation of processability and bioactivity Biofabrication 3.

[122] Urciuolo F, Imparato G, Guaccio A, Mele B, Netti PA. 2012.Novel Strategies to Engineering Biological Tissue In Vitro Methods Mol Biol 811, 223-44.

[123] Wang ZY, Teo EY, Chong MSK, Zhang QY, Lim J, Zhang ZY, et al. 2013.Biomimetic Three-Dimensional Anisotropic Geometries by Uniaxial Stretch of Poly(epsilon-Caprolactone) Films for Mesenchymal Stem Cell Proliferation, Alignment, and Myogenic Differentiation Tissue Eng Part C-Me 19, 538-49.

[124] Watari S, Hayashi K, Wood JA, Russell P, Nealey PF, Murphy CJ, et al. 2012.Modulation of osteogenic differentiation in hMSCs cells by submicron topographically-patterned ridges and grooves Biomaterials 33, 128-36.

[125] Worley K, Certo A, Wan L. 2013.Geometry–Force Control of Stem Cell Fate BioNanoSci 3, 43-51.

[126] Yin Z, Chen X, Chen JL, Shen WL, Hieu Nguyen TM, Gao L, et al. 2010.The regulation of tendon stem cell differentiation by the alignment of nanofibers Biomaterials 31, 2163-75.

[127] Zahor D, Radko A, Vago R, Gheber LA. 2007.Organization of mesenchymal stem cells is controlled by micropatterned silicon substrates Mat Sci Eng C-Bio S 27, 117-21.

[128] Zergioti I, Karaiskou A, Papazoglou DG, Fotakis C, Kapsetaki M, Kafetzopoulos D. 2005.Femtosecond laser microprinting of biomaterials Appl Phys Lett 86.

[129] Zhang WD, Chen SC. 2011.Femtosecond laser nanofabrication of hydrogel biomaterial Mrs Bull 36, 1028-33.

[130] Zhang WD, Soman P, Meggs K, Qu X, Chen SC. 2013.Tuning the Poisson's Ratio of Biomaterials for Investigating Cellular Response Adv Funct Mater 23, 3226-32.

[131] Zhang Y, Zhang X, Shi B, Miron RJ. 2013.Membranes for guided tissue and bone regeneration Annals of Oral & Maxillofacial Surgery 1, 10.

Chapter 10

TI2BioP — Topological Indices to BioPolymers. A Graphical–Numerical Approach for Bioinformatics

Guillermin Agüero-Chapin, Reinaldo Molina-Ruiz, Gisselle Pérez-Machado, Vitor Vasconcelos, Zenaida Rodríguez-Negrin and Agostinho Antunes

Additional information is available at the end of the chapter

http://dx.doi.org/10.5772/61887

Abstract

We developed a new graphical–numerical method called TI2BioP (Topological Indices to BioPolymers) to estimate topological indices (TIs) from two-dimensional (2D) graphical approaches for the natural biopolymers DNA, RNA and proteins The methodology mainly turns long biopolymeric sequences into 2D artificial graphs such as Cartesian and four-color maps but also reads other 2D graphs from the thermodynamic folding of DNA/RNA strings inferred from other programs. The topology of such 2D graphs is either encoded by node or adjacency matrixes for the calculation of the spectral moments as TIs. These numerical indices were used to build up alignment-free models to the functional classification of biosequences and to calculate alignment-free distances for phylogenetic purposes. The performance of the method was evaluated in highly diverse gene/protein classes, which represents a challenge for current bioinformatics algorithms. TI2BioP generally outperformed classical bioinformatics algorithms in the functional classification of Bacteriocins, ribonucleases III (RNases III), genomic internal transcribed spacer II (ITS2) and adenylation domains (A-domains) of nonribosomal peptide synthetases (NRPS) allowing the detection of new members in these target gene/protein classes. TI2BioP classification performance was contrasted and supported by predictions with sensitive alignment-based algorithms and experimental outcomes, respectively. The new ITS2 sequence isolated from *Petrakia* sp. was used in our graphical–numerical approach to estimate alignment-free distances for phylogenetic inferences. Despite TI2BioP having been developed for application in bioinformatics, it can be extended to predict interesting features of other biopolymers than DNA and protein sequences. TI2BioP version 2.0 is freely available from http://ti2biop.sourceforge.net/.

Keywords: 2D graphs, Topological indices, Alignment-free models, Bioinformatics

1. Introduction

Graph theory has been successfully applied in several branches of science such as mathematics, physics, chemistry, biochemistry, biology and computer science to visualize complex relationships. A graph is a collection of vertices or nodes and a compilation of edges that connect pairs of vertices. They have been deeply studied to analyse pairwise relationships in a data collection [1].

Graph theory allowed the development of chemical graph theory (CGT) to explore the chemical molecular structure by combinatorial and topological approaches that lead to the calculation of mathematical descriptors [2]. The molecular topology is simplified in graphs where its vertices and edges represent the atoms and bonds, respectively. Thus, molecular descriptors from the graph representing an approximation of the molecular structure can be estimated to carry out quantitative-structure-activity/property relationship (QSAR/QSPR). These numerical indices have been traditionally used in QSAR/QSPR studies for drug discovery and design in medicinal chemistry [3, 4].

With the arrival of the genomics and proteomics era, the CGT has been extended to characterize long biopolymeric strings such as DNA/RNA and proteins, for comparative analyses without the use of sequence alignments. The monomers (nucleotides and amino acids) of the natural biopolymers can play the role of nodes while the edges of the graph are represented by covalent bonds, hydrogen bridges, electrostatic interactions, van der Waals bonds and so on [5-7]. Thus, the structure of complex biopolymers can be simplified into the topology of a graph to provide useful insights into such molecular systems. The graphs representing molecular systems may be described using numerical descriptors like the so-called topological indices (TIs) [8]. TIs encode information about the connections between atoms in the molecule and the properties for the connected atoms [9]. In this way, they can also be applied to characterize natural biopolymers like DNA, RNA and protein sequences [10].

The use of TIs to characterize numerically biosequences to perform massive analyses without alignments is an active research topic in bioinformatics [5-7]. To determine the TIs for the natural biopolymers, we build a graph as it was described previously. There are various types of TIs depending of the dimensionality (D) of the biopolymer representation. One-dimensional (1D) representation of biosequences depicts the linear sequence order, while two-dimensional (2D) and three-dimensional (3D) representations are related to sequence arrangement or geometry into these spaces [11-13]. The 2D biopolymer graphs have grabbed special attention due to fact that they have been very effective in exploring similarities/dissimilarities among DNA and protein sequences despite not representing their real structure [2]. So far, the 2D artificial representations for DNA and protein sequences with higher potentialities in bioinformatics are the spectrum-like, star-like, Cartesian-type and four-color maps [2, 14-17]. These DNA/RNA and protein maps can generally reveal higher-order useful information contained beyond the primary structure, i.e. nucleotide/amino acid distribution into a 2D space. Such graphical features can be quantified by the TIs to easily compare a great number of sequences/maps [18-21].

http://dx.doi.org/10.5772/61887

Regardless of the biopolymer representation type, the definition of an adjacency matrix is mandatory for the calculation of any TI. There are variants of the adjacency matrix, e.g., node and edge adjacency matrix [22]. They translate the connectivity/adjacency relations between nodes or edges in the graph to a matrix arrangement [23]. Later, several algorithms can be applied on the adjacency matrix to provide different TI types such as the Winner index (*W*) [24], first defined in a chemical context; and others like Randić invariant (χ) [25], Balaban index (*J*) [26], Broto–Moreau autocorrelation (ATSd) [27] and the spectral moments introduced by Estrada [28]. The spectral moments were defined as the sum of main diagonal entries of the different powers of the bond adjacency matrix [29]. Spectral moments were implemented in the TOPS-MODE (topological substructural molecular design) program [30] and have been widely validated by many authors to encode the structure of small molecules in QSAR studies [31-33]. Despite the versatility of the spectral moments in QSAR studies, they have been poorly used to describe biopolymers structures except when they promoted the arising of the Estrada folding index (*I*3) for proteins [34, 35] or when they were redefined as stochastic spectral moments by González-Diaz et al. to numerically characterize biopolymeric systems, i.e. the protein surface of human rhinoviruses [36], Arc repressors [37], kinases [38] and different types of biological complex networks [10]. The stochastic spectral moments are implemented in the MARCH-INSIDE (Markov chain invariants for network selection and design) methodology [39] and can be estimated from star-like and Cartesian-type representations for DNA and protein sequences [6, 40, 41]. Thus, the first reported alignment-free models based on a graphical–numerical approach to annotate biological functions in biosequences were built using the MARCH-INSIDE software [40, 42, 43]. However, such predicting models were more illustrative than practical for the bioinformatics. They were built and tested on small-sized datasets and generally without considering the degree of similarity among their members and data benchmarking to evaluate the TIs as alignment-free predictors [40, 42, 43].

On the other hand, stochastic spectral moment's calculation is mathematically more complicated than the original definitions by Estrada [28]. Stochastic spectral moments rely on defining Markov chain states over the starting node adjacency matrix that are later powered at different orders, while the original spectral moments are derived directly from the powering of the bond or edge adjacency matrix weighted with some bond property [30, 39].

Considering these shared previous experiences about the potentialities of the graphical–numerical methods for bioinformatics, we aim for the development of a new methodology called TI2BioP (Topological Indices to BioPolymers) to extend the original spectral moments as simple TIs to characterize numerically 2D artificial representations for the DNA/RNA and proteins structure [5, 44]. These TIs represent alignment-free predictors to detect functional signatures in members of gene and protein classes. Its practical importance for bioinformatics consisted of dealing with gene/protein classes sharing low sequence similarity and in estimating alignment-free distances for inferring phylogenetic relationships [21].

Traditionally, the prediction of the biological function, 2D and 3D structure of a query gene or protein has relied on similarity measures provided by alignment algorithms, to other recorded members of the family. All alignment-based methods, the dynamic programming algorithms implemented by Needleman–Wunsch [45] and Smith–Waterman [46], the heuristic algorithm for basic local alignment search tool (BLAST) [47] and the probabilistic hidden Markov models (HMM) [48, 49] have a friendly interface to search structural and functional sequence classifications, but they may fail in detecting gene/protein members that share low similarity to others of the family [21, 50, 51]. There are several evidences showing a low reliability for the biological functional prediction when protein families have pairwise sequence similarities below 50% [50, 52, 53]. In addition, inaccurate alignments for proteins that share less than 30% to 40% of identity, which is commonly called the "twilight zone" for the alignment algorithms, have been reported [50, 54]. Therefore, the reliability of phylogenetic inferences is also affected by failures of the multiple sequence alignment (MSA) algorithms when the taxa represented by sequences have greatly diverged [54]. Consequently, several alignment-independent approaches have been developed to overcome this limitation for an effective functional annotation [55, 56] and for reliable phylogenetic inferences in highly diverse gene/protein families [55, 57]. Most of the alignment-free classifiers have been based on amino acid composition to annotate protein functions [51, 55, 56]. It is very likely that the most popular alignment-free approach is Chou's concept of pseudo–amino acid composition (PseAAC) that reflects the importance of the sequence order effect in addition to the amino acid composition to improve the prediction quality of protein cellular attributes [58]. On the other hand, the alignment-independent approaches reported for phylogenetic tree reconstruction have mostly been based on patterns discovered in unaligned sequences [59], amino acid composition [55] and a kernel approach for evolutionary sequence comparison [57].

While alignment methods have improved their sensitivity to detect functional and evolutionary signals in query sequences and species by using several strategies [60-62] and, on the other side, various alignment-free approaches have been reported to address the same drawback, there is still room for the development of new alignment-free biosequence descriptors. In this sense, graphical–numerical methods have been poorly explored as alignment-free tools in bioinformatics, to face current alignment algorithm limitations [2, 54]. Here, we summarize our experience in this subject through the application of TI2BioP to predict the functions of natural biopolymers (DNA and proteins) in classes representing a challenge for alignment algorithms as well as its introduction into the molecular evolutionary field.

2. Methods

2.1. TI2BioP software

TI2BioP was mainly developed from the TOPS-MODE methodology [30] for the estimation of the spectral moments series as TIs, but it takes advantage of the MARCH-INSIDE program platform [39]. It was built up on object-oriented Free Pascal IDE Tools (Lazarus) running on either a Windows or Linux operating system. TI2BioP has a friendly interface allowing users to introduce multiple fasta files containing either DNA or protein sequences to select the biopolymer 2D representation type and the calculation of TIs. We released version 2.0 of the software that can be freely downloaded from http://ti2biop.sourceforge.net/. This version contains two main types of 2D artificial representations, one based on Cartesian representation for DNA strings introduced by Nandy [63] and the other inspired by the four-color maps reported by Randic [64, 65] (Figure 1).

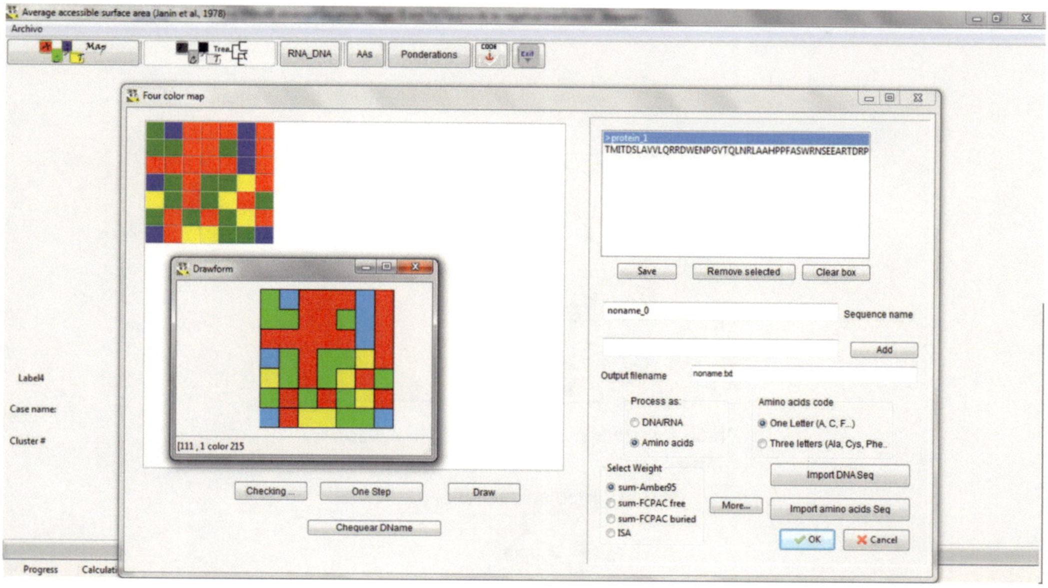

Figure 1. TI2BioP window view of the (Topological Indices to BioPolymers) software for the representation of protein four-color maps

These two 2D artificial graphs implemented in TI2BioP can be applied to nucleotide and amino acid strings as well as to the spectral moments calculations for each type of 2D DNA and protein maps [44]. It is noteworthy that the 2D Cartesian representation was extended to proteins by our group [40] and protein four-color maps were modified according to the amino acid clustering proposed in ref. [40]. Such four-color map modifications allow the speeding up of graph-building and facilitates the calculation of spectral moments as TIs [66].

TI2BioP can also import files containing 2D structures inferred by other DNA/RNA folding algorithms, e.g. Mfold implemented in the RNA structure software [67], for the calculation of the spectral moments as TIs. TI2BioP automatically represents natural biopolymers as 2D graphs and straightforward calculates spectral moments series (TIs) to be used either for statistical classification techniques in building alignment-free models for functional classification or for deriving several alignment-free distance matrices, e.g. Euclidean, Jensen–Shannon, Hamming and Minkowsk for phylogenetic purposes (Figure 2).

2.2. Database

To evaluate the performance and efficacy of our graphical–numerical approach TI2BioP to detect DNA and protein signatures and to infer phylogenetic relationships, four gene/protein families having low sequence similarity among their members were selected. The gene/protein classes targeted were:

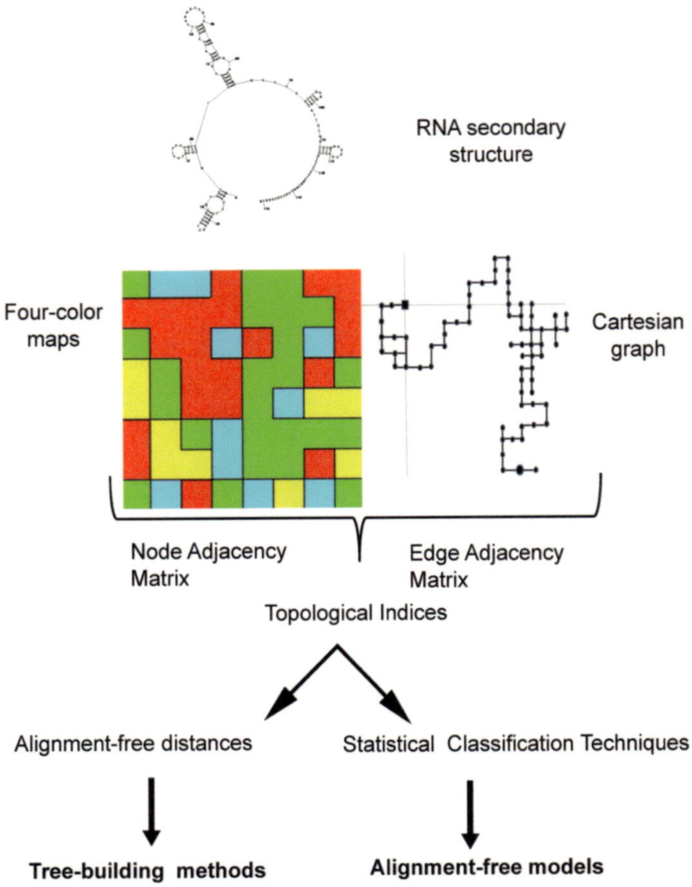

Figure 2. Workflow for the calculation of the topological indices by TI2BioP (Topological Indices to BioPolymers) from several 2D graphs for DNA, RNA and proteins.

1. *Bacteriocin protein class*: A total of 196 bacteriocin-like proteins sequences belonging to several bacterial species were collected from the two major bacteriocin databases, BAGEL [60] and BACTIBASE [68].

2. *Ribonuclease III class (RNase III)*: 206 RNase III protein sequences belonging to prokaryote and eukaryote species were downloaded from GenBank database gathering all RNAses III registered up to May of 2009.

3. *ITS2 class*: A total of 4355 ITS2 sequences from a wide variety of eukaryotic taxa (http://its2.bioapps.biozentrum.uni-wuerzburg.de) were used.

4. *Adenylation domains (A-domains)*: 138 A-domain sequences from NRPS were collected from the major NRPS–PKS database (http://www.nii. res.in/nrps-pks.html).

Because a negative set or control group to develop classification models is needed, three different control groups were selected according to some features: (1) structurally well-characterized sequences, (2) high functional diversity among its members and (3) similar sequence lengths with respect to the study case.

Protein control groups:

1. Sequences from class, architecture, topology and homology (CATH) domain database (version 3.2.0) (http://www.cathdb.info) sharing only 35% sequence similarity were selected to provide a functional representation and avoid structural redundancy. This group was used as a control to develop alignment-free models to recognize bacteriocin-like and A-domain sequences.

2. High-resolution proteins in a structurally nonredundant and representative subset from the Protein Data Bank (PDB) made up of enzymes and nonenzymes were also used. This protein subset was used as a control group to develop alignment-free models to detect RNase III enzymes [50].

3. A nonredundant subset containing both 5′- and 3′-untranslated region (UTR) sequences from the eukaryotic mRNAs database: UTRdb (http://www.ba.itb.cnr.it/UTR/). It was selected as a control group to identify ITS2 members.

3. Results

This section summarizes the main results derived from the application of TI2BioP to the functional classification of protein bacteriocins [5], RNase III [69], ITS2 [21] and NRPS A-domains [66]. All these classes show high sequence divergence among their members, which represent a handicap for the good performance of alignment algorithms. In particular, the high sequence divergence among fungal ITS2 genomic fragments has been useful for fungi identification at the genus and species level. However, such sequence diversity is not suitable for reconstructing phylogenies at a higher taxonomical level. The use of simple alignment-free classifiers, like the topological indices (TIs), containing information about the sequence/structure of the natural biopolymers may reveal a useful approach for the gene/protein functional predictions and for assessing the phylogenetic relationships at high taxonomic levels in fungal species by using the ITS2 gene class.

The TI2BioP software provides TIs (spectral moments series) that are used as input predictors for statistical classification techniques and machine-learning methods to develop alignment-free models (Figure 2). Models were statistically tested by cross- and external-validation procedures. Their usefulness was proved by identifying new members belonging to each studied gene/protein class. Such alignment-free detections were either supported by experimental evidences or by sensitive alignment methods.

Table 1 displays the alignment-free models with the best performance for the functional classification of each target gene/protein family and the procedure carried out to achieve a consensus functional prediction of new members by such models. The functional annotation of the new members resulted from the prediction agreement among the graphical–numerical based models, experimental evidences and alignment algorithms. The 2D Cartesian protein representation and its derived TIs could unravel the Cry 1Ab C-terminal domain from *Bacillus thuringiensis'* endotoxin as a bacteriocin-like protein. The bactericide action of this domain was only confirmed by experimental evidences; no alignment algorithm could anticipate such activity [5]. In addition, new ITS2 and RNase III members were registered using alignment-free models based on the same biopolymer representation [21, 69]. The predictions of these two members were verified through enzymatic assay for the new RNase III member and by evaluating both queries against profiles HMM (Table 1). The amino acid clustering strategy according to their physicochemical properties to build 2D Cartesian protein maps was extended to generate a nonclassical profile HMM with higher prediction accuracy to detect RNase III members than classical profiles [69].

The effectiveness of the presented graphical–numerical approach in bioinformatics was also demonstrated by the introduction of protein four-color maps and TIs to detect A-domains despite their sequence diversity. A DTM based on this approach was chosen as the best alignment-free model to identify the A-domain signature (Table 1).

Gene/protein class	Control group	2D-graph type	Best-reported alignment-free model	Newly detected members	Prediction procedure
Protein Bacteriocins	CATH domains	Cartesian	GDA	Cry 1Ab C-terminal domain *Bacillus thuringiensis*	1. Alignment-free prediction 2. Experimental evidences
Genomic ITS2	5′and 3′UTRs	Cartesian and RNA Secondary Structure	ANN	ITS2 genomic *Petrakia* sp.	1. Alignment-free prediction 2. Alignment-based prediction
RNase III	Nonredundan t subset PDB	Cartesian	DTM	RNase III *E. coli BL21*	1. Alignment-free prediction 2. Alignment-based prediction 3. Experimental evidences
A-domain NRPS	CATH domains	Four-color map	DTM	5 hits in the proteome of *Microcystis aeruginosa*	Pending registration

GDA, general discrimination analysis; ANN, artificial neural networks; DTM, decision tree models; ITS2, internal transcribed spacer; RNase III, ribonuclease III; A-domains, adenylation domains; CATH, class, architecture, topology and homology; PDB, Protein Data Bank; UTRs, untranslated regions.

Table 1. Best reported alignment-free models for the functional classification of each gene/protein family studied. Newly detected members of each gene/protein class and the procedure carried out for their definitive functional prediction

Its performance was contrasted to other different alignment-free approaches and homology-search methods in detecting A-domains on the same dataset. The Web server PseAAC (http://www.csbio.sjtu.edu.cn/bioinf/PseAAC/) was used to generate DTM based on other alignment-free features like amino acid composition (AAC) and pseudo amino acid composition (PseAAC) [70]. The DTM generated by four-color maps outperformed the DTM supported by AAC and PseAAC features (Table 2).

Four-color maps DTM		**Training**		**Test**
Sensitivity (Sv) (%)		100		100
Specificity (Sp) (%)		100		100
Accuracy (Acc) (%)		100		100
F-score				1.0
10-fold CV	**Sv**	**Sp**	**Acc**	
Average	98.16	99.98	99.95	
AAC DTM		**Training**		**Test**
Sensitivity (%)		53.70		3.44
Specificity (%)		100		99.68
Accuracy (%)		99.25		98.44
F-score				0.07
10-fold CV	**Sv**	**Sp**	**Acc**	
Average	21.73	100	98.78	
PseAAC DTM		**Training**		**Test**
Sensitivity (%)		67.89		20.68
Specificity (%)		99.80		99.77
Accuracy (%)		99.30		98.75
F-score				0.40
10-fold CV	**Sv**	**Sp**	**Acc**	
Average	21.73	100	98.78	

CV, Cross-validation; DTM, decision tree model.

Accuracy (Acc), Sensitivity (Sv), Specificity (Sp) and *F*-score are classification quality measures. *F*-score values range between 0 and 1.

Table 2. Classification results for alignment-free DTM based on four-color maps, amino acid composition (AAC) and pseudo–amino acid composition (PseAAC) in the A-domains detection

On the other hand, the alignment-free search of A-domains was also compared to homology-based methods such as single-template BLASTp, multitemplate BLASTp and profile HMM. These alignment-based algorithms show, by definition, different sensitivities to recognize distant homologs and therefore may provide different false-positive rates. Table 3 shows the classification results provided by different sequence-search methods including alignment-free (four-color maps, AAC and PseAAC) and homology-based (HMM, multitemplate BLASTp and BLASTp) approaches on the same dataset (138 A-domains + 8854 CATH domains). The DTM built from four-color maps TIs, HMM and multitemplate BLASTp identified all A-domains among the diversity of the dataset with no false-positives at nonstringent conditions (*E* value = 10).

Sequence search method	True positive	False positive
DTM (four-color maps)	138	0
DTM (AAC)	59	7
DTM (PseAAC)	80	18
HMM (*E* value = 10)	138	0
Multitemplate BLASTp (*E* value = 10)	138	0
BLASTp (*E* value = 10)	138	6033
BlASTp (*E* value = 0.05)	138	122
BLASTp (*E* value = 0.01)	138	24
BLASTp (*E* value = 0.001)	138	4
BLASTp (*E* value = 0.0001)	138	0

DTM, decision tree models; AAC, amino acid composition; PseAAC, pseudo–amino acid composition; HMM, hidden Markov models; BLAST, basic local alignment search tool; *E* value, a classification threshold related to the matches found by chances in alignment algorithms.

Table 3. True positives *vs.* false positives in the A-domain detection for different sequence-search methods among the overall dataset involved in the study

Considering the excellent performance of these three previous sequence-search methods, they were applied in cooperation to provide the most reliable exploration of the A-domain repertoire of NRPS in the *M. aeruginosa* proteome. DTM based on four-color map TIs detected two putative A-domain signatures among the proteome's hypothetical proteins while another three hypothetical proteins were detected as A-domains by the profile HMM. Sequence-search methods based on profiles (graphical and alignment) were able to detect five more hits than the 20 A-domains already annotated in the proteome, which were confirmed by the multitemplate BLASTp (Figure 3). These matches could reveal the presence of additional A-domain remote homologous, which would not have been detected by applying a single algorithm.

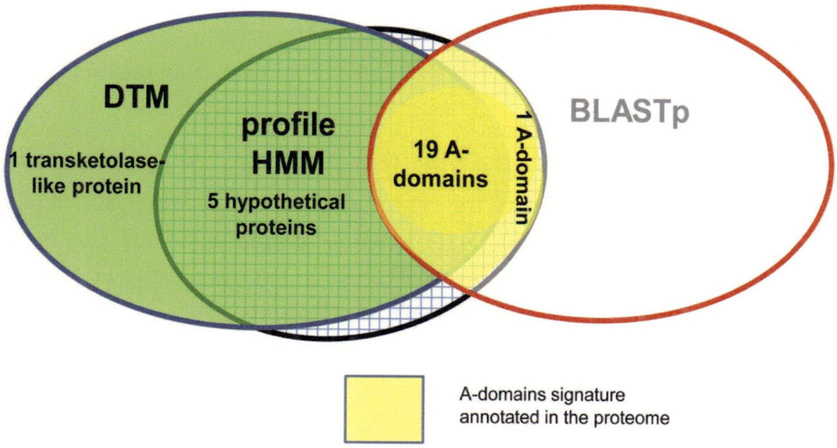

Figure 3. Reannotation of the A-domains in the proteome of *M. aeruginosa* by using an ensemble of algorithms (DTM based on four-color maps, profile HMM and multitemplate BLASTp).

The performance of DTM built from four-color map TIs was contrasted with sensitive alignment procedures like multitemplate BLASTp and HMM. Similarly, the other alignment-free models such as GDA and ANN relying on Cartesian and thermodynamic TIs were also compared to InterPro and HMM profiles for functional detection of the chosen gene/protein classes (Tables 1 and 4).

	Alignment-free models			Alignment-based procedures		
Gene/protein class	**Statistical technique**	**Sensitivity test set**	**New member prediction**	**Alignment algorithm**	**Sensitivity test set**	**New members detection**
Protein Bacteriocins	GDA	66.67%	Significant hit	InterPro	60.2%	No-hit
Genomic ITS2	ANN	92.59%	Significant hit	Profile HMM (MAFFT)	66.66%	Significant hit
RNase III	DTM	96.07%	Significant hit	Profile HMM (modified)	100%	Significant hit
A-domains NRPS	DTM	100%	Significant hits	Profiles HMM	100%	Significant hits

GDA, general discrimination analysis; ANN, artificial neural networks; DTM, decision tree models; HMM, hidden Markov models; MAFFT, multiple alignment based on fast Fourier transform; InterPro, a web resource that combines different protein signature recognition methods.

Table 4. Prediction performance measured through the sensitivity on the test set and the identification of the new member for the best reported alignment-free and alignment-based method. When alignment-based procedures achieved a sensitivity of 100%, complex algorithms were applied

The TIs supplied by TI2BioP are also used to estimate alignment-free distances that can be introduced into tree-building methods, e.g. unweighted pair group method with arithmetic mean (UPGMA), neighbor joining method (NJ) and minimum evolution (ME) to infer evolutionary relationships (Figure 2).

The newly predicted ITS2 sequence, isolated from the fungus *Petrakia* sp., was used by clustering techniques applied for the first time to the alignment-free estimation of phylogenetic inferences (Table 1). The *Petrakia* sp. fungal isolate was placed inside the *Pezizomycotina* subphylum and the *Dothideomycetes* class by the inference agreement of classical genetic distances and the alignment-free distances based on TIs (Figure 4). We concluded that our graphical–numerical approach is effective to construct distance-trees containing relevant biological information with an evolutionary significance [21].

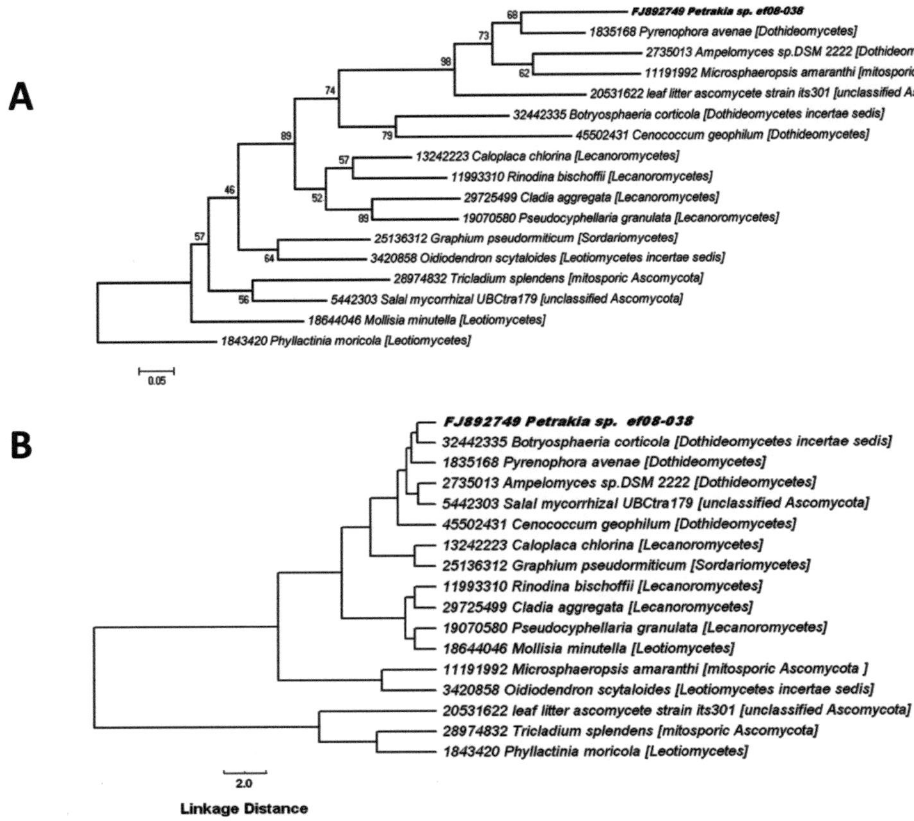

Figure 4. Higher-level phylogenetic analysis to infer the class of the fungal isolate *Petrakia* sp. (A) Neighbor-joining tree based on ITS2 sequences using the substitution Kimura 2-parameter (K2P). (B) Neighbor-joining tree clustering based on the Euclidean distance calculated from the four-color map TI values.

4. Discussion

The results shown in the previous section were motivated by related works carried out with the MARCH-INSIDE methodology. We have previously reported the 2D-Cartesian representation for proteins and its numerical characterization through stochastic sequence descriptors calculated by MARCH-INSIDE [40], to annotate biological functions in gene/protein classes. Thus, we published the first alignment-free model built up with stochastic TIs to functionally classify polygalacturonase (PG) members from plants [40]. Despite the fact that PG members were detected with high accuracy by our reported alignment-free model, classical alignment procedures also did this due to PG proteins showing a high sequence similarity, while not representing a challenge for alignment algorithms. This study opened a door to the application of graphical methods in bioinformatics for the detection of functional signatures in protein families; however, its application was more illustrative than useful to overcome the limitations of alignment algorithms [18, 40].

The 2D-Cartesian protein representation was also numerically characterized through stochastic spectral moments to detect a particular RNAse III member from *Schizosaccharomyces pombe* (Pac 1) among the diversity of this class [71]. Although alignment algorithms have demonstrated a low amino acid identity (20%–40%) between Pac 1 and other typical RNases III, the Pac 1 protein shows a remarkable ribonuclease activity [72]. This fact prompted the application of our graphical–numerical method as an alternative to traditional alignment procedures for functional annotation. Thus, an alignment-free model was developed by linear statistical techniques to successfully detect the RNase III signature among a highly diverse dataset including the Pac 1 member. The model showed a higher accuracy and a similar sensitivity in detecting the RNase III signature than that achieved using alignment procedures [71]. This report provided some clues about the potential of graphical–numerical approaches as alternative tools to detect remote homologous due to their alignment-independence essence.

Considering such promising studies, we aimed to overcome alignment handicaps to face functional detections in highly diverse gene/protein families through the creation of TI2BioP software [5, 44]. The utility of TI2BioP was proved in classifying four gene/protein classes having great sequence divergence among their members (see Section 2.2). The alignment-free model's performance was always compared to alignment algorithms since we are pursuing an alternative tool to such methods. Alignment algorithms are the most popular techniques in bioinformatics; they basically score similarity measures at a predefined biological significance between a query gene/protein against others already registered or against a family profile to predict the structural and functional class [73, 74]. Although alignment-dependent algorithms have been improved through years of use, they do not consider structural information beyond the primary sequence, e.g. long-distance interactions and also ignore the important contribution of a negative set (nonmembers of the family), especially for the alignment algorithms building a profile-based model. Another weakness of this method arises when a query sequence is similar to genes/proteins lacking functional annotations [75]. In addition, phylogenetic inferences relying on MSA methods are not reliable when gene/protein sequences show functional similarities but have greatly diverged [52]. Consequently, such handicaps motivate the arising of alignment-free methods that exploit extra-information hidden in the linearity of the sequence, e.g. amino acid pseudo amino acid composition.

To validate our graphical–numerical methodology implemented in the TI2BioP software, we applied it to detect proteinaceous bacteriocins. Bacteriocins are small proteins of bacterial origin that are lethal to bacteria other than the producing strain. They have found applications in the pharmaceutical and food industry as potential antimicrobial agents and food preservatives, respectively [76]. The bacteriocin protein family is highly diverse in terms of size, method of killing, method of production, genetics, microbial target, immunity mechanisms and release, which has contributed to its low pair-wise sequence similarity (23%–50%). These family features represent a challenge for alignment procedures in the identification of protein bacteriocins [77], demanding the implementation of complex strategies [78, 79]. However, we built an effective alignment-free model based on the 2D-Cartesian protein representation and its derived TIs to detect bacteriocin proteins among the diversity of a dataset made up of nonredundant CATH domains and bacteriocin sequences. The model retrieved 66.7% of the bacteriocin-like proteins from an external test set while the InterPro resource could just detect 60.2% (Table 4). This is the first report where an alignment-free model based on a graphical approach entirely outperforms a popular alignment-based resource for functional sequence annotation [5].

The other bioinformatics utility of our graphical–numerical method consisted of the detection of a remote bacteriocin homologous in the Cry 1Ab C-terminal domain from *Bacillus thuringiensis'* endotoxin, which had not been detected by classical alignment methods. Although the functional relationship between bacteriocins and Cry 1Ab C-terminal domain classes have been assessed by experimental procedures in previous reports [80], their sequences are completely different and consequently placed into two different protein classes by alignment procedures. TI2BioP could successfully detect the bactericide function of Cry 1Ab C-terminal, just corroborated by experimental assays, either by scoring the query sequence for the alignment-free model or by the graphical superposition of the 2D-Cartesian maps for Cry 1Ab C-terminal domain to other representative bacteriocins [5]. This graphical analysis has been useful to visualize similarities/dissimilarities between different classes of natural biopolymers [40, 71].

After having success in detecting distant homologous among the protein bacteriocin family using TIs derived from 2D-Cartesian maps, the RNase III enzymatic class was selected as the second target to evaluate TI2BioP performance. The RNase III protein class contains members having great variability regarding the primary structure and domain organization. The similarities among different RNase IIIs varies from 20% to 84%, placing many of them in the twilight zone [81]. In addition, this diversity is also influenced by differences in the domain architectures of RNase III, which have led to a subdivision of the enzymatic class into four subclasses represented by four archetypes (bacterial RNase III, fungal RNase III, Dicer and Drosha).

Spectral moments derived from the 20 amino acids clustering according to their physicochemical properties into a 2D-Cartesian space (2D Cartesian maps) were used to develop three different nonlinear approaches to detect RNase III protein signatures among the diversity of

206 RNases III and a structurally nonredundant subset of the PDB made up of enzymes and no enzymes. Two alignment-free models based on decision tree models (DTMs) and artificial neural networks (ANNs) were built from TIs provided by TI2BioP to identify RNase III members. Additionally, a nonclassical profile HMM, inspired on the graphical clustering of the amino acids was developed, to make a fair comparison among alignment-free models and alignment algorithms [82].

While machine-learning methods that use nonlinear functions like ANNs and support vector machine (SVM) have been more frequently applied to the prediction of proteins structure and function [83-86], DTMs have been poorly explored in bioinformatics despite their widespread use in other fields [87]. We reported, for the first time, a simple and interpretable DTM to identify RNase III members using spectral moments as input predictors. The reported DTM showed a high predictive power (96.07%) using just one spectral moment at different splitting values while ANNs provide a lower predictability (92.15%) (Table 4) [69, 82].

The nonclassical profile HMM showed the best performance in the classification of proteins involved in this study. It reached the highest prediction rate (100%) for the RNase III class with respect to the performance of ANN and DTM (Table 4). Amino acid clustering according to its hydrophobic/charge properties was either effective at the primary level to increase the sensitivity of the profile HMM or at the 2D level to develop highly predictive DTM. Although the nonclassical profile HMM showed a slightly better performance than the alignment-free models, its generation demands programming skills while DTM search resulted in the easiest way to detect the RNase III signature among the diversity of the dataset [69]. The usability of DTM was also shown by predicting a new bacterial RNase III class member that was isolated and subsequently enzymatically tested and registered by our group (Table 1). The efficiency of DTM as a sequence search procedure to screen a proteome in conjunction with the TIs implemented in the TI2BioP software will be seen below [20].

Up to now, the TIs generated by TI2BioP have successfully been applied as alignment-free predictors in protein families but their classification performance should also be evaluated in highly diverse gene families, as well as their ability for reconstructing phylogenies. In this sense, the original 2D Cartesian representation reported by Nandy for describing DNA sequences [63] and the secondary structure inferred by DNA/RNA folding algorithms (Mfold) [67] were used to derive two types of TIs, for the ITS2 gene class.

The ITS2 eukaryotic gene class shows a high sequence divergence among its members, which has traditionally been exploited in low-level taxonomical analyses, especially for the unequivocal classification of fungal species. However, such sequence variability has complicated the ITS2 annotation and its use in phylogenetic analyses at higher taxonomic ranks. In this sense, the ITS2 secondary structure which has been conserved among all eukaryotes has been considered in the implementation of homology-based structure modelling approaches to improve the ITS2 annotation quality and to carry out phylogenetic analyses at higher classification levels or taxonomic ranks for eukaryotes [61, 88-90]. Although alignment-based methods have been exploited to the top of its complexity to tackle the ITS2 annotation and phylogenetic inference [88, 91], no alignment-free approach has so far been able to successfully address these issues.

The use of TIs containing information about the sequence and structure of ITS2 can be an alignment-free solution to improve the ITS2 prediction and for phylogenetic reconstruction at high taxonomic levels in eukaryotes. Alignment-independent approaches are represented by two ANN-based models for ITS2 classification among a large and diverse dataset, one built with 2D-Cartesian TIs [63, 92] and the other resulting from the Mfold 2D structure TIs [67]. Although ANN models built with both TI types (Cartesian and Mfold) displayed an excellent performance to detect the ITS2 class; the Mfold graphical approach provided the best classification results. Mfold TIs contain structural information about DNA folding driven by thermodynamic rules, providing a more accurate description of the DNA/RNA structure. This is the reason why the Mfold TIs were applied as an alignment-free approach to infer phylogenetic relationship to complement the taxonomy of a fungal isolate.

The performance of both ANN models were compared to several profiles HMM generated from MSA performed with CLUSTALW [93], DIALIGN-TX [94] and MAFFT [95] to classify the test set and to identify a new fungal member of the ITS2 class. Alignment-free models outperformed profiles HMM in classifying the test set and in identifying the new fungal member of the ITS2 class, even when HMMs were built by MSA algorithms improved for sets of low overall sequence.

The new ITS2 sequence was isolated by our group (GenBank accession number FJ892749) from an endophytic fungus belonging to the genus *Petrakia.* Members of this fungal genus are potential producers of bioactive compounds but they have been hard to place taxonomically [96]. In fact, the NCBI dedicated "taxonomy" database does not have clear information about its genus and that there is no specification about its subphylum and class [97]. On the other hand, the lack of other registered ITS2 sequences from different species of the genus *Petrakia* precluded performing a phylogenetic analysis at the species level (low-level analysis). Then, assuming that our fungal isolate belongs to the *Pezizomycotina* subphylum following a recent classification found in the "The dictionary of the Fungi" [98], a higher-level phylogenetic analysis to elucidate the class of *Petrakia* sp. was carried using two different types of distance trees: (1) a traditional NJ-tree based on multiple alignments of ITS2 sequences and (2) another tree irrespective of sequence similarity built from Mfold TIs. The alignment-free distances calculated from Mfold TIs provide similar phylogenetic relationships among the different classes of the Ascomycota phylum regarding the traditional phylogenetic analysis (i.e. based on evolutionary distances derived from a multiple alignment of DNA sequences). Both phylogenetic analyses, the traditional and the alignment-free clustering, placed *Petrakia* isolate in the Dothideomycetes class (Figure 4). We concluded that our alignment-free approach was effective for constructing hierarchical distance trees containing relevant biological information with an evolutionary significance [21].

So far, the 2D Cartesian graphs have been used to derive a TI series with the aim of being applied in bioinformatics. However, there are other 2D graphical approaches reported for DNA and proteins that have been mostly unexplored in this field; such is the case of the four-color maps introduced by Randić [64, 65]. Consequently, the four-color maps for DNA and protein sequences were implemented in the latest version of TI2BioP in order estimate new alignment-free predictors that can cooperate with traditional homology search tools (e.g.

BLAST, HMMs) to carry out an exhaustive exploration of functional signatures in highly diverse gene/protein families.

The NRPS family can harbor remote homologous due to the high sequence divergence among its A-domains, ranging mostly from 10% to 40% of sequence identity. Consequently, many of them are placed in the twilight zone (20%–35% sequence identity) reported for the alignment methods [99]. In fact, A-domain members cannot be retrieved easily by BLASTp using a single template [62]. To cope with the high sequence divergence of A-domains, we propose an ensemble of homology-search methods that integrates an alignment-free model that uses TIs derived from protein four-color maps [20].

The four-color map TIs were used to develop several alignment-free models using linear and nonlinear mathematical functions. Nonlinear models outperformed linear models in classifying A-domains confirming previous outcomes. DTM was the model of choice due to its excellent performance and its simple way to detect A-domains in a highly diverse dataset [20]. The DTM built up with four-color map TIs overdid other alignment-free concepts like ACC and PseACC, providing the highest sensitivity (Table 2) and no false-positives in A-domain identification (Table 3). In addition, it showed a similar performance to sensitive alignment algorithms like profile HMM and multitemplate BLASTp (Table 3).

As a result of comparing methods to detect A-domains, we can conclude that classification results among homology-based methods agreed with the fact that multitemplate BLASTp and profile HMM are more sensitive than simple BLASTp. Both multitemplate BLASTp and profile HMM easily retrieved all A-domain members at expectation values (E value ≤ 10) without reporting any false-positive (Table 3). However, the BLASTp search using a single template provided false-positives (significant matches) among the negative set (CATH domains) at both high (E value = 10) and relatively stringent cutoffs (E values < 0.05), which is considered statistically significant and useful for filtering easily identifiable homologs pairs [47, 100].

Because of the single-template BLASTp sensitivity did not show stability in identifying the A-domain signal at different classification stringency (E value); it was considered less reliable to perform sequence searches on unknown test datasets such as an entire proteome. Therefore, the easy and reliable identification of A-domains in the proteome of the cyanobacteria *M. aeruginosa* NIES-843 [20] was carried out by the combination of multitemplate BLASTp, profile HMM and four-color maps. Profiles HMM and four-color maps found additional hits as A-domains among the hypothetical proteins, giving clues for the presence of A-domain's remote homologous in the proteome of *M. aeruginosa* (Figure 3). Hypothetical proteins have not been definitively annotated and can be reannotated by applying more sensitive strategies. The assembling of sequence-search methods encoding different features from protein sequences can provide a better description of the proteome and therefore, remote protein homologous can be detected with more confidence [20]. Thus, we are introducing a new sensitive approach to search for remote homologous by integrating graphical–numerical methods with alignment procedures.

In summary, the presented graphical–numerical method implemented in the TI2BioP software does not suffer from many of the alignment algorithm limitations. Particularly, the artificial

2D graphs and the TIs encode higher-order useful information contained beyond the primary structure of the natural biopolymers allowing the building-up of effective alignment-free models. By contrast, our graphical–numerical approach has some handicaps stemming from the artificial nature of the 2D graphs which do not represent the real secondary structure of the biopolymers. Many of these 2D graphs bear some redundancy that leads to the loss of sequence information. On the other hand, spectral moment (TIs) estimation by powering matrixes, from thousands of graphs or maps, still demands a high computational cost.

5. Conclusions

We provided several evidences of the potential use of graphical–numerical approaches to characterize DNA/RNA and proteins that can be extended to other biopolymers. This new software called TI2BioP is not in competition with currently available bioinformatics tools, but instead works in cooperation with existing methodologies, as well as with experimental procedures required to overcome hard comparative studies of the natural biopolymers.

Acknowledgements

The authors acknowledge the Portuguese Fundação para a Ciência e a Tecnologia (FCT) for financial support to GACH (SFRH/BPD/92978/2013). AA was partially supported by the Strategic Funding UID/Multi/04423/2013 through national funds provided by FCT and European Regional Development Fund (ERDF) in the framework of the programme PT2020, and the FCT projects PTDC/AAC-AMB/121301/2010 (FCOMP-01-0124-FEDER-019490) and PTDC/AAG-GLO/6887/2014. The funders had no role in the study's design, data collection and analysis, decision to publish, or preparation of the manuscript.

Author details

Guillermin Agüero-Chapin[1,2], Reinaldo Molina-Ruiz[2], Gisselle Pérez-Machado[2], Vitor Vasconcelos[1,3], Zenaida Rodríguez-Negrin[2] and Agostinho Antunes[1,3*]

*Address all correspondence to: aantunes@ciimar.up.pt

1 CIMAR/CIIMAR, Centro Interdisciplinar de Investigação Marinha e Ambiental, Universidade do Porto, Porto, Portugal

2 Centro de Bioactivos Químicos, Universidad Central "Marta Abreu" de Las Villas (UCLV), Santa Clara, Cuba

3 Faculdade de Ciências, Departamento de Biologia, Universidade do Porto, Porto, Portugal

References

[1] Biggs N, Lloyd E, Wilson R, editors. Graph Theory: Oxford University Press; 1986. 1736–936.

[2] Randic M, Zupan J, Balaban AT, Vikic-Topic D, Plavsic D. Graphical representation of proteins. Chem Rev. 2011; 111: 790-862. DOI: 10.1021/cr800198j

[3] Gonzalez-Diaz H, Vilar S, Santana L, Uriarte E. Medicinal chemistry and bioinformatics--current trends in drugs discovery with networks topological indices. Curr Top Med Chem. 2007; 7: 1015-29.

[4] Estrada E, Uriarte E. Recent advances on the role of topological indices in drug discovery research. Curr Med Chem. 2001; 8: 1573-88.

[5] Aguero-Chapin G, Perez-Machado G, Molina-Ruiz R, Perez-Castillo Y, Morales-Helguera A, Vasconcelos V, et al. TI2BioP: Topological Indices to BioPolymers. Its practical use to unravel cryptic bacteriocin-like domains. Amino Acids. 2011; 40: 431-42.

[6] Perez-Bello A, Munteanu CR, Ubeira FM, De Magalhaes AL, Uriarte E, Gonzalez-Diaz H. Alignment-free prediction of mycobacterial DNA promoters based on pseudo-folding lattice network or star-graph topological indices. J Theor Biol. 2009; 256: 458-66.

[7] Ortega-Broche SE, Marrero-Ponce Y, Diaz YE, Torrens F, Perez-Gimenez F. TOMOCOMD-CAMPS and protein bilinear indices--novel bio-macromolecular descriptors for protein research: I. Predicting protein stability effects of a complete set of alanine substitutions in the Arc repressor. FEBS J. 2010; 277: 3118-46. DOI: 10.1111/j.1742-4658.2010.07711.x

[8] Gonzalez-Diaz H, Gonzalez-Diaz Y, Santana L, Ubeira FM, Uriarte E. Proteomics, networks and connectivity indices. Proteomics. 2008; 8: 750-78. DOI: 10.1002/pmic.200700638 [doi]

[9] Estrada E, Gutman I. A Topological Index Based on Distances of Edges of Molecular Graphs. J Chem Inf Comput Sci. 1996; 36: 850-3.

[10] Riera-Fernandez P, Martin-Romalde R, Prado-Prado FJ, Escobar M, Munteanu CR, Concu R, et al. From QSAR models of drugs to complex networks: state-of-art review and introduction of new Markov-spectral moments indices. Curr Top Med Chem. 2012; 12: 927-60.

[11] Gonzalez-Diaz H, Perez-Montoto LG, Duardo-Sanchez A, Paniagua E, Vazquez-Prieto S, Vilas R, et al. Generalized lattice graphs for 2D-visualization of biological information. J Theor Biol. 2009; 261: 136-47.

[12] Concu R, Podda G, Uriarte E, Gonzalez-Diaz H. Computational chemistry study of 3D-structure-function relationships for enzymes based on Markov models for protein electrostatic, HINT, and van der Waals potentials. J Comput Chem. 2009; 30: 1510-20.

[13] Marrero-Ponce Y, Medina-Marrero R, Castillo-Garit JA, Romero-Zaldivar V, Torrens F, Castro EA. Protein linear indices of the 'macromolecular pseudograph alpha-carbon atom adjacency matrix' in bioinformatics. Part 1: prediction of protein stability effects of a complete set of alanine substitutions in Arc repressor. Bioorg Med Chem. 2005; 13: 3003-15.

[14] Randic′ M. Graphical representation of DNA as a 2-D map. Chem Phys Lett. 2004; 468–71.

[15] Randic M, Zupan J, Vikic-Topic D. On representation of proteins by star-like graphs. J Mol Graph Model. 2007; 26: 290-305.

[16] Randic M, Zupan J. Highly compact 2D graphical representation of DNA sequences. SAR QSAR Environ Res. 2004; 15: 191-205.

[17] Nandy A. Recent investigations into global characteristics of long DNA sequences. Indian J Biochem Biophys. 1994; 31: 149-55.

[18] Aguero-Chapin G, Varona-Santos J, de la Riva GA, Antunes A, Gonzalez-Vlla T, Uriarte E, et al. Alignment-free prediction of polygalacturonases with pseudofolding topological indices: experimental isolation from Coffea arabica and prediction of a new sequence. J Proteome Res. 2009; 8: 2122-8.

[19] Cruz-Monteagudo M, Gonzalez-Diaz H, Borges F, Dominguez ER, Cordeiro MN. 3D-MEDNEs: an alternative "in silico" technique for chemical research in toxicology. 2. quantitative proteome-toxicity relationships (QPTR) based on mass spectrum spiral entropy. Chem Res Toxicol. 2008; 21: 619-32. DOI: 10.1021/tx700296t [doi]

[20] Aguero-Chapin G, Molina-Ruiz R, Maldonado E, De la Riva GA, Sanchez-Rodriguez A, Vasconcelos V, et al. Exploring the Adenylation Domain Repertoire of Nonribosomal Peptide Synthetases Using an Ensemble of Sequence-Search Methods. PLoS One. 2013; 8: DOI: 10.1371/journal.pone.0065926

[21] Aguero-Chapin G, Sanchez-Rodriguez A, Hidalgo-Yanes PI, Perez-Castillo Y, Molina-Ruiz R, Marchal K, et al. An alignment-free approach for eukaryotic ITS2 annotation and phylogenetic inference. PLoS One. 2011; 6: DOI: 10.1371/journal.pone. 0026638

[22] Katritzky AR, Gordeeva EV. Traditional topological indices vs electronic, geometrical, and combined molecular descriptors in QSAR/QSPR research. J Chem Inf Comput Sci. 1993; 33: 835-57.

[23] Randic M, Zupan J. On interpretation of well-known topological indices. J Chem Inf Comput Sci. 2001; 41: 550-60.

[24] Wiener H. Structural Determination of Paraffin Boiling Points. J Am Chem Soc. 1947; 69: 17-20.

[25] Randic M. Graph theoretical approach to structure-activity studies: search for optimal antitumor compounds. Prog Clin Biol Res. 1985; 172A: 309-18.

[26] Balaban AT, Beteringhe A, Constantinescu T, Filip PA, Ivanciuc O. Four new topological indices based on the molecular path code. J Chem Inf Model. 2007; 47: 716-31. DOI: 10.1021/ci6005068 [doi]

[27] Moreau G, Broto P. The Autocorrelation of a topological structure. A new molecular descriptor. Nouv J Chim. 1980; 4: 359-60.

[28] Estrada E. Spectral Moments of the Edge Adjacency Matrix in Molecular Graphs. 1. Definition and Applications to the Prediction of Physical Properties of Alkanes. J Chem Inf Comput Sci. 1996; 36: 844-9.

[29] Estrada E. Spectral Moments of the Edge-Adjacency Matrix of Molecular Graphs. 2. Molecules Containing Heteroatoms and QSAR Applications. J Chem Inf Comput Sci. 1997; 37: 320-8.

[30] Estrada E. On the topological sub-structural molecular design (TOSS-MODE) in QSPR/QSAR and drug design research. SAR QSAR Environ Res. 2000; 11: 55-73.

[31] Markovic S, Markovic Z, McCrindle RI. Spectral moments of phenylenes. J Chem Inf Comput Sci. 2001; 41: 112-9.

[32] González MP, Teran C, Teijeira M. A topological function based on spectral moments for predicting affinity toward A_3 adenosine receptors. Bioorg Med Chem Lett. 2006; 16: 1291-6.

[33] Morales AH, González MP, Briones JR. TOPS-MODE approach to predict mutagenicity in dental monomers. Polymer. 2004; 45: 2045-50.

[34] Estrada E. Characterization of the folding degree of proteins. Bioinformatics. 2002; 18: 697-704.

[35] Estrada E, Hatano N. A Tight-Binding "Dihedral Orbitals" Approach to Electronic Communicability in Protein Chains. Chemical Physics Letters. 2007; 449: 216-20.

[36] Gonzalez-Diaz H, Uriarte E. Biopolymer stochastic moments. I. Modeling human rhinovirus cellular recognition with protein surface electrostatic moments. Biopolymers. 2005; 77: 296-303.

[37] Gonzalez-Diaz H, Uriarte E, Ramos de Armas R. Predicting stability of Arc repressor mutants with protein stochastic moments. Bioorg Med Chem. 2005; 13: 323-31.

[38] Gonzalez-Diaz H, Saiz-Urra L, Molina R, Gonzalez-Diaz Y, Sanchez-Gonzalez A. Computational chemistry approach to protein kinase recognition using 3D stochastic

van der Waals spectral moments. J Comput Chem. 2007; 28: 1042-8. DOI: 10.1002/jcc. 20649 [doi]

[39] González-Díaz H, Molina-Ruiz R, Hernandez I. MARCH-INSIDE v3.0 (MARkov CHains INvariants for SImulation & DEsign). 3.0 ed2007. p. Windows supported version under request to the main author contact email: gonzalezdiazh@yahoo.es.

[40] Aguero-Chapin G, Gonzalez-Diaz H, Molina R, Varona-Santos J, Uriarte E, Gonzalez-Diaz Y. Novel 2D maps and coupling numbers for protein sequences. The first QSAR study of polygalacturonases; isolation and prediction of a novel sequence from Psidium guajava L. FEBS Lett. 2006; 580: 723-30. DOI: 10.1016/j.febslet. 2005.12.072 [doi]

[41] Aguero-Chapin G, Antunes A, Ubeira FM, Chou KC, Gonzalez-Diaz H. Comparative study of topological indices of macro/supramolecular RNA complex networks. J Chem Inf Model. 2008; 48: 2265-77. DOI: 10.1021/ci8001809 [doi]

[42] Gonzalez-Diaz H, Aguero-Chapin G, Varona-Santos J, Molina R, de la Riva G, Uriarte E. 2D RNA-QSAR: assigning ACC oxidase family membership with stochastic molecular descriptors; isolation and prediction of a sequence from Psidium guajava L. Bioorg Med Chem Lett. 2005; 15: 2932-7.

[43] Gonzalez-Diaz H, Aguero-Chapin G, Varona J, Molina R, Delogu G, Santana L, et al. 2D-RNA-coupling numbers: a new computational chemistry approach to link secondary structure topology with biological function. J Comput Chem. 2007; 28: 1049-56. DOI: 10.1002/jcc.20576 [doi]

[44] Molina R, Agüero-Chapin G, Pérez-González MP. TI2BioP (Topological Indices to BioPolymers) *version 2.0.*: Molecular Simulation and Drug Design (MSDD), Chemical Bioactives Center, Central University of Las Villas, Cuba; 2011

[45] Needleman SB, Wunsch CD. A general method applicable to the search for similarities in the amino acid sequence of two proteins. J Mol Biol. 1970; 48: 443-53.

[46] Smith TF, Waterman MS. Identification of common molecular subsequences. J Mol Biol. 1981; 147: 195-7. DOI: 0022-2836(81)90087-5 [pii]

[47] Altschul SF, Gish W, Miller W, Myers EW, Lipman DJ. Basic Local Alignment Search Tool. J Mol Biol. 1990; 215: 403-10.

[48] Krogh AB, M.; Mian, I. S.; Sjeander, K.; Haussler, D. Hidden Markov models in computational biology. Applications to protein modeling. J Mol Biol. 1994; 235: 1501-31.

[49] Eddy SR. A new generation of homology search tools based on probabilistic inference. Genome Inform. 2009; 23: 205-11.

[50] Dobson PD, Doig AJ. Distinguishing Enzyme Structures from Non-enzymes Without Alignments. J Mol Biol. 2003; 330: 771–83.

[51] Strope PK, Moriyama EN. Simple alignment-free methods for protein classification: a case study from G-protein-coupled receptors. Genomics. 2007; 89: 602-12. DOI: 10.1016/j.ygeno.2007.01.008

[52] Schwarz RF, Fletcher W, Förster F, Merget B, Wolf M, Schultz J, et al. Evolutionary Distances in the Twilight Zone—A Rational Kernel Approach. PLoS ONE. 2010; 5:

[53] Rost B. Enzyme function less conserved than anticipated. J Mol Biol. 2002; 318: 595-608.

[54] Pearson WR, Sierk ML. The limits of protein sequence comparison? Current Opinion in Strctural Biology. 2005; 15: 254-60.

[55] Kumar M, Thakur V, Raghava GP. COPid: composition based protein identification. In Silico Biol. 2008; 8: 121-8.

[56] Deshmukh S, Khaitan S, Das D, Gupta M, Wangikar PP. An alignment-free method for classification of protein sequences. Protein Pept Lett. 2007; 14: 647-57.

[57] Schwarz RF, Fletcher W, Forster F, Merget B, Wolf M, Schultz J, et al. Evolutionary distances in the twilight zone--a rational kernel approach. PLoS One. 2010; 5: e15788. DOI: 10.1371/journal.pone.0015788

[58] Chou KC. Prediction of protein cellular attributes using pseudo-amino acid composition. Proteins. 2001; 43: 246-55.

[59] Hohl M, Rigoutsos I, Ragan MA. Pattern-based phylogenetic distance estimation and tree reconstruction. Evol Bioinform Online. 2006; 2: 359-75.

[60] de Jong A, van Hijum SA, Bijlsma JJ, Kok J, Kuipers OP. BAGEL: a web-based bacteriocin genome mining tool. Nucleic Acids Res. 2006; 34: W273-9.

[61] Selig C, Wolf M, Muller T, Dandekar T, Schultz J. The ITS2 Database II: homology modelling RNA structure for molecular systematics. Nucleic Acids Res. 2008; 36: D377-80.

[62] Ansari MZ, Yadav G, Gokhale RS, Mohanty D. NRPS-PKS: a knowledge-based resource for analysis of NRPS/PKS megasynthases. Nucleic Acids Res. 2004; 32: W405-13. DOI: 10.1093/nar/gkh359

[63] Nandy A. Two-dimensional graphical representation of DNA sequences and intron-exon discrimination in intron-rich sequences. Comput Appl Biosci. 1996; 12: 55-62.

[64] Randic M, Mehulic K, Vukicevic D, Pisanski T, Vikic-Topic D, Plavsic D. Graphical representation of proteins as four-color maps and their numerical characterization. J Mol Graph Model. 2009; 27: 637-41. DOI: 10.1016/j.jmgm.2008.10.004

[65] Randic M, Lers N, Plavšić D, Basak S, Balaban A. Four-color map representation of DNA or RNA sequences and their numerical characterization. Chemical Physics Letters 2005; 407: 205-8.

[66] Aguero-Chapin G, Molina-Ruiz R, Maldonado E, de la Riva G, Sanchez-Rodriguez A, Vasconcelos V, et al. Exploring the adenylation domain repertoire of nonribosomal peptide synthetases using an ensemble of sequence-search methods. PLoS One. 2013; 8: e65926. DOI: 10.1371/journal.pone.0065926

[67] Mathews DH. RNA secondary structure analysis using RNAstructure. Curr Protoc Bioinformatics. 2006; Chapter 12: Unit 12 6. DOI: 10.1002/0471250953.bi1206s13 [doi]

[68] Hammami R, Zouhir A, Hamida JB, Fliss I. BACTIBASE: a new web-accessible database for bacteriocin characterization. BMC Microbiology 2007; 7: 89 DOI: 10.1186/1471-2180-7-89

[69] Aguero-Chapin G, de la Riva GA, Molina-Ruiz R, Sanchez-Rodriguez A, Perez-Machado G, Vasconcelos V, et al. Non-linear models based on simple topological indices to identify RNase III protein members. J Theor Biol. 2011; 273: 167-78.

[70] Shen HB, Chou KC. PseAAC: a flexible web server for generating various kinds of protein pseudo amino acid composition. Anal Biochem. 2008; 373: 386-8.

[71] Aguero-Chapin G, Gonzalez-Diaz H, de la Riva G, Rodriguez E, Sanchez-Rodriguez A, Podda G, et al. MMM-QSAR recognition of ribonucleases without alignment: comparison with an HMM model and isolation from Schizosaccharomyces pombe, prediction, and experimental assay of a new sequence. J Chem Inf Model. 2008; 48: 434-48. DOI: 10.1021/ci7003225 [doi]

[72] Lamontagne B, Elela SA. Evaluation of the RNA determinants for bacterial and yeast RNase III binding and cleavage. J Biol Chem. 2004; 279: 2231-41. DOI: 10.1074/jbc.M309324200

[73] Altschul SF, Madden, T.L., Schaffer, A.A., Zhang, J., Zhang, Z., Miller, W. and Lipman, D.J. Gapped BLAST and PSI-BLAST: a new generation of protein database search programs. Nucl Acids Res. 1997; 25: 3389-402.

[74] Finn RD, Clements J, Eddy SR. HMMER web server: interactive sequence similarity searching. Nucleic Acids Res. 2011; 39: W29-37. DOI: 10.1093/nar/gkr367

[75] Davies MN, Secker A, Freitas AA, Timmis J, Clark E, Flower DR. Alignment-Independent Techniques for Protein Classification. Current Proteomics. 2008; 5: 217-23.

[76] Gillor O, Etzion A, Riley MA. The dual role of bacteriocins as anti- and probiotics. Appl Microbiol Biotechnol. 2008; 81: 591-606.

[77] Cotter P, Hill C, Ross R. What's in a name? Class distinction for bacteriocins. Nature Reviews Microbiology. 2006; 4:

[78] Dirix G, Monsieurs P, Dombrecht B, Daniels R, Marchal K, Vanderleyden J, et al. Peptide signal molecules and bacteriocins in Gram-negative bacteria: a genome-wide in silico screening for peptides containing a double-glycine leader sequence and their cognate transporters. Peptides. 2004; 25: 1425-40.

[79] Stein T. Bacillus subtilis antibiotics: structures, syntheses and specific functions. Mol Microbiol. 2005; 56: 845-57.

[80] Vazquez-Padron RI, de la Riva G, Aguero G, Silva Y, Pham SM, Soberon M, et al. Cryptic endotoxic nature of Bacillus thuringiensis Cry1Ab insecticidal crystal protein. FEBS Lett. 2004; 570: 30-6. DOI: 10.1016/j.febslet.2004.06.021

[81] Lamontagne B, Elela S.A. Evaluation of the RNA determinants for bacterial and yeast RNase III binding and cleavage. J Biol Chem. 2004; 279: 2231-41.

[82] Chapin GA. A Graphical and Numerical Approach for Functional Annotation and Phylogenetic Inference: University of Porto; 2013.

[83] Punta M, Rost B. Neural networks predict protein structure and function. Methods Mol Biol. 2008; 458: 203-30.

[84] Nair R, Rost B. Protein subcellular localization prediction using artificial intelligence technology. Methods Mol Biol. 2008; 484: 435-63. DOI: 10.1007/978-1-59745-398-1_27 [doi]

[85] Cai YD, Ricardo PW, Jen CH, Chou KC. Application of SVM to predict membrane protein types. J Theor Biol. 2004; 226: 373-6.

[86] Fernandez M, Caballero J, Fernandez L, Abreu JI, Garriga M. Protein radial distribution function (P-RDF) and Bayesian-Regularized Genetic Neural Networks for modeling protein conformational stability: chymotrypsin inhibitor 2 mutants. J Mol Graph Model. 2007; 26: 748-59. DOI: 10.1016/j.jmgm.2007.04.011

[87] Ripley B. Pattern Recognition and Neural Networks. Cambridge, UK: Cambridge University Press; 1996

[88] Koetschan C, Forster F, Keller A, Schleicher T, Ruderisch B, Schwarz R, et al. The ITS2 Database III--sequences and structures for phylogeny. Nucleic Acids Res. 2009;

[89] Schultz J, Maisel S, Gerlach D, Müller T, and Wolf M. A common core of secondary structure of the internal transcribed spacer 2 (ITS2) throughout the Eukaryota. RNA 2005; 11: 361-4.

[90] Schultz J, Muller T, Achtziger M, Seibel PN, Dandekar T, Wolf M. The internal transcribed spacer 2 database--a web server for (not only) low level phylogenetic analyses. Nucleic Acids Res. 2006; 34: W704-7.

[91] Schultz J, Müller T, Achtziger M, Seibel P, Dandekar T, Wolf M. The internal transcribed spacer 2 database--a web server for (not only) low level phylogenetic analyses. Nucleic Acids Research. 2006; 34: DOI: 10.1093/nar/gkl129

[92] Nandy A. Empirical relationship between intra-purine and intra-pyrimidine differences in conserved gene sequences. PLoS One. 2009; 4: e6829. DOI: 10.1371/journal.pone.0006829

[93] Thompson JD, Higgins DG, Gibson TJ. CLUSTAL W: improving the sensitivity of progressive multiple sequence alignment through sequence weighting, position-specific gap penalties and weight matrix choice. Nucleic Acids Res. 1994; 22: 4673-80.

[94] Subramanian AR, Kaufmann M, Morgenstern B. DIALIGN-TX: greedy and progressive approaches for segment-based multiple sequence alignment. Algorithms Mol Biol. 2008; 3: 6. DOI: 10.1186/1748-7188-3-6

[95] Katoh K, Kuma K, Miyata T, Toh H. Improvement in the accuracy of multiple sequence alignment program MAFFT. Genome Inform. 2005; 16: 22-33.

[96] Qi FH, Jing TZ, Wang ZX, Zhan YG. Fungal endophytes from Acer ginnala Maxim: isolation, identification and their yield of gallic acid. Lett Appl Microbiol. 2009; 49: 98-104.

[97] Bisby F, Roskov Y, Ruggiero M, Orrell T, Paglinawan L, Brewer P, et al. Species 2000 & ITIS Catalogue of Life: 2007 Annual Checklist Taxonomic Classification. CD-ROM; Species 2000: Reading, U.K. 2007;

[98] Kirk PM, Cannon PF, Stalpers JA. The dictionary of the Fungi. 10th ed. UK: CABI; 2008. 784.

[99] Rost B. Twilight zone of protein sequence alignments. Protein Eng. 1999; 12: 85-94.

[100] Boekhorst J, Snel B. Identification of homologs in insignificant blast hits by exploiting extrinsic gene properties. BMC Bioinformatics. 2007; 8.